Lecture Notes in Mathematics

Edited by A. Dold and B. Eckmann

1068

Number Theory Noordwijkerhout 1983

Proceedings of the Journées Arithmétiques
held at Noordwijkerhout, The Netherlands
July 11-15, 1983

Edited by H. Jager

Springer-Verlag
Berlin Heidelberg New York Tokyo 1984

Editor

Hendrik Jager
Mathematisch Instituut, Universiteit van Amsterdam
Roetersstraat 15, 1018 WB Amsterdam, The Netherlands

AMS Subject Classification (1980): 10-02

ISBN 3-540-13356-9 Springer-Verlag Berlin Heidelberg New York Tokyo
ISBN 0-387-13356-9 Springer-Verlag New York Heidelberg Berlin Tokyo

Printed in Germany

Printing and binding: Beltz Offsetdruck, Hemsbach/Bergstr.
2146/3140-543210

P R E F A C E

In 1961, French number theorists gathered during a weekend in Grenoble, where six lectures on the subject were delivered. This proved a success and subsequent *Journées Arithmétiques*, as these now five-day -conferences became generally known, were held in Lille (1963), Besançon (1965), Grenoble (1967), Bordeaux (1969), Marseille Saint-Jérôme (1971), Grenoble (1973), Bordeaux (1974), Caen (1976) and Marseille-Luminy (1978). In return for the hospitality of the French to their colleagues from abroad, the 1980 conference was organized by the London Mathematical Society in Exeter, U.K. and after the conference in Metz (1981) the Netherlands became the host country for the 13th Journées Arithmétiques.

These Journées Arithmétiques took place from 11th to 15th July, 1983 at the conference centre "de Leeuwenhorst" at Noordwijkerhout near Leiden. Organizers were Prof. H.W. Lenstra, Jr. of the University of Amsterdam and Prof. R. Tijdeman of the University of Leiden. The conference was supported by the Mathematical Centre, Amsterdam, the University of Amsterdam, the University of Leiden, the Royal Netherlands Academy of Sciences, the Dutch Mathematical Society, the Société Mathématique de France / C.N.R.S., I.B.M. Netherlands and Shell Netherlands. The participants came from thirteen different countries and the approximately eighty lectures covered almost all aspects of number theory.

For various reasons, not all these lectures could be collected in the present volume. However, the contributions presented here in the Proceedings give a fair impression of the variety and high level of the conference. They deal with many branches of number theory, e.g. algebraic number theory, Riemann's zeta function and prime numbers, transcendental number theory and uniform distribution. Some contributions are self-contained articles giving all the proofs, others are surveys or give heuristically and numerically supported conjectures. Hence, every number theorist will find some papers here which are of particular interest to him.

H. Jager

TABLE OF CONTENTS

NEW RESULTS IN THE THEORY OF IRREGULARITIES OF POINT DISTRIBUTIONS

J. Beck
Mathematical Institute
of the Hungarian Academy of Sciences
Budapest, 1053 Reáltanoda u. 13-15, HUNGARY

I.1. Let $\xi = (z_1, z_2, z_3, \ldots)$ be an infinite sequence of real numbers in the unit interval $U = [0,1)$. Given an x in U and a positive integer N, we write $Z_N(\xi; x)$ for the number of integers i with $1 \le i \le N$ and $0 \le z_i < x$ and we put

$$\Delta_N(\xi; x) = |Z_N(\xi; x) - N \cdot x| .$$

The sequence ξ is called uniformly distributed if $\Delta_N(\xi)/N \to 0$ as $N \to +\infty$, where $\Delta_N(\xi)$ is the supremum of $\Delta_N(\xi; x)$ over all numbers x in U. The theory of irregularities of distribution was initiated by a conjecture of van der Corput that $\Delta_N(\xi)$ cannot remain bounded as $N \to +\infty$. It was proved by Mrs. T. van Aardenne - Ehrenfest [1] in 1945. Later this beautiful theorem was improved and extended in various directions by the basic works of K.F.Roth and W.M.Schmidt. There now is a large literature of this subject, we refer the reader to Schmidt's book [13].

This paper is above all a review of the author's recent research in the topic, nevertheless it contains a complete proof of Theorem 1.13 (see Section I.7).

I.2. We introduce the concept of discrepancy concerned with a set of points on the unit torus ("static" case). Let U^2 be the unit square consisting of points $\underline{x} = (x_1, x_2)$ with $0 \le x_1 < 1$, $0 \le x_2 < 1$, and let $\underline{z}_1, \underline{z}_2, \ldots, \underline{z}_N$ be N points (not necessarily distinct) in U^2. Further let

$$Q = Q(\underline{z}_1, \ldots, \underline{z}_N) = \{\underline{z} + \underline{w} : \underline{z} \in \{\underline{z}_1, \ldots, \underline{z}_N\}, \underline{w} \in Z^2\}.$$

Given a bounded and measurable set A of area $\mu(A)$, write $Z(Q;A)$ for the number (counted with multiplicities) of points of Q in A, and put

$$\Delta_N(A) = \Delta_N(Q;A) = |Z(Q;A) - N \cdot \mu(A)|.$$

The quality $\Delta_N(Q;A)$ is called discrepancy of A, and measures the deviation of the distribution of $\underline{z}_1,\ldots,\underline{z}_N$ from the uniform distribution (i.e., Lebesgue measure).

More than ten years ago W.M.Schmidt (see [11] and [12]) proved the following two remarkable lower bounds in the theory of irregularities of distribution.

THEOREM A (W.M.Schmidt): *Let there be given* N *points* $\underline{z}_1,\ldots,\underline{z}_N$ *on the unit torus.*

(α) *There exists a rectangle* R *contained in* U^2 *and with sides parallel to the coordinate axes (aligned rectangle) such that*

$$\Delta_N(R) \gg \log N.$$

(β) *Suppose* $N\cdot\delta^2 > \varepsilon > 0$. *There exists a circular disc* D *of diameter* $\leq \delta$ *and with*

$$\Delta_N(D) \gg (N\cdot\delta^2)^{\frac{1}{4}-\varepsilon} .$$

(All the constants of this paper indicated by $\gg$ are positive and absolute.)

Note that statement (α) is sharp apart from the value of the constant factor, since already in 1922 A. Ostrowski [8] constructed a sequence which yields the inequality $\Delta_N(R) \ll \log N$ for any aligned rectangle R. This "static" result of Schmidt is equivalent with the solution of the long-standing "dynamic" conjecture that for any infinite sequence $\xi = (z_1,z_2,\ldots)$ of real numbers, $\Delta_N(\xi) \gg \log N$ infinitely often (see Section I.1). Moreover, the exponent ¼ in statement (β) is also the best possible (see [2]).

In the proofs of these surprisingly sharp results Schmidt used completely different ideas depending heavily on the special shapes of aligned rectangles and circular discs, respectively. This led him to the question of understanding the astonishing phenomenon that circular discs have much greater discrepancy than aligned rectangles (see [13]). We announce the following results which give a partial answer to this question.

I.3. Let us consider the following more general problem. Let $S = \{\underline{z}_1,\underline{z}_2,\ldots\}$ be a completely arbitrary infinite discrete set of points in Euclidean plane $\mathbb{R}^2$. Given a bounded and measurable set $A \subset \mathbb{R}^2$, write $Z(S;A)$ for the number of points $\underline{z}_i \in A$ and set

$$\Delta(A) = \Delta(S;A) = |Z(S;A)-\mu(A)|$$

where $\mu(A)$ denotes the usual area of A (i.e.,Lebesgue measure of A).

Let B be an arbitrary bounded convex domain in the plane. If λ is a real number and $\underline{x} \in \mathbb{R}^2$ is a vector, write

$$B(\lambda,\underline{x}) = \{\lambda\cdot\underline{y} + \underline{x} : \underline{y} \in B\}$$

Clearly $B(\lambda,\underline{x})$ is <u>homothetic</u> to B. Now Schmidt's question can be formulated as follows: Estimate the quantity

$$\Delta^*(B) = \inf_S \sup_{B(\lambda,\underline{x})} \Delta(S;B(\lambda,\underline{x}))$$

where the supremum is taken over all sets $B(\lambda,\underline{x})$ with $0 < \lambda \le 1$, $\underline{x} \in \mathbb{R}^2$ and the infimum is extended over all infinite discrete sets $S \subset \mathbb{R}^2$ (i.e., we may contract and translate, but rotation is forbidden). We need the following basic definition which tells us how well can the convex set B be approximated by an inscribed polygon of few sides.

For any bounded convex domain B let $B_k \subset B$ denote the inscribed k-gon (i.e., polygon of k sides) of largest area. Denote by app(B) the smallest integer $k \ge 3$ such that $\mu(B \setminus B_k) < k$, that is, the area of B minus B_k is less than the number of sides of B_k. We call app(B) the <u>approximation number</u> of the convex domain B. Note that in general the approximation number of B depends on both the shape and the size of B. For instance, if D_r is the circular disc of radius $r \ge 1$ then, as elementary calculation shows, $\mathrm{app}(D_r)$ is about $c_1 \cdot r^{2/3}$. By a well-known result in discrete geometry we know that $\mu(B_k) \ge \mu(B) \cdot \frac{k}{2\pi} \cdot \sin\left(\frac{2\pi}{k}\right)$ and equality holds only for the ellipse (see [5] p.38). This implies $\mathrm{app}(B) \ll n^{1/3}$ where $n = \mu(B)$. Furthermore, if P_k is a polygon of k sides then $\mathrm{app}(P_k) \le k$ and if λ is sufficiently large depending on the shape of P_k then $\mathrm{app}(\lambda \cdot P_k) = k$.

The following two results illustrate the importance of the concept of approximation number. Roughly speaking they state that the quantity $\Delta^*(B)$ can be estimated by two powers of $\mathrm{app}(B) \cdot \log n$ where $n = \mu(B)$, but we cannot determine the exact constant in the exponent.

THEOREM 1.1: Let there be given an infinite discrete set $S = \{\underline{z}_1, \underline{z}_2, \underline{z}_3, \ldots\} \subset \mathbb{R}^2$ and a bounded convex domain B with $\mu(B) \ge 1$. Then there exist a real λ_o, $0 < \lambda_o \le 1$ and a vector $\underline{x}_0 \in \mathbb{R}^2$ such that

$$\Delta(B(\lambda_0,\underline{x}_0)) = \Delta(S;B(\lambda_0,\underline{x}_0)) \gg \max\{(\mathrm{app}(B))^{\frac{1}{2}}, (\log n)^{\frac{1}{2}}\}$$

where $n = \mu(B)$ denotes the area of B.

For the sake of brevity, we write $B \approx B_0$ if the domains B and B_0 are homothetic to each other.

In the opposite direction we have

THEOREM 1.2 : Let B_0 be an arbitrary but fixed bounded convex domain. Then there exists an infinite discrete set $S_0 = S_0(B_0)$ such that for <u>all</u> B with $B \approx B_0$ and $\mu(B) \ge 2$, $\Delta(S_0;B) \ll (\mathrm{app}(B))^2 \cdot (\log n)^5$ where $n = \mu(B)$.

Unfortunately, Theorem 1.1-2 are not sharp. For example, if D_r

is the circular disc of radius $r \geq 1$, then Theorem 1.1 yields the lower bound $r^{1/3}$, but a theorem of Schmidt (see Theorem A 4 in [11]) states the lower bound $r^{1/2-\varepsilon}$. However, for a class of smooth domains we have the following nearly best possible result.

THEOREM 1.3: Let S be an arbitrary infinite discrete set. Let B be a bounded convex domain and let $M > 1$ be a real number. Suppose the boundary $\Gamma = \Gamma(B)$ of B is smooth and the ratio

$$\frac{\text{maximum curvature of } \Gamma}{\text{minimum curvature of } \Gamma} \leq M.$$

Then there exist a real λ_o, $0 < \lambda_o \leq 1$ and a vector $\underline{x}_0 \in \mathbb{R}^2$ such that $\Delta(B(\lambda_0, \underline{x}_0)) = \Delta(S; B(\lambda_0, \underline{x}_0)) > c_2(M) \cdot n^{1/4}$ where $n = \mu(B) \geq 1$ and the constant $c_2(M)$ depends only on M.

We remark that here the exponent $1/4$ of n is the best possible (for a stronger result, see Theorem 1.9 later).

I.4. From Theorem 1.1 and 1.3 one can immediately obtain results concerning points on the unit torus (see Section I.2). Given a bounded convex domain B, let $\text{app}_N(B)$ denote the smallest integer $k \geq 3$ such that $N \cdot \mu(B \setminus B_k) < k$. If we blow up the periodic set $Q(\underline{z}_1, \ldots, \underline{z}_N)$ in ratio $\sqrt{N}:1$, and apply Theorem 1.1 we obtain

COROLLARY 1.4 : Let there be given N points $\underline{z}_1, \underline{z}_2, \ldots, \underline{z}_N$ on the unit torus and let B be a convex domain with diameter less than one. Then there exist a real λ_o, $0 < \lambda_o \leq 1$ and a vector $\underline{x}_0 \in \mathbb{R}^2$ such that

$$\Delta_N(B(\lambda_0, \underline{x}_0)) \gg (\text{app}_N(B))^{1/2}.$$

The proof of Theorem 1.1 easily yields

THEOREM 1.5: Under the hypothesis of Corollary 1.4 there exist a real λ_o, $0 < \lambda_o \leq 1$ and a vector $\underline{x}_0 \in \mathbb{R}^2$ such that $B(\lambda_0, \underline{x}_0)$ <u>is contained in the unit square</u> U^2 and

$$\Delta_N(B(\lambda_0, \underline{x}_0)) \gg (\log(N \cdot \mu(B)))^{1/2}$$

This result seems to be new already in the simplest case B = "square" (the problem is due to Vera T. Sós). A few days ago G.Halász informed me that for aligned squares he can prove the sharp (apart from constant factor) lower bound logN.

In the opposite direction we have the following "dynamic" result.

THEOREM 1.6: Let B_0 be an arbitrary but fixed convex domain on the unit torus. Then there exists an infinite <u>sequence</u> $(\underline{z}_1, \underline{z}_2, \ldots)$ of points such that for <u>all</u> B with $B \approx B_0$ and diameter$(B) < 1$, the inequality

$$\Delta_N(\underline{z}_1, \underline{z}_2, \ldots, \underline{z}_N; B) \ll (\text{app}_N(B))^2 \cdot (\log N)^6$$

holds for each $N \geq 2$.

Our method is very insensitive to the Lebesgue measure μ, and all the results above remain true for arbitrary nonnegative normed Borel measure ν such that the Radon-Nikodym derivative $d\nu/d\mu$ is bounded above.

In higher dimensions we state only a result concerned with aligned cubes (i.e., the sides of the cube are parallel to the coordinate axes) and general measures.

THEOREM 1.7: Let there be given N points in the ℓ-dimensional unit cube $U^{\ell} = [0,1)^{\ell}$. Let ν_{ℓ} be an arbitrary nonnegative normed Borel measure in U^{ℓ} such that the Radon-Nikodym derivative $d\nu_{\ell}/d\mu_{\ell}$ is bounded above, where μ_{ℓ} denotes the usual ℓ-dimensional volume. Then there exists an aligned cube B contained in U^{ℓ} such that the ν_{ℓ}-discrepancy

$$\Delta_N(B;\nu_{\ell}) = |Z(Q;B)-N\nu_{\ell}(B)| > c_3(\ell)(\log N)^{(\ell-1)/2}.$$

Here the constant $c_3(\ell)$ depends only on the dimension ℓ.

Note that Roth's old theorem [9] guarantees the existence of an aligned box (i.e., product of intervals; not necessarily a cube) with the same order of discrepancy in the case of the usual volume μ_{ℓ}. But Roth's "orthogonal function" method heavily depends on the homogenity of μ_{ℓ}, and it is not clear how to modify the method for general measures. Probably Theorem 1.7 remains true if we replace the exponent $(\ell-1)/2$ of log N by the twice as large value $(\ell-1)$, but we are unable to handle this very hard and basic conjecture.

I.5. The situation is completely different if we allow rotations. Our starting point is again a surprising and deep result of W.M.Schmidt [11].

THEOREM B (W.M.Schmidt): *Let there be given* N *points* $\underline{z}_1,\dots,\underline{z}_N$ *on the unit torus. Suppose* $N\cdot\delta^2 > N^{\varepsilon} > 0$ *then there exists a tilted rectangle* T *with diameter* $\le \delta$ *and with*

$$\Delta_N(T) \gg (N\cdot\delta^2)^{1/4-\varepsilon}.$$

That is, if rotations are allowed then the lower bound here involves a power of N rather than log N (compare Theorem B with the sharp estimate (α) in Theorem A, see Section I.2). Note that using a technical refinement due to G. Harman [7] the lower bound $N^{1/4-\varepsilon}$ can be replaced by the sharper $N^{1/4}(\log N)^{-1/2}$.

We can prove an analogous "large discrepancy" phenomenon for an arbitrary convex domain. Let rot(ϑ) denote the rotation of the plane with angle ϑ, $0 \le \vartheta < 2\pi$. For arbitrary angle ϑ, real λ and vector $\underline{x} \in \mathbb{R}^2$ set

$$B(\vartheta,\lambda,\underline{x}) = \{\mathrm{rot}(\vartheta)(\lambda\underline{y}+\underline{x}) : y \in B\},$$

that is, $B(\vartheta,\lambda,\underline{x})$ can be obtained from B by a similarity transformation with parameters ϑ, λ and $\underline{x}$.

Let per(B) denote the <u>perimeter</u> of the bounded convex domain B, and we recall that $\mu(B)$ denotes the area of B. The next two results say that the quantity

$$\Delta^{**}(B) = \inf_{S} \sup_{B(\vartheta,\lambda,\underline{x})} \Delta(S;B(\vartheta,\lambda,\underline{x})) \quad ,$$

where $0 \le \vartheta < 2\pi$, $0 < \lambda \le 1$, $\underline{x} \in \mathbb{R}^2$ and S is an arbitrary infinite discrete subset of the plane, roughly equals $\min\{(\mathrm{per}(B))^{1/2}, (\mu(B))^{1/2}\}$.

THEOREM 1.8: Let there be given an infinite discrete set $S \subset \mathbb{R}^2$ and a bounded convex domain $B \subset \mathbb{R}^2$. Then there exist an angle ϑ_0, a real λ_o, $0 < \lambda_o \le 1$ and a vector $\underline{x}_0 \in \mathbb{R}^2$ such that

$$\Delta(B(\vartheta_0,\lambda_0,\underline{x}_0)) = \Delta(S;B(\vartheta_0,\lambda_0,\underline{x}_0)) \gg \max\{1,(m(B))^{1/2}\}$$

where $m(B) = \min\{\mathrm{per}(B),\mu(B)\}$.

Observe that, by the isoperimetric inequality, $m(B) \ge (\mu(B))^{1/2}$ if $\mu(B) \ge 1$. Moreover, $m(\lambda B) = \mathrm{per}(\lambda B)$ if λ is sufficiently large depending only on the shape of B.

Note that in the particular case B = "rectangle of size $n \times 1$" Theorem 1.8 yields the existence of a tilted rectangle with discrepancy of "random error" size, i.e., there is a tilted rectangle T with $\mu(T) \le n$ and $\Delta(T) \gg \sqrt{n}$.

Next we state that Theorem 1.8 is nearly best possible. We write $B \sim B_0$ if the convex domains B, B_0 are similar to each other.

THEOREM 1.9: Let B_0 be a given convex domain and $\varepsilon > 0$ be an arbitrarily small real. Then there exists an infinite discrete set $S = \{\underline{z}_1,\underline{z}_2,\ldots\}$ in the plane and a threshold $c_4(\varepsilon)$ such that for <u>every</u> B with $B \sim B_0$ and diameter $(B) > c_4(\varepsilon)$,

$$\Delta(S;B) \ll \max\{1,(m(B))^{1/2}\}(\mathrm{per}(B))^{\varepsilon}$$

where $m(B) = \min\{\mathrm{per}(B),\mu(B)\}$.

I.6. Again we return to the case of the unit torus. Theorem 1.8 immediately yields.

COROLLARY 1.10: Let $\underline{z}_1,\ldots,\underline{z}_N$ be N points on the unit torus and let B be a convex domain of diameter less than one. Then there exist an angle ϑ_0, a real λ_o, $0 < \lambda_o \le 1$ and a vector $\underline{x}_0 \in \mathbb{R}^2$ such that

$$\Delta_N(B(\vartheta_0,\lambda_0,\underline{x}_0)) \gg \max\{1,\min\{N^{1/4}(\mathrm{per}(B))^{1/2},N^{1/2}(\mu(B))^{1/2}\}\}.$$

Note that in the particular case B = "square" we obtain a slight improvement on Schmidt's estimate (see Theorem B), namely the ε in the exponent can be cancelled.

As a converse we have the following "dynamic" result.

THEOREM 1.11: Let B_0 be a bounded convex domain and $\varepsilon > 0$ be an arbitrarily small real. Then there exists an infinite sequence $(\underline{z}_1, \underline{z}_2, \underline{z}_3, \ldots)$ of points on the unit torus and a threshold $c_5(\varepsilon)$ such that for every B with $B \sim B_0$ and diameter(B)<1.

$$\Delta_N(B) \ll \max\{1, \min\{N^{1/4+\varepsilon}(\mathrm{per}(B))^{1/2}, N^{1/2+\varepsilon}(\mu(B))^{1/2}\}\}$$

holds for each $N > c_5(\varepsilon)$.

In higher dimensions we mention a particular result only.

THEOREM 1.12: Let there be given N points in the ℓ-dimensional unit torus. Then there is a tilted cube A of diameter less than one such that

$$\Delta_N(A) > c_6(\ell) N^{1/2-1/(2\ell)} .$$

Note that the cases $\ell=2$ and 3 are essentially due to W.M.Schmidt [11] (apart from an ε in the exponent), but for $\ell > 3$ the result is new. Here the exponent $1/2-1/(2\ell)$ of N is the best possible, see [2].

I.7. The proofs of our lower bounds are based on the Fourier transform technique (Parseval's identity, Fejér kernel, etc.). This method enables us to handle the case of an arbitrary convex domain. The first appearance of this method can be found in K.F.Roth's paper [10] concerning irregularities of integer sequences relative to arithmetic progressions. As an illustration of the method, see the proof of Theorem 1.13 in Section II.

To prove the upper bounds we use probabilistic and combinatorial arguments.

The following "size localization" conjecture, due to P.Erdős, is an example of a question which probably cannot be attacked by this method: Does there exist a universal function f(r) tending to infinity as $r \to +\infty$ such that for every discrete set of points on the plane and for every real $r > 1$ one can find a circular disc of radius r with discrepancy greater than f(r)?

We shall prove, however, the following result of the same spirit.

THEOREM 1.13: *Let there be given a completely arbitrary discrete set of points on the plane. Then for every real* $r>1$ *there is a tilted square of size* $r \times r$ *with discrepancy* $\gg r^{1/4}$.

Section II is devoted to the proof of this theorem.

Note that for aligned squares of fixed size one cannot guarantee unbounded discrepancy. Indeed, using the well-known van der Corput sequence it is not hard to construct an infinite discrete set on the plane such that any aligned square of size $2^n \times 2^n$ (n is an arbitrary natural number) has discrepancy < 1.

I.8. Now we mention two related results which will appear in [3] and [4]. Suppose that S is the set of the N points $\underline{z}_1,\dots,\underline{z}_N$ in the circular disc D of unit area. For each segment G (i.e., an intersection of D with a half-plane) let $Z(S;G)$ denote the number of points among $\underline{z}_1,\dots,\underline{z}_N$ which lie in G, and write

$$\Delta(S;G) = |Z(S;G)-N\mu(G)|$$

(we recall that $\mu(\cdot)$ denotes the Lebesgue measure).

Set

$$\Delta_o(N) = \inf_S \sup_G \Delta(S;G)$$

where the supremum is extended over all segments G of D and the infimum is taken over all N-element subsets S of D. Many years ago K.F. Roth suspected that $\Delta_o(N)$ cannot be bounded as $N \to +\infty$. We have succeeded in proving the following sharper form of Roth's conjecture (see [3]).

THEOREM 1.14: If N is sufficiently large depending only on $\varepsilon>0$ then

$$\Delta_o(N) > N^{1/4-\varepsilon} .$$

It has been shown earlier by the author [2] that

$$\Delta_o(N) = O(N^{1/4}(\log N)^{1/2}) ,$$

that is, the exponent 1/4 of N in the theorem is the best possible.

Next we consider the following interesting problem of a geometrical nature: For what set of N points on the sphere is the sum of all $\binom{N}{2}$ distances between points maximal, and what is the maximum?

Write S^ℓ for the surface of the unit sphere in $(\ell+1)$-dimensional Euclidean space $\mathbb{R}^{\ell+1}$. Let P be a set of N points on S^ℓ. Let $\rho(\cdot,\cdot)$ denote the usual Euclidean distance. We define

$$\Sigma(P,\ell) = \sum_{1\le i<j\le N} \rho(\underline{p}_i,\underline{p}_j) \quad \text{where } P = \{\underline{p}_1,\underline{p}_2,\dots,\underline{p}_N\} ,$$

and

$$\Sigma(N,\ell) = \max_{\substack{P\subset S^\ell \\ |P|=N}} \Sigma(P,\ell)$$

($|P|$ denotes, as usual, the number of elements of the set P.)

The determination of $\Sigma(N,\ell)$ is a long-standing open problem in discrete geometry. For $\ell=1$ the solution is given by the regular N-gon (see L. Fejes Tóth [6]). It is also known [6] that for $N = \ell+2$, the regular simplex is optimal. For $N > \ell+2$ and $\ell > 1$ the exact value of $\Sigma(N,\ell)$ is unknown.

It is a natural idea to compare the discrete sum $\Sigma(P,\ell)$ with the integral

$$\frac{N^2}{2}\cdot\frac{1}{\sigma(S^\ell)}\int_{S^\ell}\rho(\underline{x}_0,\underline{x})\,d\sigma(\underline{x})$$

where $\sigma(S^\ell)$ denotes the surface area of S^ℓ, $d\sigma(\underline{x})$ represents an element of surface area on S^ℓ and $\underline{x}_0 = (1,0,\ldots,0) \in \mathbb{R}^{\ell+1}$. K.B.Stolarsky [14] has discovered the following beautiful identity.

THEOREM C (K.B.Stolarsky): *For any* $P \subset S^\ell$ *with* $|P| = N$,

$$\Sigma(P,\ell) + \int_{-1}^{+1}\Big(\frac{1}{\sigma(S^\ell)}\int_{S^\ell}(Z(P,\underline{x},t) - N\cdot\sigma^*(t))^2 d\sigma(\underline{x})\,dt = \tag{1}$$

$$= \frac{N^2}{2}\cdot\frac{1}{\sigma(S^\ell)}\int_{S^\ell}\rho(\underline{x}_0,\underline{x})\,d\sigma(\underline{x}) .$$

Here $Z(P,\underline{x},t)$ *is the number of points of* P *in the spherical cap* $\{y \in S^\ell : \underline{x}\cdot\underline{y} \le t\}$ *and* $\sigma^*(t)$ *denotes the normalized surface area of a spherical cap* $\{y \in S^\ell : \underline{x}\cdot\underline{y} \le t\}$.

The second term on (1) is clearly a measure of the discrepancy of P. The sum of distances is thus maximised by a well distributed set of points.

Set $C_0(\ell) = \frac{1}{2\sigma(S^\ell)}\int_{S^\ell}\rho(\underline{x}_0,\underline{x})\,d\sigma(\underline{x})$. Note that the constant $C_0(\ell)$ can be calculated explicitly . Identity (1) immediately yields $C_0(\ell)N^2 - \Sigma(N,\ell) > 0$. Using a result of W.M.Schmidt on the discrepancy of spherical caps (see Theorem D), Stolarsky [14] has shown the lower bound

$$C_0(\ell)N^2 - \Sigma(N,\ell) > c_7(\ell)N^{1-3/\ell-2/\ell^2-\varepsilon} .$$

Later it was improved by G.Harman [7] to

$$C_0(\ell)N^2 - \Sigma(N,\ell) > c_8(\ell)N^{1-2/\ell-\varepsilon}$$

for ℓ even.

Note that these estimates gave non-trivial results only for $\ell \ge 3$. In the classical case $\ell = 2$ these results do not imply that $C_0(2)N^2 - \Sigma(N,2) \to +\infty$ as $N \to +\infty$. We could determine the correct order of magnitude of this difference ([4]).

THEOREM 1.15 : For any $\ell \ge 1$

$$c_9(\ell)N^{1-1/\ell} < C_0(\ell)N^2 - \Sigma(N,\ell) < c_{10}(\ell)N^{1-1/\ell} .$$

The upper bound here is established in Stolarsky [14].

As we promised we recall Schmidt's theorem concerning spherical caps. Let P be a set of N points on S^ℓ. We write $Cap(r,\underline{x})$ for the spherical cap of all points on S^ℓ whose spherical distance from

$x \in S^\ell$ is no more than r. Put $Z(P;\mathrm{Cap}(r,\underline{x}))$ for the number of points of P in $\mathrm{Cap}(r,\underline{x})$ and

$$\Delta(P;r,\underline{x}) = |Z(\mathrm{Cap}(r,\underline{x}))-N\sigma^*(\mathrm{Cap}(r,\underline{x}))|$$

(we recall that σ^* denotes the normalized surface area, so $\sigma^*(S^\ell)=$ $=1$).

We write

$$V(P;r) = \int_{S^\ell} \Delta^2(P;r,\underline{x})d\sigma^*(\underline{x}).$$

Using his remarkable "integral equation method", Schmidt [11] established

THEOREM D (W.M.Schmidt): *If* N *is sufficiently large depending on* $\varepsilon > 0$ *and if* $N\delta^\ell \geq N^\varepsilon$ *then*

$$\int_0^\delta r^{-1}\cdot V(P,r)\,dr < (N\cdot\delta^\ell)^{1-1/\ell-\varepsilon} \tag{2}$$

We emphasize that in (2) the integration goes on the spherical distance r, but in the second term of (1) the integration goes on the "inner product distance" (i.e., we call $1-\underline{x}\cdot\underline{y}$ the "inner-product distance" of $\underline{x},\underline{y} \in S^\ell$). These two distances have entirely different speeds of tending to zero. This is the reason why Schmidt's estimate (2) was somewhat "inconvenient" for estimating the difference $C_o(\ell)N^2 - \Sigma(N,\ell)$.

From (2) the following result easily follows by averaging argument

COROLLARY E (W.M.Schmidt): *Let there be given* N *points on the sphere* S^ℓ. *If* N *is sufficiently large depending only on* $\varepsilon > 0$ *then there is a spherical cap* $\mathrm{Cap}(r,\underline{x}) \subset S^\ell$ *with large discrepancy* $\Delta(r,\underline{x}) > N^{1/2-1/(2\ell)-\varepsilon}$.

We remark that (1) and Theorem 1.14 immediately yield the slightly sharper lower bound $\Delta(r,\underline{x}) \gg N^{1/2-1/(2\ell)}$. In the opposite direction, using probabilistic ideas it is not hard to show the existence of an N-element subset of S^ℓ such that any spherical cap has discrepancy $\ll$ $\ll N^{1/2-1/(2\ell)}(\log N)^{1/2}$ (for the idea, see [2]).

II.1. Proof of Theorem 1.13. First we introduce two measures. For any $H \subset \mathbb{R}^2$ let

$$S_o(H) = \sum_{\underline{z}_j\in H\cap[-M,M]^2} 1 ,$$

i.e., $S_o(\cdot)$ denotes the discrete measure generated by the set $\{\underline{z}_1,\underline{z}_2,\dots\} \cap [-M,M]^2$ (we recall that $\underline{z}_1,\underline{z}_2,\dots$ are the given points; $M = M(r)$ is a parameter fixed later). For any measurable $H \subset \mathbb{R}^2$ let

$\mu_o(H) = \mu(H \cap [-M,M]^2)$, i.e., $\mu_o(\cdot)$ denotes the restriction of the usual area to the square $[-M,M]^2$.

Let $\chi_{r,\vartheta}$ denote the characteristic function of the tilted square

$$H(r,\vartheta)=\{\underline{x}=(x_1,x_2) : |x_1\cos\vartheta-x_2\sin\vartheta|\le\tfrac{r}{2} , |x_1\sin\vartheta+x_2\cos\vartheta|\le\tfrac{r}{2}\}$$

Moreover, let $H(r,\vartheta,\underline{y}) = H(r,\vartheta)+\underline{y}$ (i.e., a translate of $H(r,\vartheta)$).

Consider now the function

$$F_\vartheta = \chi_{r,\vartheta} * (dS_o-d\mu_o) \tag{3}$$

where $*$ denotes the <u>convolution</u> operation.

More explicitly,

$$\begin{aligned} F_\vartheta(\underline{x}) &= \int_{\mathbb{R}^2} \chi_{r,\vartheta}(\underline{x}-\underline{y})\,(dS_o(\underline{y})-d\mu_o(\underline{y})) = \\ &= \sum_{\underline{z}_j\in H(r,\vartheta,\underline{x})\cap[-M,M]^2} 1 \quad - \mu(H(r,\vartheta,\underline{x})\cap[-M,M]^2). \end{aligned} \tag{4}$$

Therefore, if $H(r,\vartheta,\underline{x}) \subset [-M,M]^2$ then $|F_\vartheta(\underline{x})|$ equals the discrepancy of the tilted square $H(r,\vartheta,\underline{x})$ of size $r \times r$. Since the function F_ϑ was defined as a convolution, we shall employ the theory of <u>Fourier transformation</u>. We recall some well-known facts.

Let

$$\hat{f}(\underline{t}) = \frac{1}{\pi}\int_{\mathbb{R}^2} e^{-i\underline{x}\underline{t}} f(\underline{x})\,d\underline{x}$$

be the Fourier transform of the function $f(\underline{x})$, $x \in \mathbb{R}^2$. It is known that (see any textbook)

$$\widehat{f*g} = \hat{f}\cdot\hat{g} \tag{5}$$

$$\widehat{f\cdot g} = \hat{f}*\hat{g} \tag{6}$$

$$\int_{\mathbb{R}^2} |f(\underline{x})|^2 d\underline{x} = \int_{\mathbb{R}^2} |\hat{f}(\underline{t})|^2 d\underline{t} \quad \text{(Parseval-Plancherel identity)} \tag{7}$$

By (7), (3) and (5) we have

$$\int_0^{2\pi} \Big(\int_{\mathbb{R}^2} F_\vartheta^2(\underline{x})\,d\underline{x}\Big)d\vartheta = \int_0^{2\pi} \Big(\int_{\mathbb{R}^2} |\hat{F}_\vartheta(\underline{t})|^2 d\underline{t}\Big)d\vartheta =$$

$$= \int_0^{2\pi} \Big(\int_{\mathbb{R}^2} |\hat{\chi}_{r,\vartheta}(\underline{t})|^2 |\widehat{(dS_o-d\mu_o)}(\underline{t})|^2 d\underline{t}\Big)d\vartheta .$$

For the sake of brevity, let $\varphi(\underline{t}) = \widehat{(dS_o-d\mu_o)}(\underline{t})$, i.e.,

$$\varphi(\underline{t}) = \frac{1}{\pi} \int_{\mathbb{R}^2} e^{-i\underline{x}\underline{t}} (dS_o(\underline{x}) - d\mu_o(\underline{x})) .$$

Thus we can write

$$\int_0^{2\pi} \left(\int_{\mathbb{R}^2} F_\vartheta^2(\underline{x}) d\underline{x} \right) d\vartheta = \int_{\mathbb{R}^2} |\varphi(\underline{t})|^2 \left(\int_0^{2\pi} |\hat{\chi}_{r,\vartheta}(\underline{t})|^2 d\vartheta \right) d\underline{t} . \tag{8}$$

We may assume that

$$\sum_{\underline{z}_j \in [-M,M]^2} 1 \geq \frac{1}{2}(2M)^2 . \tag{9}$$

Indeed, M will be specified so as $M > 4r$, and thus if (9) does not hold then by some averaging argument we obtain the existence of a translate $[-r/2,r/2]^2+\underline{x}$ of the square $[-r/2,r/2]^2$ of size $r \times r$ such that it is contained in $[-M,M]^2$ and has less than $8r^2/9$ points $\underline{z}_1, \underline{z}_2, \ldots$. Consequently, this square has discrepancy $\gg r^2$.

II.2. We need

LEMMA 2.1:

$$\int_{[-100,100]^2} |\varphi(\underline{t})|^2 d\underline{t} \gg M^2$$

<u>Proof</u>. We introduce the following auxiliary function (<u>Fejér kernel</u>)

$$g(\underline{x}) = \pi^2 \left(2\frac{\sin 50x_1}{x_1}\right)^2 \cdot \left(2\frac{\sin 50x_2}{x_2}\right)^2 .$$

Since the Fourier transform of $\sqrt{\pi} \cdot 2 \cdot \frac{\sin y \cdot x}{x}$ (x is the variable) is the characteristic function of the interval $[-y,y]$, by (6) we know that $\hat{g}$ equals the convolution of the characteristic function of the square $[-50,50]^2$ with itself, that is,

$$\hat{g}(\underline{t}) = (100-|t_1|)^+ (100-|t_2|)^+ \quad \text{where} \quad (y)^+ = \begin{cases} y & \text{if } y > 0 \\ 0 & \text{otherwise} \end{cases} .$$

Let

$$G(\underline{x}) = \int_{\mathbb{R}^2} g(\underline{x}-\underline{y}) (dS_o(\underline{y}) - d\mu_o(\underline{y})) .$$

Then by (5) we have

$$\hat{G}(\underline{t}) = \hat{g}(\underline{t}) \varphi(\underline{t}) \qquad \text{(we recall that } \varphi = \widehat{dS_o - d\mu_o}\text{)} .$$

By (7),

$$\int_{\mathbb{R}^2} G^2(\underline{x}) d\underline{x} = \int_{\mathbb{R}^2} |\hat{g}(\underline{t})|^2 |\varphi(\underline{t})|^2 d\underline{t} \ll \int_{[-100,100]^2} |\varphi(\underline{t})|^2 d\underline{t} \tag{10}$$

since $\hat{g}(\underline{t}) = 0$ whenever $\underline{t} \notin [-100,100]^2$.

On the other hand, since $g(\underline{x})$ is a <u>positive</u> function and, as elementary calculation shows, the inequality

$$g(\underline{x}-\underline{z}_j) - \int_{\mathbb{R}^2} g(\underline{x}-\underline{y})d\underline{y} \geq 1$$

is certainly true for $\underline{z}_j \in [-1/100,1/100]^2+\underline{x}$, we obtain

$$|G(\underline{x})| = |\int_{\mathbb{R}^2} g(\underline{x}-\underline{y})(dS_o-d\mu_o)(\underline{y})| =$$

$$= |\sum_{\underline{z}_j\in[-M,M]^2} g(\underline{x}-\underline{z}_j) - \int_{[-M,M]^2} g(\underline{x}-\underline{y})d\underline{y}| \geq$$

$$\geq \sum_{\underline{z}_j\in([-1/100,1/100]^2+\underline{x})\cap[-M,M]^2} 1 \quad \stackrel{\text{def}}{=} w(\underline{x})$$

Since $w^2(\underline{x}) \geq w(\underline{x})$, we get

$$\int_{\mathbb{R}^2} G^2(\underline{x})d\underline{x} \geq \int_{\mathbb{R}^2} w^2(\underline{x})d\underline{x} \geq \int_{\mathbb{R}^2} w(\underline{x})d\underline{x} \gg \sum_{\underline{z}_j\in[-M,M]^2} 1 \tag{11}$$

Now the lemma follows from (9), (10) and (11).□

II.3. Next we estimate the integral

$$\int_0^{2\pi} |\hat{\chi}_{r,\vartheta}(\underline{t})|^2 d\vartheta$$

from below simultaneously for all $\underline{t} \in [-100,100]^2$ (see (8)).
Since

$$\hat{\chi}_{[-y,y]}(t) = \frac{2}{\sqrt{\pi}} \frac{\sin yt}{t} ,$$

the function $\hat{\chi}_{r,\vartheta}$ can be written in the following explicit form

$$\hat{\chi}_{r,\vartheta}(\underline{t}) = \left(\frac{2}{\sqrt{\pi}} \frac{\sin(\frac{r}{2}\underline{t}\cdot\underline{e}_1(\vartheta))}{\underline{t}\cdot\underline{e}_1(\vartheta)}\right)\cdot\left(\frac{2}{\sqrt{\pi}} \frac{\sin(\frac{r}{2}\underline{t}\cdot\underline{e}_2(\vartheta))}{\underline{t}\cdot\underline{e}_2(\vartheta)}\right)$$

where $\underline{e}_1(\vartheta) = (\cos\vartheta,-\sin\vartheta)$ and $\underline{e}_2(\vartheta) = (\sin\vartheta,\cos\vartheta)$.
And so

$$\int_0^{2\pi} |\hat{\chi}_{r,\vartheta}(\underline{t})|^2 d\vartheta = \frac{16}{\pi^2} \int_0^{2\pi} \left(\frac{\sin(\frac{r}{2}\underline{t}\cdot\underline{e}_1(\vartheta))}{\underline{t}\cdot\underline{e}_1(\vartheta)}\right)^2 \cdot \left(\frac{\sin(\frac{r}{2}\underline{t}\cdot\underline{e}_2(\vartheta))}{\underline{t}\cdot\underline{e}_2(\vartheta)}\right)^2 d\vartheta \tag{12}$$

Now assume that $\underline{t} \in [-100,100]^2$ and let $\alpha = \alpha(\underline{t})$ be defined by the equation

$$\underline{t}\cdot\underline{e}_1(\alpha)=0, \text{ and to have a unique solution, let}$$
$$\underline{t}\cdot\underline{e}_2(\alpha) = |\underline{t}| . \tag{13}$$

Consequently,

$$(\underline{t}\cdot\underline{e}_2(\vartheta))^2 = |\underline{t}|^2\cdot(1-\sin^2(\vartheta-\alpha)).$$

From this it follows that the value of the function $f(\vartheta)=\frac{r}{2}\underline{t}\cdot\underline{e}_2(\vartheta)$ goes from $\frac{r}{2}|\underline{t}|-c_{11}|\underline{t}|$ $(c_{11}>0)$ to $\frac{r}{2}|\underline{t}|$ when the argument ϑ goes from $\alpha-\frac{1}{\sqrt{r}}$ to α. Thus there is a subinterval $I(\underline{t})\subset[\alpha-\frac{1}{\sqrt{r}},\alpha]$ with length $\gg\frac{1}{\sqrt{r}}$ such that $|\sin(\frac{r}{2}\underline{t}\cdot\underline{e}_2(\vartheta))|\geq c_{12}|\underline{t}|$ $(c_{12}>0)$ for all $\vartheta\in I(\underline{t})$. Hence

$$\left(\frac{\sin(\frac{r}{2}\underline{t}\cdot\underline{e}_2(\vartheta))}{\underline{t}\cdot\underline{e}_2(\vartheta)}\right)^2\geq c_{12}^2\gg 1 \quad \text{for all} \quad \vartheta\in I(\underline{t}) \tag{14}$$

Furthermore, there is a subset $J(\underline{t})$ of $I(\underline{t})$ such that its total length $\gg\frac{1}{\sqrt{r}}$ and $|\sin(\frac{r}{2}\underline{t}\cdot\underline{e}_1(\vartheta))|\geq 1/2$ for each $\vartheta\in J(\underline{t})$. Since $J(\underline{t})\subset I(\underline{t})\subset[\alpha-\frac{1}{\sqrt{r}},\alpha]$, by (13) we know (we recall that $\underline{t}\in[-100,100]$

$$|\underline{t}\cdot\underline{e}_1(\vartheta)|\leq\frac{|\underline{t}|}{\sqrt{r}}\leq 100\sqrt{2}/\sqrt{r} \quad \text{for} \quad \vartheta\in J(\underline{t}).$$

Therefore, we obtain

$$\left(\frac{\sin(\frac{r}{2}\underline{t}\cdot\underline{e}_1(\vartheta))}{\underline{t}\cdot\underline{e}_1(\vartheta)}\right)^2\gg r \quad \text{whenever} \quad \vartheta\in J(\underline{t}). \tag{15}$$

Combining (12), (14) and (15) we get that for every $\underline{t}\in[-100,100]^2$,

$$\int_0^{2\pi}|\hat{\chi}_{r,\vartheta}(\underline{t})|^2 d\vartheta > \int_{J(\underline{t})}\left(\frac{\sin(\frac{r}{2}\underline{t}\cdot\underline{e}_1(\vartheta))}{\underline{t}\cdot\underline{e}_1(\vartheta)}\right)^2\left(\frac{\sin(\frac{r}{2}\underline{t}\cdot\underline{e}_2(\vartheta))}{\underline{t}\cdot\underline{e}_2(\vartheta)}\right)^2 d\vartheta \gg \tag{16}$$

$$\gg r\int_{J(\underline{t})}d\vartheta\gg\sqrt{r}.$$

II.4. By (8), Lemma 2.1 and (16) we see

$$\int_0^{2\pi}\left(\int_{\mathbb{R}^2}F_\vartheta^2(\underline{x})d\underline{x}\right)d\vartheta\geq\int_{[-100,100]^2}|\varphi(\underline{t})|^2\left(\int_0^{2\pi}|\hat{\chi}_{r,\vartheta}(\underline{t})|^2 d\vartheta\right)d\underline{t}\geq \tag{17}$$

$$\geq\min_{\underline{t}\in[-100,100]^2}\int_0^{2\pi}|\hat{\chi}_{r,\vartheta}(\underline{t})|^2 d\vartheta\cdot\int_{[-100,100]^2}|\varphi(\underline{t})|^2 d\underline{t}\gg\sqrt{r}\cdot M^2$$

Now we are in the position to end the proof of Theorem 1.13. Let $M=r^5$. If for some $\vartheta\in[0,2\pi)$ and $\underline{x}\in\mathbb{R}^2$ the inequality

$$|F_\vartheta(\underline{x})|=\left|\sum_{\underline{z}_j\in H(r,\vartheta,\underline{x})\cap[-M,M]^2}1-\mu(H(r,\vartheta,\underline{x})\cap[-M,M]^2)\right|>2r^2$$

holds, then obviously the tilted square $H(r,\vartheta,\underline{x})$ has a huge discrepancy (greater than r^2), and we are done.

Thus we may assume

$$\max_{(\vartheta,\underline{x})} |F_\vartheta(\underline{x})| \le 2r^2. \tag{18}$$

By (4) and (17) (we recall that $\Delta(H)$ denotes the discrepancy of the square H and that $F_\vartheta(\underline{x})=0$ for all $(\vartheta,\underline{x})$ with $H(r,\vartheta,\underline{x})\cap[-M,M]^2=\phi$)

$$(2M)^2 \cdot \max_{H(r,\vartheta,\underline{x})\subset[-M,M]^2} \Delta^2(H(r,\vartheta,\underline{x}) +$$

$$+ c_{13}Mr \cdot \max_{H(r,\vartheta,\underline{x})\not\subset[-M,M]^2} F_\vartheta^2(\underline{x}) \tag{19}$$

$$\ge \frac{1}{2\pi}\int_0^{2\pi}\Big(\int_{\mathbb{R}^2} F_\vartheta^2(\underline{x})\,d\underline{x}\Big)d\vartheta \gg \sqrt{r}M^2 .$$

Finally, from (18) and (19) we see

$$\max_{H(r,\vartheta,\underline{x})\subset[-M,M]^2} \Delta^2(H(r,\vartheta,\underline{x})) \gg \sqrt{r} - O\Big(\frac{r^5}{M}\Big) = \sqrt{r} - O(1) .$$

Theorem 1.13 follows, since we may assume that r is sufficiently large.

REFERENCES

[1] AARDENNE-EHRENFEST,T.van, *Proof of the impossibility of a just distribution,* Indag. Math. 7(1945), 71-76.
[2] BECK,J., *Some upper bounds in the theory of irregularities of distribution,* to appear in Acta Arithmetica.
[3] BECK,J., *On a problem of K.F.Roth concerning irregularities of point distribution,* to appear in Inventiones Math.
[4] BECK,J., *On the sum of distances between N points on a sphere - an application of the theory of irregularities of distribution to discrete geometry,* to appear in Mathematika.
[5] FEJES TÓTH,L., Lagerungen in der Ebene auf der Kugel und im Raum, Springer, Berlin, 1953.
[6] FEJES TÓTH,L., *On the sum of distances determined by a pointset,* Acta Math. Acad. Sci. Hungar. 7(1956) 397-401.
[7] HARMAN,G., *Sums of distances between points of a sphere,* Internat.J.Math. & Math. Sci. 5(1982) 707-714.
[8] OSTROWSKI,A., *Bemerkungen zur Theorie der diophantischen Approximationen,* Abh. Hamburg Sem. I(1922), 77-98.
[9] ROTH, K.F., *On irregularities of distribution,* Mathematika 7(1954), 73-79.

[10] ROTH,K.F., *Remark concerning integer sequences*, Acta Arithmetica 9(1964) 257-260.

[11] SCHMIDT, W.M., *Irregularities of distribution IV*. Inventiones Math. 7(1969) 55-82.

[12] SCHMIDT, W.M., *Irregularities of distribution VII*. Acta Arithmetica 21(1972) 45-50.

[13] SCHMIDT, W.M., Lectures on irregularities of distribution, Tata Institute, Bombay, 1977.

[14] STOLARSKY, K.B., *Sums of distances between points on a sphere II*. Proc. Amer. Math. Soc. 41(1973) 575-582.

Finally, we mention two hand-written manuscripts which contain the ("rough" and "clumsy") proofs of Theorem 1.1 and 1.8, resp.

BECK,J. Discrepancy relative to sets which are *homothetic* to a given convex domain, 1983.

BECK,J., Discrepancy relative to sets which are *similar* to a given convex domain, 1983.

PROPRIÉTÉS ARITHMÉTIQUES DE FONCTIONS THÊTA À PLUSIEURS VARIABLES

par Daniel BERTRAND

Soit E une courbe elliptique définie sur un corps de nombres F plongé dans $\mathbb{C}$, admettant des multiplications complexes par un sous-corps quadratique imaginaire K de F, et munie d'une forme différentielle de première espèce définie sur F, dont on note $\mathcal{L}$ le réseau des périodes. Si $\sigma_{\mathcal{L}}$ désigne la fonction sigma de Weierstrass associée à $\mathcal{L}$, et N la norme relative à l'extension $K/\mathbb{Q}$, on sait (voir [2]) qu'il existe un unique élément $s_2(\mathcal{L})$ de F tel que la fonction

$$\theta_E(z) = \sigma_{\mathcal{L}}(z) \exp(-s_2(\mathcal{L})\, z^2/2)$$

vérifie la propriété suivante : pour tout endomorphisme α de E, le carré de $\theta_E(\alpha z)/\theta_E(z)^{N\alpha}$ s'identifie à une fonction F-rationnelle sur E. Ce sont les produits de valeurs de cette fonction thêta que nous étudions ici, généralisant ainsi les résultats de [3]. On place tout d'abord cette étude (§ 1) dans le cadre des variétés abéliennes de type (K). Les propriétés de transcendance de leurs fonctions thêta sont données au § 2, le rôle du théorème de Masser et Wüstholz [7] utilisé dans [3] étant maintenant joué par le récent théorème de Wüstholz [9] sur la méthode de Baker. Enfin, le § 3 est consacré aux applications d'un analogue ultramétrique de ces résultats à l'étude des hauteurs p-adiques sur la courbe elliptique E. On trouvera le détail des démonstrations dans [5].

§ 1 Fonctions thêta à multiplications complexes

Soient A une variété abélienne de dimension g définie sur la clôture algébrique $\bar{\mathbb{Q}}$ de $\mathbb{Q}$ dans $\mathbb{C}$, d'espace tangent à l'origine t_A, et $e_A : t_A(\mathbb{C}) \longrightarrow A(\mathbb{C})$ l'application exponentielle sur le groupe des points complexes de A. Nous identifierons $t_A(\mathbb{C})$ à $\mathbb{C}^g$ au moyen d'une base de t_A définie sur $\bar{\mathbb{Q}}$.

Soit par ailleurs X un diviseur sur A défini sur $\bar{\mathbb{Q}}$, non dégénéré, et dont le support évite l'origine de A, de sorte que les fonctions thêta de diviseur $e_A^* X$ sont analytiques en O. Comme me l'a indiqué L. Breen, on dispose de plus de l'information suivante, établie par Barsotti [1] et, par une voie différente, par Candelera et Cristante ([6], théorème A.4).

LEMME 1 : <u>Il existe une fonction thêta sur</u> $t_A(\mathbb{C})$ <u>de diviseur</u> $e_A^* X$, <u>dont le développement de Taylor en</u> 0 <u>a tous ses coefficients algébriques</u>.

Pour définir plus précisément une fonction thêta associée à X, nous supposerons désormais que A et X vérifient les hypothèses suivantes, où sont reprises les notations K et N de l'introduction :

(C_1) <u>l'anneau des endomorphismes de</u> A <u>contient un ordre</u> $\mathcal{O}$ <u>du corps quadratique imaginaire</u> K.

(C_2) <u>le diviseur</u> X <u>est symétrique</u> ($(-1)^* X = X$) <u>et, pour tout élément</u> α <u>de</u> $\mathcal{O}$, <u>les classes d'équivalence linéaire de</u> $\alpha^* X$ <u>et de</u> $(N\alpha).X$ <u>coïncident</u>.

Du fait de (C_1), il existe un entier ν compris entre 0 et g et une base de t_A pour laquelle l'action des éléments α de K est représentée par :

$$\alpha(z_1,\dots,z_g) = (\alpha z_1,\dots \alpha z_\nu, \bar{\alpha} z_{\nu+1},\dots,\bar{\alpha} z_g).$$

Soient alors θ une fonction thêta répondant aux conditions du lemme 1 et, pour $i=1,\dots,g$, H_i la dérivée logarithmique de θ relativement à $\partial/\partial z_i$. L'hypothèse (C_2) entraîne que, pour $i \leq \nu$ (resp. $i > \nu$) et pour α dans K, les formes différentielles

$$\alpha^* dH_i - \bar{\alpha} dH_i \quad (\text{resp. } \alpha^* dH_i - \alpha\, dH_i)$$

s'identifient, à l'addition de formes exactes près, à des formes de première espèce sur A. La considération de l'action de K sur les classes de cohomologie de formes de deuxième espèce (voir [4], lemme 2) conduit alors à l'énoncé suivant :

LEMME 2 : <u>Sous les hypothèses</u> (C_1) et (C_2), <u>il existe une fonction thêta</u> θ <u>sur</u> $\mathbb{C}^g$ <u>de diviseur</u> $e_A^* X$, <u>paire, valant</u> 1 <u>en</u> 0, <u>dont le développement de Taylor en</u> 0 <u>a tous ses coefficients algébriques, et telle que, pour tout élément</u> α <u>de</u> $\mathcal{O}$, <u>la fonction</u> $\theta(\alpha z)/\theta(z)^{N\alpha}$ <u>définit une fonction</u> $\bar{\mathbb{Q}}$<u>-rationnelle sur</u> A. <u>De plus, toute fonction thêta vérifiant ces conditions est multiple de</u> θ <u>par l'exponentielle d'un polynôme de la forme</u> $\sum_{i\leq\nu}\sum_{j>\nu} a_{ij} z_i z_j$ <u>à coefficients</u> a_{ij} <u>algébriques</u>.

Le cas le plus intéressant pour les applications est celui où $\nu = g$, c'est-à-dire où

(C_3) K <u>agit de façon scalaire sur</u> t_A.

Il existe alors une unique fonction thêta, que nous noterons $\theta_{A,X}$ répondant aux conditions du lemme 2. Comme l'a remarqué D. Masser, A est alors en fait isogène à une puissance d'une courbe elliptique à multiplications complexes par K, et l'étude de $\theta_{A,X}$ équivaut à celle de produits de fonctions $\theta_{E_i}(z_i)$ correspondant aux différents facteurs de A.

§ 2 Résultats de transcendance

Nous supposons que A et X vérifient les conditions (C_1), (C_2) et (C_3) du § 1, et nous notons θ la fonction $\theta_{A,X}$ associée canoniquement à X par le lemme 2.

THEOREME 1 : Soient u *un élément de* $t_A(\mathbb{C})$, *non situé sur le support de* $e_A^* X$, *dont l'image* P *par* e_A *soit un point d'ordre infini de* $A(\bar{\mathbb{Q}})$, *et* B *la plus petite sous-variété abélienne de* A *telle que* $t_B(\mathbb{C})$ *contienne* u. *Si* $\theta(u)$ *est un nombre algébrique, il existe un entier* N *non nul tel que la restriction de* θ^N *à* $t_B(\mathbb{C})$ *s'identifie à une fonction rationnelle sur* B.

La démonstration du théorème 1, qui repose sur le théorème fondamental de Wüstholz [9], reprend les arguments de [3]. En voici le principe.

Notons φ_X l'isogénie canonique de A sur $\mathrm{Pic}^{\circ}(A)$ définie par le diviseur non dégénéré X. En vertu de (C_2), l'involution de Rosati associée à X induit sur K la conjugaison complexe. Soient alors β un élément de $\mathcal{O}$ non rationnel, et G une extension de A par le tore déployé $\mathbb{G}_m^2$, dont la classe d'isomorphisme est paramétrée par le couple $\{\varphi_X(P), \varphi_X(\bar{\beta}P)\}$ de $(\mathrm{Pic}^{\circ}A)^2$. D'après la proposition 1 de [4], les éléments de l'ordre $\mathbb{Z} \oplus \mathbb{Z}\beta$ de K se relèvent de façon unique en des endomorphismes de G . La représentation analytique de End G sur l'espace tangent à l'origine t_G de G , jointe à l'hypothèse (C_3), fournit dans ces conditions un hyperplan t^+ de t_G , défini sur $\bar{\mathbb{Q}}$, sur lequel K agit de façon scalaire.

Supposons alors $\theta(u)$ algébrique. En explicitant l'écriture de l'application exponentielle e_G sur $G(\mathbb{C})$ dans une base convenable de $t_G(\mathbb{C})$ (voir [4], § 9), on vérifie comme dans [3] qu'il existe un élément ℓ de $t^+(\mathbb{C})$ dont l'image par la projection canonique de $t_G(\mathbb{C})$ sur $t_A(\mathbb{C})$ soit u et tel que $e_G(\ell)$ appartienne à $G(\bar{\mathbb{Q}})$. D'après le théorème 1 de [9], il existe donc un sous-groupe algébrique H de G , se projetant sur la sous-variété abélienne B de A , et tel que t_H soit contenu dans t^+. Mais puisque $t^+ \cap t_{\mathbb{G}_m^2}$ n'est pas défini sur $\mathbb{Q}$, l'intersection de H avec $\mathbb{G}_m^2$ est finie, et H est isogène à B. On déduit alors de la remarque de [4], § 2, en désignant par Y la restriction à B du diviseur X, que l'application φ_Y s'annule en tout multiple suffisamment grand de P. Par conséquent, la classe d'équivalence linéaire de Y est un point de torsion de Pic(B), et donc un élément de $\mathrm{Pic}^o(B)$. Or B est, du fait de (C_3), stable sur $\mathcal{O}$, et on peut répéter le raisonnement précédent, en considérant cette fois l'extension de B par $\mathbb{G}_m^2$ paramètrée par le couple $\{Y, \beta^* Y\}$ de $(\mathrm{Pic}^o(B))^2$. Il s'ensuit que la classe d'un multiple non nul N.Y de Y est nulle dans $\mathrm{Pic}^o(B)$, ce qui, en vertu de l'unicité de θ, équivaut à l'énoncé du théorème 1. (On pourrait d'ailleurs, plus directement, invoquer le caractère symétrique du diviseur Y pour conclure.)

Bien entendu, une telle conclusion ne peut avoir lieu si le diviseur non dégénéré X est effectif ; mais les applications que nous avons en vue interdisent de se limiter à cette situation. Enonçons néanmoins :

COROLLAIRE : <u>On reprend les hypothèses du théorème 1, et on suppose de plus</u> X <u>ample. Alors,</u> θ(u) <u>est transcendant.</u>

En particulier, avec les notations de l'introduction : si $u_1, \dots, u_g$ désignent des nombres complexes dont les images par e_E soient des points d'ordre infini de $E(\bar{\mathbb{Q}})$, et si $n_1, \dots, n_g$ sont des entiers rationnels > 0, alors, le nombre $\prod_{i=1}^{g} \theta_E(u_i)^{n_i}$ est transcendant.

§ 3 <u>Hauteurs p-adiques</u>

On reprend les notations de l'introduction concernant la courbe elliptique E, et on fixe un nombre premier p au dessus duquel E a bonne réduction ordinaire, ainsi qu'un plongement de $\bar{\mathbb{Q}}$ dans le corps p-adique $\mathbb{C}_p$. On désigne par $\mathfrak{p}$ et $\bar{\mathfrak{p}}$ les idéaux de K au dessus de p, par S l'ensemble des plongements de F dans $\bar{\mathbb{Q}}$, et par S^+ le sous-ensemble de S correspondant aux places de F au dessus de $\mathfrak{p}$. Pour tout élément σ de S, soient θ_σ la fonction analytique sur un sous-groupe $\mathfrak{D}_\sigma$ de $\mathbb{C}_p$ que définit la série de Taylor en 0 de la fonction thêta canonique θ_{E^σ} associée à l'image E^σ de E par σ, et e_σ une exponentielle p-adique de $\mathfrak{D}_\sigma$ vers le groupe $E^\sigma(\mathbb{C}_p)$. Les courbes E^σ étant isogènes sur $\bar{\mathbb{Q}}$, on peut sans restreindre la généralité supposer que les corps $\bar{\mathbb{Q}}(e_\sigma(z))$ définissent des extensions algébriques d'un corps de fonctions elliptiques indépendant de σ.

Le théorème 1 admet une version p-adique, dont la démonstration est identique à celle du § 2, et qu'on peut énoncer de la façon suivante.

THEOREME 2 : <u>Pour tout élément</u> σ <u>de</u> S, <u>soient</u> u_σ <u>un point non nul de</u> $\mathfrak{D}_\sigma$ <u>dont l'image par</u> e_σ <u>appartienne à</u> $E^\sigma(\bar{\mathbb{Q}})$, <u>et</u> n_σ <u>un entier rationnel. On</u>

note T *le plus petit sous-espace vectoriel de* $\prod_{\sigma \in S} (\mathbb{Q} \otimes \mathcal{D}_\sigma)$ *défini sur* K *et contenant le point* $(u_\sigma ; \sigma \in S)$, *et on suppose que* $\prod_{\sigma \in S} \theta_\sigma(u_\sigma)^{n_\sigma}$ *est algébrique. Alors, une puissance non nulle de la restriction à* T *de la fonction* $\prod_{\sigma \in S} \theta_\sigma(z_\sigma)^{n_\sigma}$ *coïncide avec la restriction à* T *d'une fonction rationnelle en* $\{e_\sigma(z_\sigma) ; \sigma \in S\}$.

Considérons dans ces conditions la hauteur p-adique usuelle $h_\wp$ sur le groupe E(F) associée à l'idéal $\wp$. A l'addition de logarithmes ($\wp$-adiques) d'expressions algébriques près, elle est donnée (voir [2], et [8], chapitre III) par le logarithme de la fonction

$$\prod_{\sigma \in S^+} \theta_\sigma \circ \ell_\sigma$$

où ℓ_σ désigne le logarithme du groupe $E^\sigma(\mathbb{C}_p)$. Si donc $h_\wp$ s'annule en un point P de E(F), un produit de puissances de fonctions $\theta_\sigma \circ \ell_\sigma$, affectées d'exposants $\geqslant 0$ et non tous nuls, prend en chaque multiple entier de P où il est défini une valeur algébrique. D'après le théorème 2 (ou, plus directement, l'analogue p-adique du corollaire au théorème 1), P est alors nécessairement d'ordre fini. Quitte à étendre le corps F, on a ainsi démontré :

COROLLAIRE 1 : Soit P un point de $E(\bar{\mathbb{Q}})$ de hauteur $h_\wp(P)$ nulle. Alors, P est un point de torsion de E.

L'application $h_\wp$, et son homologue $h_{\bar{\wp}}$, s'interprètent également comme les hauteurs associées aux caractères κ et $\bar{\kappa}$ du groupe $Gal(\bar{\mathbb{Q}}/F)$ définis par les composantes $\wp$-primaires et $\bar{\wp}$-primaires du sous-groupe de torsion de E (voir [8]). Le raisonnement précédent s'étend sans difficulté à l'étude des hauteurs $h_{a,b}$ associées aux caractères $\kappa^a \bar{\kappa}^b$, pourvu que a et b soient deux entiers rationnels > 0 : leurs seuls vecteurs isotropes sur $E(\bar{\mathbb{Q}})$ sont les points de torsion. En revanche, des dégénérescences peuvent apparaître lorsque a et b sont de signes opposés, le diviseur correspondant sur $\prod_{\sigma \in S} E^\sigma$ n'étant alors plus effectif. Le théorème 2 permet encore de les décrire. On obtient ainsi, en se restreignant, pour alléger l'énoncé, au cas où K est principal :

COROLLAIRE 2 : On suppose E définie sur $\mathbb{Q}$. Soient (a,b) un élément de $\mathbb{Z}^2$ non nul et P un point d'ordre infini de E(K) tel que $h_{a,b}(P) = 0$. Alors, a+b est nul, et le conjugué complexe de P est égal à un multiple de P par une racine de l'unité de K.

Le théorème 2, appliqué aux fonctions de la forme $\theta(z_1+z_2)/\theta(z_1)\theta(z_2)$, fournit également des renseignements sur la forme bilinéaire $\langle , \rangle_{\wp}$ associée à la hauteur $h_{\wp}$. En conservant l'hypothèse K principal, on peut ainsi énoncer :

COROLLAIRE 3 : <u>On suppose</u> E <u>définie sur</u> $\mathbb{Q}$. <u>Soient</u> P_1 <u>et</u> P_2 <u>deux points d'ordre infini de</u> E(K) <u>tels que</u> $\langle P_1, P_2\rangle_{\wp}$ <u>soit nul</u>. <u>Alors</u>, P_2 <u>est égal à l'image de</u> P_1 <u>sous l'action d'un élément de</u> K <u>de trace nulle</u>.

Mais la méthode exposée ici ne permet bien entendu d'étudier ni les hauteurs $h_{a,b}$ associées à des couples entiers p-adiques de rapport irrationnel, ni la non-dégénérescence de la forme $\langle , \rangle_{\wp}$ sur le groupe $E(\overline{\mathbb{Q}}) \otimes \mathbb{Z}_p$.

BIBLIOGRAPHIE

[1] I. BARSOTTI : Considerazioni sulle funzione theta ; Symp. Math., 3, 1970, 247-277.

[2] D. BERNARDI : Hauteur p-adique sur les courbes elliptiques ; Birkhäuser Prog. Math., 12, 1981, 1-14.

[3] D. BERTRAND : Valeurs de fonctions thêta et hauteurs p-adiques ; Birkhäuser Prog. Math., 22, 1982, 1-11.

[4] D. BERTRAND : Endomorphismes de groupes algébriques ; applications arithmétiques ; Birkhäuser Prog. Math., 31, 1983, 1-45.

[5] D. BERTRAND : Fonctions thêta à multiplications complexes, en préparation.

[6] M. CANDELERA - V. CRISTANTE : Biextensions associated to divisors an abelian varieties and theta functions ; preprint, Univ. Padova, 1983.

[7] D. MASSER - G. WÜSTHOLZ : Zero estimates on group varieties I ; Invent. Math., 64, 1981, 489-516.

[8] B. PERRIN-RIOU : Arithmétique des courbes elliptiques et théorie d'Iwasawa ; thèse, Univ. Orsay, 1983.

[9] G. WÜSTHOLZ : Some remarks on a conjecture of Waldschmidt, Birkhäuser Prog. Math., 31, 1983, 329-336.

Université de Paris VI
Mathématiques, T.46
4, Place Jussieu
75230 Paris - cedex 05
(France)

MULTIPLICATIVE GALOIS STRUCTURE

Ted Chinburg

Department of Mathematics

University of Pennsylvania

I. Introduction.

Let N/K be a finite Galois extension of number fields with group G = Gal(N/K). Define $\mathcal{O}_N$ to be the integers of N. E. Noether proved that $\mathcal{O}_N$ is a projective G-module if and only if N/K is at most tamely ramified. One then has a class $\Omega_a(N/K) = (\mathcal{O}_N) - [K:Q](Z[G])$ in the Grothendieck group $K_0(Z[G])$ of finitely generated G-modules of finite projective dimension. Motivated by work of A. Fröhlich [8], H. Stark [16] and J. Tate [17], we began in [5] a unified theory of the G-structure of $\mathcal{O}_N$ and of the group $U_{N,S}$ of S-units of N when S is a sufficiently large finite set of places of N stable under the action of G.

In this theory, one has for all N/K, not only those which are tamely ramified, a class $\Omega_m(N/K)$ in $K_0(Z[G])$ which measures the G-structure of $U_{N,S}$ for all sufficiently large S stable under G. The theory is based on a parallel between Galois Gauss sums, which Fröhlich has related in $\Omega_a(N/K)$, and the leading terms in the expansions at s = 0 of Artin L-functions, which are the subject of Stark's conjectures.

In this note we recall the definition of $\Omega_m(N/K)$ and discuss a conjecture which would give an exact expression for it in terms of the root numbers of the symplectic rep-presentations of G. This conjecture, and M. Taylor's recent proof of Fröhlich's conjecture ([19]) about $\Omega_a(N/K)$, have the suprising implication that $\Omega_m(N/K) = \Omega_a(N/K)$ if N/K is tame. We will report on a proof of our conjecture for some tame extensions N of K = Q in which G is the quaternion group H_8 of order eight. One of these extensions provides the first known example in which $\Omega_m(N/K)$ is non-trivial.

These results were found using a general algorithm for computing $\Omega_m(N/K)$ which we will sketch. Further details and applications of this algorithm will be given in subsequent papers.

II. Definition of $\Omega_m(N/K)$.

Let S be a finite set of places of N stable under G. Unless otherwise specified, we will assume that S satisfies the following two conditions:

(2.1) S contains the archimedean places of N and those places which are ramified over K, and

(2.2) the S-class number of every subfield of N containing K is 1.

Given N/K, one can always find a set of places S satisfying these conditions.

Let $Y = Y_{N,S}$ be the free abelian group on S, and let $X = X_{N,S}$ be the kernel of the homomorphism $Y \longrightarrow Z$ which sends each $v \in S$ to 1. Let U denote $U_{N,S}$. For S as above, J. Tate proved in [18] that there is a 'canonical class' $\alpha(N/K,S)$ in $Ext_G^2(X,U)$ which is represented by an exact sequence

$$(2.3) \qquad 0 \longrightarrow U \longrightarrow A \longrightarrow B \longrightarrow X \longrightarrow 0$$

in which A and B are finitely generated G-modules of finite projective dimension.

In [5] we showed that $\Omega_m = (A) - (B)$ is a torsion class in $K_0(Z[G])$, and that Ω_m is independent of the choice of an S as above and of the choice of a sequence (2.3) with class $\alpha(N/K,S)$. Thus on defining $\Omega_m(N/K) = \Omega_m$, one arrives at an invariant which depends only on N/K.

III. A conjecture.

In [4] we recalled the definition of a class $q(W'(N/K))$ in $K_0(Z[G])$ which depends only on the root numbers of the symplectic representations of G. This class was first defined for tamely ramified N/K by Ph. Cassou-Noguès in [2] (c.f. [9, p. 16]). It was first defined for arbitrary N/K by A. Frohlich in [10]. The class appears also in work of J. Queyrut ([13], and more recently, [14]). The class $q(W'(N/K))$ has order one or two, and is trivial if all of the symplectic root numbers above are 1.

Conjecture 1 (c.f. [4, 'Question']): $\Omega_m(N/K) = q(W'(N/K))$.

Let T be a subgroup of $K_0(Z[G])$. It is natural to consider also the following possibly weaker form of Conjecture 1.

Conjecture 1 mod T : $\Omega_m(N/K) \equiv q(W'(N/K)) \mod T$ in $K_0(Z[G])$.

The evidence that we now have for Conjecture 1, and for Conjecture 1 mod T for various T, is as follows.

A. Let D(Z[G]) be the kernel subgroup of $K_0(Z[G])$ (c.f. [20]). In [5] we showed that Conjecture 1 mod D(Z[G]) is a consequence of a strong form of Stark's conjecture on Artin L-functions at s = 0. This form of Stark's conjecture has been proved by J. Tate in [17] for the L-functions of representations with rational character, and is analogous to conjectures proposed by S. Lichtenbaum in [11].

B. Let $G_0(Z[G])$ be the Grothendieck group of all finitely generated G-modules. Let $h : K_0(Z[G]) \longrightarrow G_0(Z[G])$ be the Cartan ('forgetful') homomorphism. Suppose that $\mathcal{L}$ is a finite set of rational primes. Define $G_0^{\mathcal{L}}(Z[G])$ to be the subgroup of $G_0(Z[G])$ generated by the classes of finite modules supported on $\mathcal{L}$. Define $h_{\mathcal{L}} : K_0(Z[G]) \longrightarrow G_0(Z[G])/G_0^{\mathcal{L}}(Z[G$ to be the composition of h with the natural quotient map.

Theorem 1 ([6], [5]): *Suppose* K = Q, G *is abelian*, N *is either real or cyclotomic, and that* $\mathcal{L}$ *contains* 2 *and the prime divisors of the order of* G. *Then Conjecture* 1 *mod* ker($h_{\mathcal{L}}$) *is true for* N/K.

This result follows from work of B. Mazur and A. Wiles in [12]. If one assumes that the 'Gras Conjecture' is true at the prime 2, which will almost surely be proved by the methods of Mazur and Wiles, then one can remove the hypothesis that $\mathcal{L}$ contains 2.

Conjecture 1 states $\Omega_m(N/K) = 0$ if G is abelian (compare

[5, Question 3.1]). J. Queyrut has shown (c.f. [5, Corollary 3.4]) that Conjecture 1 mod ker($h_{\mathcal{L}}$) is equivalent to $h_{\mathcal{L}}(\Omega_m(N/K)) = 0$ for all N/K and all $\mathcal{L}$.

C. We showed in [4] that Conjecture 1 is functorial with respect to restriction and coinflation. This implies that if L/F is a finite Galois extension such that $F \subseteq K \subseteq N \subseteq L$, and Conjecture 1 is true for L/F, then Conjecture 1 is true for N/K. This fact, together with relations between the $K_0(Z[H])$ of subquotients H of G, allows one to conclude that Conjecture 1 is true for N/K in certain non-trivial cases provided N/K may be embedded in a suitably large extension L/F. It also shows that conjectures weaker than Conjecture 1 for L/F may imply that Conjecture 1 is true for N/K. For further details and examples, see [4] and [6].

D. Conjecture 1 makes the following prediction when G is the quaternion group H_8 of order eight. Let V be an irreducible two-dimensional representation of G. Then V is symplectic and is unique up to isomorphism. The root number W(V) of V is ± 1. The torsion subgroup of $K_0(Z[G])$ has order two, and may hence be identified with $\{\pm 1\}$.

<u>Conjecture</u> H_8 : <u>For</u> N/K, G <u>and</u> V <u>as</u> <u>above</u>, $\Omega_m(N/K) = W(V)$.

If N/K is at most tamely ramified, A. Fröhlich proved in [7] for K = Q, and in [8] for all K, that $\Omega_a(N/K) = W(V)$.

For r = 7 and r = 43, define N_r to be the unique H_8 extension of Q which contains $Q(\sqrt{5})$ and which is ramified at exactly 3, 5, r and ∞ . Define V_r to be the representation

V for $N/K = N_r/Q$. Notice that N_r/Q is tamely ramified, so that $\Omega_a(N/K) = \pm 1$ is well-defined.

<u>Theorem</u> 2 : $\Omega_m(N_7/Q) = \Omega_a(N_7/Q) = W(V_7) = -1$ <u>and</u>

$$\Omega_m(N_{43}/Q) = \Omega_a(N_{43}/Q) = W(V_{43}) = 1.$$

By the functorality of $\Omega_m(N/K)$ with respect to inflation, $\Omega_m(N/Q) \neq 0$ for all finite Galois extensions N of Q which contain N_7.

IV. <u>An algorithm for computing</u> $\Omega_m(N/K)$.

Let J be the group of S-ideles of N, and let C be the group of idele classes of N. In [18], J. Tate defines classes $\alpha_2(N/K,S) \in H^2(G,\mathrm{Hom}(Y,J))$ and $\alpha_1(N/K) \in H^2(G,\mathrm{Hom}(Z,C)) = H^2(G,C)$ The class $\alpha_2(N/K,S)$ results from the local canonical classes of places in S, while $\alpha_1(N/K)$ is the global canonical class of N/K.

One may construct a diagram of finitely generated G-modules

$$\begin{array}{ccccccccc}
 & & 0 & & 0 & & 0 & & 0 & & \\
 & & \downarrow & & \downarrow & & \downarrow & & \downarrow & & \\
0 & \longrightarrow & X(2) & \xrightarrow{\partial_X} & F_X & \longrightarrow & F'_X & \longrightarrow & X & \longrightarrow & 0 \\
 & & \downarrow & & \downarrow & & \downarrow & & \downarrow & & \\
0 & \longrightarrow & Y(2) & \xrightarrow{\partial_Y} & F_Y & \longrightarrow & F'_Y & \longrightarrow & Y & \longrightarrow & 0 \\
 & & \downarrow & & \downarrow & & \downarrow & & \downarrow & & \\
0 & \longrightarrow & Z(2) & \xrightarrow{\partial_Z} & F_Z & \longrightarrow & F'_Z & \longrightarrow & Z & \longrightarrow & 0 \\
 & & \downarrow & & \downarrow & & \downarrow & & \downarrow & & \\
 & & 0 & & 0 & & 0 & & 0 & &
\end{array}$$

in which all the modules in the middle two columns are free. The theory of canonical classes implies that there is a diagram

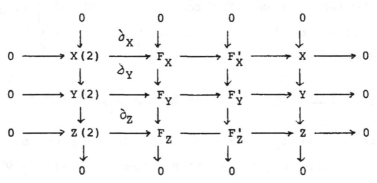

$$
\begin{array}{ccccccccc}
0 & \longrightarrow & X(2) & \longrightarrow & Y(2) & \longrightarrow & Z(2) & \longrightarrow & 0 \\
 & & \downarrow f_3 & & \downarrow f_2 & & \downarrow f_1 & & \\
0 & \longrightarrow & U & \longrightarrow & J & \longrightarrow & C & \longrightarrow & 0
\end{array} \tag{4.1}
$$

in which f_2 (resp. f_1) represents $\alpha_2(N/K,S) \in \mathrm{Ext}^2_G(Y,J)$ (resp. $\alpha_1(N/K) \in \mathrm{Ext}^2_G(Z,C)$). These conditions imply that f_3 represents $\alpha(N/K,S)$.

To construct explicitly such f_3, f_2 and f_1, one begins by finding $f_2' \in \mathrm{Hom}_G(Y(2),J)$ (resp. $f_1' \in \mathrm{Hom}_G(Z(2),C)$) representing $\alpha_2(N/K,S)$ (resp. $\alpha_1(N/K)$). An f_2' may be found from the explicit representatives for local canonical classes given by Serre in [15, p. 210]. An f_1' may be found by the methods of Artin and Tate in [1]. This is greatly simplified if (as in our examples) G is the decomposition group of some place v in S. Then one can compute a representative two-cocycle for the global canonical class from a two-cocycle for the local class at v. One then finds $g_2 \in \mathrm{Hom}_G(F_Y,J)$ and $g_1 \in \mathrm{Hom}_G(F_Z,C)$ such that when $f_2 = f_2' + \partial_Y g_2$ and $f_1 = f_1' + \partial_Z g_1$, one has a diagram (4.1) for some f_3.

Adding a large, finitely generated free module F to X(2), and defining f_3 on F appropriately, one may find an exact sequence

$$0 \longrightarrow P \longrightarrow X(2) \oplus F \xrightarrow{f_3} U \longrightarrow 0$$

This P must be projective; let n be its rank. One finds $n(Z[G]) - (P) = \Omega_m(N/K)$ in $K_0(Z[G])$. The class $\Omega_m(N/K)$ may then be computed by the methods of [20] or [8].

In practice, it is necessary to have only that the cokernel of $f_3 : X(2) \oplus F \longrightarrow U$ is cohomologically trivial.

The class of the cokernel in $K_0(Z[G])$ then enters into $\Omega_m(N/K)$. Furthermore, one may replace condition (2.2) by the weaker condition

(4.2) <u>The</u> S-<u>class number of every subfield of</u> N <u>containing</u> K <u>is relatively prime to the order of</u> G.

The class in $K_0(Z[G])$ of the S-class group $Cl_S(N)$ of N then enters into $\Omega_m(N/K)$. As we will show in subsequent papers, this leads to formulas for $\Omega_m(N/K)$ which contain the leading terms in the expansions at s = 0 of Artin L-functions. This in turn leads to a 'method of congruences' in the multiplicative theory which is analogous to the corresponding method in the additive theory (c.f. [9, p. 16]). In the multiplicative (resp. additive) case, one studies $\Omega_m(N/K)$ (resp. $\Omega_a(N/K)$) via congruences satisfied by the algebraic constants in Stark's conjectures (resp. satisfied by Galois Gauss sums). These methods are also available to study the counterpart of $\Omega_m(N/K)$ for finite Galois extensions of global function fields ([3]).

We will conclude with a table of numerical data concerning the examples N_7/Q and N_{43}/Q discussed in §III. In each case, let S be the set of places of N_r which are infinite or ramified over K = Q. The S-units of N_r are generated by -1, ε_5, ε_{3r}, ε_{15r}, a_3, a_r, a_q, a_q^y and $\sqrt{\xi}$, where

$$\xi = -\varepsilon_5\varepsilon_{3r}\varepsilon_{15r}\sqrt{15\cdot r}.$$

The field N_7 satisfies (2.2), while N_{43} satisfies (4.2).

TABLE : S-units, class number and S-class number of N_r

	N_7	N_{43}
ε_5	$(1+\sqrt{5})/2$	$(1+\sqrt{5})/2$
ε_{3r}	$(5+\sqrt{21})/2$	$16855-1484\sqrt{129}$
ε_{15r}	$\sqrt{21}+2\sqrt{5}$	$(5\sqrt{5}+\sqrt{129})/2$
a_3	$(3+\sqrt{21})/2$	$-159+14\sqrt{129}$
a_r	$(7+\sqrt{21})/2$	$602-53\sqrt{129}$
a_q	$(1+\sqrt{21}+2\sqrt{5})/2$	$(91+8\sqrt{129}+34\sqrt{5}+3\sqrt{645})/2$
a_q^y	$(1-\sqrt{21}+2\sqrt{5})/2$	$(91-8\sqrt{129}+34\sqrt{5}-3\sqrt{645})/2$
$\sqrt{\xi}$	$\sqrt{-\varepsilon_5\,\varepsilon_{15r}\,a_p\,a_r\,a_q\,a_q^y}$	$\sqrt{\varepsilon_5\,\varepsilon_{15r}\,a_p\,a_r\,a_q\,a_q^y}$
class number	8	$8\cdot 7^2$
S-class number	1	7^2

REFERENCES

1. Artin, E. and Tate, J.: Class Field Theory. Benjamin : New York (1967).
2. Cassou-Nogues, Ph.: Quelques theorèmes de base normale d'entiers. Ann. Inst. Fourier 28 (1978), 1-33.
3. Chinburg, T.: Letter to D. Hayes, 9/15/1983.
4. Chinburg, T.: Multiplicative Galois Module Structure. to appear in the J. of the London Math. Soc.
5. Chinburg, T.: On the Galois structure of algebraic integers and S-units. Inv. Math. (in press).

6. Chinburg, T.: The Galois structure of S-units. to appear in the Sem. de Theorie de Nombres de Bordeaux.

7. Fröhlich, A.: Artin root numbers and normal integral bases for quaternion fields. Inv. Math. 17(1972), 143-166.

8. Frohlich, A.: Galois module structure, in "Algebraic Number Fields," Proceedings of the Durham Symposium 1975. Academic Press : London 1977, 133-191.

9. Fröhlich, A.: Galois module structure of algebraic integers. Springer-Verlag : Berlin, Heidelberg, New York, Tokyo (1983).

10. Fröhlich, A.: Some problems of Galois module structure for wild extensions. Proc. London Math. Soc. 37(1978), 193-212.

11. Lichtenbaum, S.: Letter to J. Tate (1978).

12. Mazur, B. and Wiles, A.: Class fields of abelian extensions of Q. Inv. Math. (in press).

13. Queyrut, J.: Sommes de Gauss et structure Galoisienne des anneaux d'entiers. Sem. de Theorie de Nombres de Bordeaux, exp. 20, (1981-1982).

14. Queyrut, J.: Structure Galoisienne des groupes d'unites et des group des classes d'un anneau d'entiers. Sem. de Theorie de Nombres de Bordeaux, exp. 2, (1982-1983).

15. Serre, J. P.: Corps Locaux. 2nd ed. Paris : Hermann (1968).

16. Stark, H. M.: L-functions at s = 1. I, II, III, IV. Advances in Math. 7, 301-343(1971); 17, 60-92(1975); 22, 64-84(1976); 35, 197-235(1980).

17. Tate, J.: Les conjectures de Stark sur les fonctions L d'Artin an s = 0; notes d'un cours a Orsay redigees par D. Bernardi et N. Schappacher. to appear.

18. Tate, J.: The cohomology groups of tori in finite Galois extensions of number fields. Nagoya Math. J. 27, 709-719 (1966).

19. Taylor, M. J.: On Frohlich's conjecture for rings of integers of tame extensions. Inv. Math. 63(1981), 41-79.

20. Ullom, S. V.: A survey of class groups of integral group rings, in "Algebraic Number Fields (L-functions and Galois properties)," 709-719. New York : Academic Press (1977).

<u>Note added in proof</u> (3/84) : It has now been shown that Conjecture H_8 is true for an infinite number of examples (to appear).

HEURISTICS ON CLASS GROUPS OF NUMBER FIELDS

by

H. COHEN and H. W. LENSTRA, Jr.

-:-:-:-

§ 1. - Motivations

The motivation for this work came from the desire to understand heuristically (since proofs seem out of reach at present) a number of experimental observations about class groups of number fields, and in particular imaginary and real quadratic fields. In turn the heuristic explanations that we obtain may help to find the way towards a proof.

Three of these observations are as follows :

A/ The odd part of the class group of an imaginary quadratic field seems to be quite rarely non cyclic.

B/ If p is a small odd prime, the proportion of imaginary quadratic fields whose class number is divisible by p seems to be significantly greater than $1/p$ (for instance 43 % for $p = 3$, 23.5 % for $p = 5$).

C/ It seems that a definite non zero proportion of real quadratic fields of prime discriminant (close to 76 %) has class number 1 , although it is not even known whether there are infinitely many.

The main idea, due to the second author, is that the scarcity of noncyclic groups can be attributed to the fact that they have many automorphisms. This naturally leads to the heuristic assumption that isomorphism classes G of abelian groups should be weighted with a weight proportional to $1/\#\,\mathrm{Aut}\,G$. This is a very natural and common weighting factor, and it is the purpose of this paper to show that the assumption above, plus another one to take into account the units, is sufficient to give very satisfactory heuristic answers of quantitative type to most natural questions about class groups. For example we find that the class number of an imaginary quadratic field should be divisible by 3 with probability close to 43.987 %, and that the proportion of real quadratic fields with class number one (having prime discriminant) should be close to 75.446 %.

To distinguish clearly between theorem and conjectural statements, this paper can be considered as having two parts. In the first part (§ 2 to § 7) we give theorems about finite modules over certain Dedekind domains. The second part (§ 8 to § 10) explains in detail the heuristic assumptions that we make, and gives a large sample of conjectures which follow from these heuristic assumptions using the theory developed in the first part.

§ 2. - Notations

In what follows, A will be the ring of integers of a number field. It will be seen that more general Dedekind domains can be used, and also direct products of such, but for simplicity we will assume that A is as above. The special case $A = \mathbb{Z}$ is of particular importance. We denote by $\mathcal{P}$ the set of non zero prime ideals of A, and if $\mathfrak{p} \in \mathcal{P}$, the norm of $\mathfrak{p}$ is by definition $N\mathfrak{p} = \#(A/\mathfrak{p})$. The letter $\mathfrak{p}$ will be used only for elements of $\mathcal{P}$.

. If G_1 and G_2 are A-modules, we write $G_1 \leq G_2$ to mean that G_1 is a submodule of G_2.

. If $\mathfrak{p} \in \mathcal{P}$ and G is a finite A-module, then we write $r_{\mathfrak{p}}(G)$ for the $\mathfrak{p}$-rank of G, i.e. the dimension of $G/\mathfrak{p}G$ as an $A/\mathfrak{p}$-vector space.

. k will be a non negative integer or ∞. If $k \neq \infty$ and G is a finite A-module $s_k(G)$ (or $s_k^A(G)$ when the ring A must be specified) will be the number of surjective A-homomorphisms from A^k to G.

. If G is a finite A-module we define the k-weight $w_k(G)$ of G, and the weight $w(G)$ of G as follows :

$$w_k(G) = s_k(G)\,(\# G)^{-k}\,(\# \operatorname{Aut} G)^{-1}$$

$$w(G) = w_\infty(G) = (\# \operatorname{Aut} G)^{-1}$$

where $\operatorname{Aut} G = \operatorname{Aut}_A G$ is the group of A-automorphisms of G.

. For $\mathfrak{p} \in \mathcal{P}$ we set $\eta_k(\mathfrak{p}) = \prod_{1 \leq i \leq k} (1 - N\mathfrak{p}^{-i})$, $\eta_\infty(\mathfrak{p}) = \prod_{1 \leq i} (1 - N\mathfrak{p}^{-i})$.

. If $0 \le b \le a$ with a, b integers we define

$$\left[\begin{smallmatrix} a \\ b \end{smallmatrix}\right]_p = \eta_a(p)/(\eta_{a-b}(p)\,\eta_b(p))$$

and if $a \ge 0$ but $b<0$ or $b>a$, we set $\left[\begin{smallmatrix} a \\ b \end{smallmatrix}\right]_p = 0$.

. Let $\zeta_A(s)$ be the Dedekind zeta function of the ring A. Then we set

$$C_k = \varkappa \prod_{2 \le i \le k} \zeta_A(i) \quad , \quad C_\infty = \varkappa \prod_{2 \le i} \zeta_A(i)$$

where $\varkappa = \varkappa_A$ is the residue at $s=1$ of the function $\zeta_A(s)$ (see also section 7).

. We will need the well known notion which generalises to finite A-modules G the notion of cardinality for finite $\mathbb{Z}$-modules. This has several names in the literature (1st Fitting ideal [10], 0th determinantial ideal [2] for example). We will call it the A-cardinal of G, and write $\chi_A(G)$ or $\chi(G)$ as in [12]. It is an ideal of A which can be defined as follows : every finite A-module G can be written in a non canonical way

$$G = \bigoplus_i A/\mathfrak{a}_i \qquad (\mathfrak{a}_i \text{ ideals in } A) .$$

Then we set $\chi_A(G) = \prod_i \mathfrak{a}_i$, and this <u>is</u> canonical and does not depend on the decomposition. In the case $A = \mathbb{Z}$, $\chi_{\mathbb{Z}}(G) = n\mathbb{Z}$ where $n = \# G$. In the general case, $\# G = N(\chi_A(G))$.

. We shall use the notations

$\sum_{G(\mathfrak{a})}$ as an abbreviation for $\sum_{G \text{ up to A-isomorphism}, \ \chi_A(G) = \mathfrak{a}}$

$\sum_{G(\mathfrak{a}),\, \varphi_u}$ as an abbreviation for $\sum_{G \text{ up to A-isomorphism}, \ \chi_A(G) = \mathfrak{a},\ \varphi \in \mathrm{Hom}_A(A^u, G)}$.

. We define the k-<u>weight</u> $w_k(\mathfrak{a})$ of an integral ideal $\mathfrak{a}$ as follows :

$$w_k(\mathfrak{a}) = \sum_{G(\mathfrak{a})} w_k(G)$$

and we set $w(\mathfrak{a}) = w_\infty(\mathfrak{a})$.

. The letter u will be used to denote a non negative integer, and it will be in our applications the A-rank of a certain group of units (see section 8).

§ 3. - Fundamental properties of the functions $w_k(G)$ and $w_k(\mathfrak{a})$

We first show :

PROPOSITION 3.1. - Let J be a projective A-module of rank k and G a finite A-module. Set $\chi_A(G) = \mathfrak{a}$

i) The number of surjective A-homomorphisms from J to G is equal to $s_k(G)$

ii) $s_k(G) = (N\mathfrak{a})^k \prod_{\mathfrak{p}|\mathfrak{a}} \eta_k(\mathfrak{p})/\eta_{k-r_\mathfrak{p}(G)}(\mathfrak{p})$

$w_k(G) = \left(\prod_{\mathfrak{p}|\mathfrak{a}} \eta_k(\mathfrak{p})/\eta_{k-r_\mathfrak{p}(G)}(\mathfrak{p})\right)(\# \operatorname{Aut} G)^{-1}$

iii) $\# \{H \le J : J/H \simeq G\} = (N\mathfrak{a})^k w_k(G)$

iv) $\lim_{k\to+\infty} w_k(G) = w(G)$.

Proof. - i) By inverting all the prime ideals which are not in $\mathfrak{a}$, one easily sees that G is unchanged and A becomes a semilocal Dedekind ring and in particular is a principal ideal ring. In that case i) is trivial since $J \simeq A^k$ as an A-module.

ii) It is easy to check that $s_k(G) = \prod_{\mathfrak{p}|\mathfrak{a}} s_k(G_\mathfrak{p})$ where $G_\mathfrak{p}$ is the $\mathfrak{p}$-component of G (note that $G_\mathfrak{p}$ is non trivial if and only if $\mathfrak{p}|\mathfrak{a}$). Hence we may assume that G is a $\mathfrak{p}$-group. Then we know (e.g. see [1], § 3, prop. 11) that if $\varphi \in \operatorname{Hom}_A(A^k, G)$, φ is surjective if and only if $\bar{\varphi}$ is surjective, where $\bar{\varphi} \in \operatorname{Hom}_{A/\mathfrak{p}}((A/\mathfrak{p})^k, G/\mathfrak{p}G)$ is obtained by reduction mod. $\mathfrak{p}$. Hence it follows that :

$$s_k^A(G) = s_k^{A/\mathfrak{p}}(G/\mathfrak{p}G)\, \# \{\varphi \in \operatorname{Hom}_A(A^k, G) / \bar{\varphi} = 0\} .$$

If we set $r = r_\mathfrak{p}(G)$, then $s_k^{A/\mathfrak{p}}(G/\mathfrak{p}G)$ is equal to the number of $k \times r$ matrices of rank r over $A/\mathfrak{p}$, i.e. to the number of $A/\mathfrak{p}$-linearly independent r-tuples $(v_1, \dots, v_r)$, where $v_i \in (A/\mathfrak{p})^k$. Since a vector space of dimension i over $A/\mathfrak{p}$ has $(N\mathfrak{p})^i$ elements, we obtain :

$$s_k^{A/\mathfrak{p}}(G/\mathfrak{p}G) = \prod_{0 \le i < r} (N\mathfrak{p}^k - N\mathfrak{p}^i) = (N\mathfrak{p})^{kr}\, \eta_k(\mathfrak{p})/\eta_{k-r}(\mathfrak{p}) .$$

On the other hand $\bar{\varphi} = 0 \Leftrightarrow \forall v \in A^k,\ \varphi(v) \in \mathfrak{p}G$, and so

$$\# \{\varphi \in \operatorname{Hom}_A(A^k, G) / \bar{\varphi} = 0\} = (\# \mathfrak{p}G)^k = \left(\frac{\# G}{\# G/\mathfrak{p}G}\right)^k = \frac{(N\mathfrak{a})^k}{(N\mathfrak{p})^{kr}}$$

and proposition 3.1 ii) follows.

iii) Follows from the fact that if M and N are two A-modules and $\varphi_1, \varphi_2 \in \mathrm{Hom}_A(M,N)$ are surjective then

$$\mathrm{Ker}\,\varphi_1 = \mathrm{Ker}\,\varphi_2 \Leftrightarrow \exists \sigma \in \mathrm{Aut}\,N \text{ such that } \varphi_2 = \sigma \circ \varphi_1 .$$

Finally iv) is a trivial consequence of ii).

PROPOSITION 3.2. - For $k_1, k_2 \neq \infty$ and G a finite A-module :

$$s_{k_1+k_2}(G) = \sum_{G_1 \leq G} s_{k_1}(G_1)\, s_{k_2}(G/G_1)\, (\# G_1)^{k_2} .$$

Proof. - Write $A^{k_1+k_2}$ as $A^{k_1} \times A^{k_2}$. It is clear that :

$$s_{k_1+k_2}(G) = \sum_{G_1 \leq G} \# \{\varphi \in \mathrm{Hom}_A(A^{k_1+k_2}, G) / \varphi \text{ surjective, } \varphi(A^{k_1}) = G_1\}$$

$$= \sum_{G_1 \leq G} \sum_{\substack{\varphi_1 \in \mathrm{Hom}(A^{k_1}, G_1) \\ \varphi_1 \text{ surjective}}} \# \{\varphi \in \mathrm{Hom}_A(A^{k_1+k_2}, G) / \varphi \text{ surjective, } \varphi|_{A^{k_1}} = \varphi_1\} .$$

It will thus suffice to prove the following lemma :

LEMMA 3.3. - If $\varphi_1 \in \mathrm{Hom}_A(A^{k_1}, G_1)$, φ_1 surjective, then

$$\# \{\varphi \in \mathrm{Hom}_A(A^{k_1+k_2}, G) / \varphi \text{ surjective, } \varphi|_{A^{k_1}} = \varphi_1\} = s_{k_2}(G/G_1)(\# G_1)^{k_2} .$$

Proof. - Put $E = \{\varphi \in \mathrm{Hom}_A(A^{k_1+k_2}, G) / \varphi \text{ surjective, } \varphi|_{A^{k_1}} = \varphi_1\}$ and $F = \{\overline{\varphi}_2 \in \mathrm{Hom}_A(A^{k_2}, G/G_1),\ \overline{\varphi}_2 \text{ surjective}\}$.
Then it is not difficult to check that the natural map obtained by restricting to A^{k_2} and then reducing mod. G_1 is indeed a map from E to F . Furthermore, by writing down explicitly a set of representatives in G of G/G_1, one can also easily see that every $\varphi_2 \in F$ has exactly $(\# G_1)^{k_2}$ preimages. Hence

$$\# E = \# F \,.\, (\# G_1)^{k_2} ,$$

thus proving lemma 3.3 and proposition 3.2.

COROLLARY 3.4. - If $k_1 \neq \infty$:

$$w_{k_1+k_2}(G) = (\# \mathrm{Aut}\ G)^{-1} \sum_{G_1 \leq G} (\# G/G_1)^{-k_1} (\# \mathrm{Aut}\, G_1)(\# \mathrm{Aut}\, G/G_1)\, w_{k_1}(G_1)\, w_{k_2}(G/G_1) .$$

Proof. - For $k_2 \neq \infty$ this is just a restatement of proposition 3.2, and the case $k_2 = \infty$ follows by letting $k_2 \to +\infty$ and using proposition 3.1, iv).

The following theorem, although not difficult to prove, will be very important in the sequel :

THEOREM 3.5. - Let K and C be finite A-modules. Then for all k :

$$\sum_{G \text{ up to A-isomorphism}} w_k(G)\ \#\{G_1 \leq G : G_1 \simeq K \text{ and } G/G_1 \simeq C\} = w_k(K)\, w_k(C) \ .$$

Proof. - We consider only k finite since the case $k = \infty$ follows by making $k \to \infty$. We shall count the number of pairs (H, J) of A-modules such that $H \subset J \subset A^k$, $A^k/J \simeq C$, $J/H \simeq K$. Note that H and J are necessarily projective modules of rank k.

If we write $m = \# K$ and $n = \# K \# C$, then the number of J is $(n/m)^k w_k(C)$ (proposition 3.1 (iii)), while for a given J the number of H is equal to $m^k w_k(K)$ by the same proposition. Hence the number of pairs (H, J) as above is equal to $n^k w_k(K)\, w_k(C)$.

Now let H be fixed and set $G = A^k/H$. Then every submodule of G can be written uniquely in the form J/H for some J such that $H \subset J \subset A^k$, hence the number of J is equal to

$$\#\{G_1 \leq G : G_1 \simeq K \text{ and } G/G_1 \simeq C\}$$

where we have set $G_1 = J/H$.
Finally, using again proposition 3.1, we see that for a given G the number of H such that $A^k/H \simeq G$ is equal to $n^k w_k(G)$, and Theorem 3.5 follows. (Note that $m = N(\chi_A(K))$, $(n/m) = N(\chi_A(C))$ and that if $0 \to G_1 \to G \to G/G_1 \to 0$ is an exact sequence then $\chi_A(G) = \chi_A(G_1)\, \chi_A(G/G_1)$.)

THEOREM 3.6. - Let $\mathfrak{a}$ be a non zero ideal of A

i) For any $k_2 \neq \infty$:

$$w_{k_1+k_2}(\mathfrak{a}) = \sum_{\mathfrak{b} | \mathfrak{a}} (N\mathfrak{b})^{-k_2} w_{k_1}(\mathfrak{b})\, w_{k_2}(\mathfrak{a}\mathfrak{b}^{-1}) \ .$$

ii) For any k, $\sum_{\mathfrak{b}|\mathfrak{a}} w_k(\mathfrak{b}) = (N\mathfrak{a})\, w_{k+1}(\mathfrak{a})$. In particular

$$\sum_{\mathfrak{b}|\mathfrak{a}} w(\mathfrak{b}) = (N\mathfrak{a})\, w(\mathfrak{a}) \ .$$

Proof. - i) By corollary 3.4 we have, setting $\mathfrak{b} = \chi_A(G/G_1)$:

$$w_{k_1+k_2}(\mathfrak{a}) = \sum_{G(\mathfrak{a})} w_{k_1+k_2}(G) = \sum_{\mathfrak{b}|\mathfrak{a}} (N\mathfrak{b})^{-k_1} \sum_{C(\mathfrak{b})} (\# \mathrm{Aut}\, C)\, w_{k_2}(C) \times$$

$$\times \sum_{K(\mathfrak{a}\mathfrak{b}^{-1})} (\# \mathrm{Aut}\, K)\, w_{k_1}(K) \sum_{G(\mathfrak{a})} w(G) \# \{G_1 \leq G : G_1 \simeq K \text{ and } G/G_1 \simeq C\}$$

so using Theorem 3.5 with $k = \infty$:

$$w_{k_1+k_2}(\mathfrak{a}) = \sum_{\mathfrak{b}|\mathfrak{a}} (N\mathfrak{b})^{-k_1} \sum_{C(\mathfrak{b})} w_{k_2}(C) \sum_{K(\mathfrak{a}\mathfrak{b}^{-1})} w_{k_1}(K)$$

$$= \sum_{\mathfrak{b}|\mathfrak{a}} (N\mathfrak{b})^{-k_2} w_{k_2}(\mathfrak{b})\, w_{k_1}(\mathfrak{a}\mathfrak{b}^{-1})$$

and (i) follows after interchanging k_1 and k_2 .

For (ii) we apply (i) with $k_1 = k$, $k_2 = 1$. Note that $s_1(G) \neq 0$ if and only if $G \simeq A/\mathfrak{a}$, where $\mathfrak{a}$ is a non zero ideal of A , and $s_1(A/\mathfrak{a}) = \#(A/\mathfrak{a})^* = \# \mathrm{Aut}(A/\mathfrak{a})$

Since $\chi_A(A/\mathfrak{a}) = \mathfrak{a}$ and that $A/\mathfrak{a} \simeq A/\mathfrak{b}$ if and only if $\mathfrak{a} = \mathfrak{b}$, it easy follows that

$$w_1(\mathfrak{a}) = 1/N\mathfrak{a}$$

and ii) follows.

This theorem is best expressed in terms of Dirichlet series as follows :

COROLLARY 3.7. - (i) *Let* $\mathfrak{p} \in \mathcal{P}$. *Then for* $\mathrm{Re}\, s > -1$

$$\sum_{\alpha \geq 0} w_k(\mathfrak{p}^{\alpha})(N\mathfrak{p})^{-\alpha s} = \prod_{1 \leq j \leq k} (1 - N\mathfrak{p}^{-j-s})^{-1} .$$

(ii) *If we set* $\zeta_{k,A}(s) = \zeta_k(s) = \sum w_k(\mathfrak{a})(N\mathfrak{a})^{-s}$ *for* $\mathrm{Re}\, s > 0$, *then*

$$\zeta_{k,A}(s) = \prod_{1 \leq j \leq k} \zeta_A(s+j)$$

where $\zeta_A(s)$ *is the Dedekind zeta function of* A .

Proof. - Clear by induction on k .

Note that theorem 3.6 (i) follows from the identity

$$\zeta_{k_1+k_2}(s) = \zeta_{k_1}(s+k_2)\, \zeta_{k_2}(s) .$$

COROLLARY 3.8. - For every $k \geq 1$:

$$w_k(\mathfrak{a}) = \frac{1}{N\mathfrak{a}} \prod_{\mathfrak{p}^\alpha \| \mathfrak{a}} \begin{bmatrix} \alpha+k-1 \\ \alpha \end{bmatrix}_{\mathfrak{p}} \quad \text{and in particular}$$

$$w(\mathfrak{a}) = \frac{1}{N\mathfrak{a}} \prod_{\mathfrak{p}^\alpha \| \mathfrak{a}} (\eta_\alpha(\mathfrak{p}))^{-1} \ . \ (^*)$$

(See section 2 for notations ; $\mathfrak{p}^\alpha \| \mathfrak{a}$ means that α is the exact exponent of $\mathfrak{p}$ in the prime ideal decomposition of $\mathfrak{a}$.)

Proof. - We use induction on k . For $k=1$, $w_1(\mathfrak{a}) = 1/N\mathfrak{a}$ as we have seen so the formula is true. Assume that it is true for some $k \geq 1$ and let us prove it for $k+1$. First note that both sides of the formula are multiplicative functions of $\mathfrak{a}$, hence it suffices to prove it for $\mathfrak{a} = \mathfrak{p}^\alpha$. Now by theorem 3.6 (ii) and our induction hypothesis :

$$w_{k+1}(\mathfrak{p}^\alpha) = (N\mathfrak{p})^{-\alpha} \sum_{0 \leq \beta \leq \alpha} (N\mathfrak{p})^{-\beta} \begin{bmatrix} \beta+k-1 \\ \beta \end{bmatrix}_{\mathfrak{p}} \ .$$

Now the following lemma is well known and straightforward to prove (it is the q-analogue of the formula

$\binom{\beta+k}{k} = \binom{\beta+k-1}{k} + \binom{\beta+k-1}{k-1}$ for binomial coefficients, with $q = (N\mathfrak{p})^{-1}$) :

LEMMA 3.9.- $$\begin{bmatrix} \beta+k \\ k \end{bmatrix}_{\mathfrak{p}} = \begin{bmatrix} \beta+k-1 \\ k \end{bmatrix}_{\mathfrak{p}} + (N\mathfrak{p})^{-\beta} \begin{bmatrix} \beta+k-1 \\ k-1 \end{bmatrix}_{\mathfrak{p}}$$

Hence

$$w_{k+1}(\mathfrak{p}^\alpha) = (N\mathfrak{p})^{-\alpha} \sum_{0 \leq \beta \leq \alpha} \left(\begin{bmatrix} \beta+k \\ k \end{bmatrix}_{\mathfrak{p}} - \begin{bmatrix} \beta+k-1 \\ k \end{bmatrix}_{\mathfrak{p}} \right) = (N\mathfrak{p})^{-\alpha} \begin{bmatrix} \alpha+k \\ k \end{bmatrix}_{\mathfrak{p}} = (N\mathfrak{p})^{-\alpha} \begin{bmatrix} \alpha+k \\ \alpha \end{bmatrix}_{\mathfrak{p}}$$

and so corollary 3.8 follows by induction and then letting $k \to \infty$ since $\lim_{k\to\infty} \begin{bmatrix} \alpha+k-1 \\ \alpha \end{bmatrix}_{\mathfrak{p}} = \eta_\alpha(\mathfrak{p})^{-1}$.

(*) In the case $A = \mathbb{Z}$, this last formula was proved by a completely different method by P. Hall [8] .

§ 4. - Some consequences of theorems 3.5 and 3.6

In this section we collect a number of almost direct consequences of theorems 3.5 and 3.6 which will be useful to us later on.

PROPOSITION 4.1. - Let $\mathfrak{a}$; $\mathfrak{b}$ be (non zero) ideals of A, such that $\mathfrak{b} | \mathfrak{a}$, and K a finite A-module such that $\mathfrak{b} = \chi_A(K)$. Then for all k :

(i) $\sum_{G(\mathfrak{a})} w_k(G) \# \{G_1 \leq G : G_1 \simeq K\} = w_k(\mathfrak{a}\mathfrak{b}^{-1}) \, w_k(K)$

(ii) $\sum_{G(\mathfrak{a})} w_k(G) \# \{G_1 \leq G : G/G_1 \simeq K\} = w_k(\mathfrak{a}\mathfrak{b}^{-1}) \, w_k(K)$

(iii) $\sum_{G(\mathfrak{a})} w_k(G) \# \{G_1 \leq G : \chi_A(G_1) = \mathfrak{b}\} = w_k(\mathfrak{a}\mathfrak{b}^{-1}) \, w_k(\mathfrak{b})$.

Proof. - Clear from theorem 3.5 by summing over suitable isomorphism classes.

We now want to generalize theorems 3.5 and its consequences to the case where G is replaced by $G/\mathrm{Im}\,\varphi$, where $\varphi \in \mathrm{Hom}_A(A^u, G)$. For this we need a new definition :

DEFINITION 4.2. - For u, k arbitrary and $\mathfrak{a}$ non zero ideal, we set :

$$w_{k,u}(\mathfrak{a}) = \sum_{G(\mathfrak{a})} w_k(G) \, w_u(G) \# \mathrm{Aut}\, G \quad \text{and} \quad \zeta_{k,u}(s) = \sum_{\mathfrak{a}} \frac{w_{k,u}(\mathfrak{a})}{(N\mathfrak{a})^s} .$$

Note that $w_{k,u} = w_{u,k}$ and that $w_{\infty,u} = w_u$, and similarly for $\zeta_{k,u}$. The first result that we need is the following :

PROPOSITION 4.3. - Let $\mathfrak{a}$, $\mathfrak{b}$ be (non zero) ideals of A with $\mathfrak{b} | \mathfrak{a}$, and K a finite A-module such that $\chi_A(K) = \mathfrak{b}$. Then

$$\sum_{G(\mathfrak{a})} w_k(G) \# \{\varphi \in \mathrm{Hom}_A(A^u, G) : G/\mathrm{Im}\,\varphi \simeq K\} = (N(\mathfrak{a}\mathfrak{b}^{-1}))^u \, w_{k,u}(\mathfrak{a}\mathfrak{b}^{-1}) \, w_k(K) .$$

Proof. - The left hand side clearly equals

$$\sum_{L(\mathfrak{a}\mathfrak{b}^{-1})} s_u(L) \sum_{G(\mathfrak{a})} w_k(G) \# \{G_1 \leq G : G_1 \simeq L \text{ and } G/G_1 \simeq K\}$$

$$= \sum_{L(\mathfrak{a}\mathfrak{b}^{-1})} (N(\mathfrak{a}\mathfrak{b}^{-1}))^u \# \mathrm{Aut}\, L \, w_u(L) \, w_k(L) \, w_k(K) \quad \text{by theorem 3.5}$$

$$= (N(\mathfrak{a}\mathfrak{b}^{-1}))^u \, w_{k,u}(\mathfrak{a}\mathfrak{b}^{-1}) \, w_k(K), \text{ and the proposition is proved.}$$

COROLLARY 4.4. - Let $\mathfrak{a}, \mathfrak{b}, \mathfrak{c}$ be (non zero) ideals of A with $\mathfrak{b}\mathfrak{c} \mid \mathfrak{a}$ and K, C finite A-modules such that $\chi_A(K) = \mathfrak{b}$, $\chi_A(C) = \mathfrak{c}$. Then

$$\sum_{G(\mathfrak{a}), \varphi_u} w_k(G) \# \{G_1 \leq G/\mathrm{Im}\,\varphi : G_1 \simeq K \text{ and } (G/\mathrm{Im}\,\varphi)/G_1 \simeq C\}$$
$$= N(\mathfrak{a}\mathfrak{b}^{-1}\mathfrak{c}^{-1})^u \, w_{k,u}(\mathfrak{a}\mathfrak{b}^{-1}\mathfrak{c}^{-1}) \, w_k(C) \, w_k(K).$$

Proof. - The left hand side clearly equals

$$\sum_{L(\mathfrak{b}\mathfrak{c})} \# \{G_1 \leq L : G_1 \simeq K \text{ and } L/G_1 \simeq C\} \sum_{G(\mathfrak{a})} w_k(G) \# \{\varphi \in \mathrm{Hom}(A^u, G) : (G/\mathrm{Im}\,\varphi) \simeq L\}$$
$$= \sum_{L(\mathfrak{b}\mathfrak{c})} \# \{G_1 \leq L : G_1 \simeq K \text{ and } L/G_1 \simeq C\} N(\mathfrak{a}\mathfrak{b}^{-1}\mathfrak{c}^{-1})^u \, w_{k,u}(\mathfrak{a}\mathfrak{b}^{-1}\mathfrak{c}^{-1}) \, w_k(L)$$

by proposition 4.3, and the result follows from theorem 3.5.

PROPOSITION 4.5. - Let $\mathfrak{a}, \mathfrak{b}$ be (non zero) ideals of A with $\mathfrak{b} \mid \mathfrak{a}$, and K a finite A-module such that $\chi_A(K) = \mathfrak{b}$. Then :

(i) $\displaystyle\sum_{G(\mathfrak{a}), \varphi_u} w_k(G) \# \{G_1 \leq G/\mathrm{Im}\,\varphi : G_1 \simeq K\} = N(\mathfrak{a}\mathfrak{b}^{-1})^u \, w_k(\mathfrak{a}\mathfrak{b}^{-1}) \, w_k(K)$

(ii) $\displaystyle\sum_{G(\mathfrak{a}), \varphi_u} w_k(G) \# \{G_1 \leq G/\mathrm{Im}\,\varphi : (G/\mathrm{Im}\,\varphi)/G_1 \simeq K\} = N(\mathfrak{a}\mathfrak{b}^{-1})^u \, w_k(\mathfrak{a}\mathfrak{b}^{-1}) \, w_k(K)$

(iii) $\displaystyle\sum_{G(\mathfrak{a}), \varphi_u} w_k(G) \# \{G_1 \leq G/\mathrm{Im}\,\varphi : \chi_A(G_1) = \mathfrak{b}\} = N(\mathfrak{a}\mathfrak{b}^{-1})^u \, w_k(\mathfrak{a}\mathfrak{b}^{-1}) \, w_k(\mathfrak{b})$.

THEOREM 4.6. - We have for $\mathrm{Re}\, s > 0$:

$$\zeta_{k,u}(s) = \sum_{\mathfrak{a}} \frac{w_{k,u}(\mathfrak{a})}{(N\mathfrak{a})^s} = \frac{\zeta_k(s)}{\zeta_k(s+u)} = \frac{\zeta_k(s)\,\zeta_u(s)}{\zeta_{k+u}(s)} = \prod_{1 \leq j \leq k} \frac{\zeta_A(s+j)}{\zeta_A(u+s+j)} .$$

Proofs. - We prove proposition 4.5 (i) and theorem 4.6 simultaneously.

If we sum the formula in corollary 4.4 over $C(\mathfrak{c})$ and then over all $\mathfrak{c} \mid \mathfrak{a}\mathfrak{b}^{-1}$, it is clear that we obtain :

$$\sum_{G(\mathfrak{a}), \varphi_u} w_k(G) \# \{G_1 \leq G/\mathrm{Im}\,\varphi : G_1 \simeq K\} = f(\mathfrak{a}\mathfrak{b}^{-1}) \, w_k(K)$$

$$\text{where } \quad f(\mathfrak{a}) = \sum_{\mathfrak{c} \mid \mathfrak{a}} N(\mathfrak{a}\mathfrak{c}^{-1})^u \, w_{k,u}(\mathfrak{a}\mathfrak{c}^{-1}) \, w_k(\mathfrak{c}) .$$

Now the important point is that $f(\mathfrak{a}\mathfrak{b}^{-1})$ depends only on the ideal $\mathfrak{a}\mathfrak{b}^{-1}$. Hence if we take $K = \{0\}$ the trivial A-module, we have $w_k(K) = 1$ and $\chi_A(K) = \mathfrak{b} = A$ hence :

$$f(\mathfrak{a}) = \sum_{G(\mathfrak{a}), \varphi_u} w_k(G) = (N\mathfrak{a})^u \, w_k(\mathfrak{a}), \quad \text{so } f(\mathfrak{a}\mathfrak{b}^{-1}) = N(\mathfrak{a}\mathfrak{b}^{-1})^u \, w_k(\mathfrak{a}\mathfrak{b}^{-1})$$

and proposition 4.5 (i) follows.

Now we have just proven the identity

$$f(\mathfrak{a}) = \sum_{\mathfrak{c} \mid \mathfrak{a}} N(\mathfrak{a}\mathfrak{c}^{-1})^u \, w_{k,u}(\mathfrak{a}\mathfrak{c}^{-1}) \, w_k(\mathfrak{c}) = (N\mathfrak{a})^u w_k(\mathfrak{a}) \, .$$

In terms of Dirichlet series, this gives :

$$\zeta_{k,u}(s-u) \, \zeta_k(s) = \zeta_k(s-u) \, ,$$

and theorem 4.6 follows immediately.

The proofs of (ii) and (iii) in proposition 4.5 are now trivial and left to the reader

§ 5. - Some u-probabilistics and u-averages

In the beginning of this section, f will be a complex-valued function defined on isomorphism classes of finite A-modules.

DEFINITION 5.1. - We set

$$w_k(f;\mathfrak{a}) = \sum_{G(\mathfrak{a})} w_k(G) \, f(G) \, , \qquad \zeta_k(f;s) = \sum_{\mathfrak{a}} w_k(f;\mathfrak{a}) (N\mathfrak{a})^{-s}$$

and we define the (k,u)-average $M_{k,u}(f)$ of f as follows :

$$M_{k,u}(f) = \lim_{x \to \infty} \frac{\sum_{N\mathfrak{a} \le x} (N\mathfrak{a})^{-u} \sum_{G(\mathfrak{a}),\varphi_u} f(G/\mathrm{Im}\,\varphi) \, w_k(G)}{\sum_{N\mathfrak{a} \le x} (N\mathfrak{a})^{-u} \sum_{G(\mathfrak{a}),\varphi_u} w_k(G)} \, .$$

If $k = \infty$ we will simply speak of u-average of f and write $M_u(f)$ instead of $M_{\infty,u}(f)$.

Remarks. - 1) The (k,u)-average of f may not exist if the expression after the lim does not tend to a limit when $x \to \infty$.

2) The denominator in the definition of $M_{k,u}(f)$ is equal to

$$\sum_{N\mathfrak{a} \le x} w_k(\mathfrak{a})$$

but we have written it in the above manner to make it clear that we are dealing with an average (i.e. the (k,u)-average of a function which is constant is that constant).

3) When f is the characteristic function of a property P (i.e. $f = 1$ if P is true, $f = 0$ if P is false) we will speak of (k,u)-probability or u-probability of P instead of (k,u)-average or u-average of f.

4) If $u = 0$ and $k = \infty$ we will speak of the average of f, or the probability of P. It should be noted that this is only a finitely additive measure, hence the word

probability should be taken to mean exactly that in our context.

The aim of this section is to show how, in many cases, one can easily compute (k,u)-averages and probabilities.

The most direct way is by using the following Tauberian theorem :

LEMMA 5.2. - *If* $D(s)=\sum_{\mathfrak{a}} c(\mathfrak{a})(N\mathfrak{a})^{-s}$ *converges for* $\mathrm{Re}\, s>0$ *and if* $D(s)-C/s$ *can be analytically continued for* $\mathrm{Re}\, s\geq 0$ *then if the* $c(\mathfrak{a})$ *are non negative we have*

$$\sum_{N\mathfrak{a}\leq x} c(\mathfrak{a}) \sim C \operatorname{Log} x \quad \text{as} \quad x\to+\infty .$$

Note that this follows from a classical Tauberian theorem (see e.g. [15]) by writing

$$\sum_{\mathfrak{a}} c(\mathfrak{a}) N\mathfrak{a}^{-s} = \sum_{n\geq 1} n \Big(\sum_{N\mathfrak{a}=n} c(\mathfrak{a})\Big) n^{-s-1}$$

and then using partial summation.

Applying this to $w_k(\mathfrak{a})=c(\mathfrak{a})$ and using corollary 3.7 we obtain

LEMMA 5.3. -

$$\sum_{N\mathfrak{a}\leq x} w_k(\mathfrak{a}) \sim C_k \operatorname{Log} x \quad (x\to+\infty)$$

(see notations for C_k).

In fact one can obtain a more precise estimate, but this asymptotic equality will be sufficient for us since we only want a limit.

PROPOSITION 5.4. - *Write*

$$\zeta_k(f\,;s+u)\,\zeta_k(s)/\zeta_k(s+u) = \sum_{\mathfrak{a}} a_{k,u}(f\,;\mathfrak{a})(N\mathfrak{a})^{-s} .$$

Then

$$M_{k,u}(f) = \lim_{x\to\infty} \frac{\sum_{N\mathfrak{a}\leq x} a_{k,u}(f;\mathfrak{a})}{C_k \operatorname{Log} x} .$$

Proof. - We have

$$\sum_{G(\mathfrak{a}),\varphi_u} f(G/\mathrm{Im}\,\varphi)\, w_k(G) = \sum_{\mathfrak{b}|\mathfrak{a}} \sum_{L(\mathfrak{b})} f(L) \sum_{G(\mathfrak{a})} w_k(G)\,\#\{\varphi\in \mathrm{Hom}(A^u,G) : G/\mathrm{Im}\,\varphi\simeq L\}$$

$$= \sum_{\mathfrak{b}|\mathfrak{a}} \sum_{L(\mathfrak{b})} f(L)\, N(\mathfrak{a}\mathfrak{b}^{-1})^u\, w_{k,u}(\mathfrak{a}\mathfrak{b}^{-1})\, w_k(L)$$

by proposition 4.3

$$= \sum_{\mathfrak{b}|\mathfrak{a}} N(\mathfrak{a}\mathfrak{b}^{-1})^u\, w_{k,u}(\mathfrak{a}\mathfrak{b}^{-1})\, w_k(f\,;\mathfrak{b}) .$$

Hence

$$\sum_{\mathfrak{a}} (N\mathfrak{a})^{-s-u} \sum_{G(\mathfrak{a}),\varphi_u} f(G/\mathrm{Im}\,\varphi)\, w_k(G) = \zeta_{k,u}(s)\, \zeta_k(f\,;\, s+u)$$

$$= \zeta_k(s)\, \zeta_k(f\,;\, s+u)/\zeta_k(s+u) = \sum_{\mathfrak{a}} a_{k,u}(f;\mathfrak{a})(N\mathfrak{a})^{-s}$$

where we have used theorem 4.6. Hence

$$\sum_{N\mathfrak{a}\leq x} (N\mathfrak{a})^{-u} \sum_{G(\mathfrak{a}),\varphi_u} f(G/\mathrm{Im}\,\varphi)\, w_k(G) = \sum_{N\mathfrak{a}\leq x} a_{k,u}(f\,;\,\mathfrak{a})$$

and the proposition follows from remark 2 and lemma 5.3.

COROLLARY 5.5. - <u>Assume that</u> f <u>is a non-negative valued function on the set of is isomorphism classes of finite</u> A-<u>modules</u>. <u>Assume further that</u> $\zeta_k(f\,;\,s)$ <u>converges for</u> $\mathrm{Re}\, s > 0$ <u>and that</u> $\zeta_k(f\,;\,s) - C/s$ <u>can be analytically continued to</u> $\mathrm{Re}\, s \geq 0$. <u>Then</u> :

<u>For</u> $u \neq 0$, $M_{k,u}(f) = \zeta_k(f\,;\,u)\, C_u/C_{u+k} = \zeta_k(f\,;\,u)/\zeta_k(u)$

<u>For</u> $u = 0$, $M_{k,0}(f) = C/C_k = \lim_{s\to 0} (\zeta_k(f\,;\,s)/\zeta_k(s))$.

<u>Proof</u>. - From proposition 5.4 it is clear that $\sum_{\mathfrak{a}} a_{k,u}(f\,;\,\mathfrak{a})\, N\mathfrak{a}^{-s}$ converges for $\mathrm{Re}\, s > 0$ and is asymptotic to $(\zeta_k(f\,;\,u)/\zeta_k(u)) \times C_k/s$ if $u > 0$ and to C/s if $u = 0$.

Since $\zeta_k(u) = C_{u+k}/C_u$ the corollary follows from proposition 5.4 and our Tauberian lemma 5.2.

For our applications, we need to be able to restrict our attention to A-modules having only certain $\mathfrak{p}$-components. More precisely, in what follows we let $\mathcal{P}_1 \subset \mathcal{P}$ and we call an A-module G a $\mathcal{P}_1$-A-module if $G = G_{\mathcal{P}_1}$, with an evident notation ($G_{\mathcal{P}_1} = \bigoplus_{\mathfrak{p}\in\mathcal{P}_1} G_{\mathfrak{p}}$). Then in a straightforward way one can define the notion of (k,u)-average of a function f restricted to $\mathcal{P}_1$-A-modules. The following proposition is easy and left to the reader :

PROPOSITION 5.6. - <u>The</u> (k, u)-<u>average of a function</u> f <u>restricted to</u> $\mathcal{P}_1$-A-<u>modules is the same as the</u> (k,u)-<u>average of the function</u> $f \circ \mathcal{P}_1$ <u>defined by</u>

$$f \circ \mathcal{P}_1(G) = f(G_{\mathcal{P}_1}) .$$

Essentially this proposition says that the $\mathfrak{p}$-components of a finite A-module behave independently.

The last information we need about $\mathcal{P}_1$-A-modules is the following :

PROPOSITION 5.7. - With the notations of proposition 5.6, we have :

$$w_k(f \circ \mathcal{P}_1 ; \mathfrak{a}) = w_k(f ; \mathfrak{a}_1)\, w_k(\mathfrak{a}_2)$$

where $\mathfrak{a}_1$ is the $\mathcal{P}_1$-part of $\mathfrak{a}$, and $\mathfrak{a}_2 = \mathfrak{a}\,\mathfrak{a}_1^{-1}$, and consequently

$$\zeta_k(f \circ \mathcal{P}_1 ; s) = \Big(\sum_{\substack{\mathfrak{a} \\ \mathfrak{p} | \mathfrak{a} \Rightarrow \mathfrak{p} \in \mathcal{P}_1}} w_k(f ; \mathfrak{a})(N\mathfrak{a})^{-s}\Big) \prod_{\mathfrak{p} \notin \mathcal{P}_1} \prod_{1 \leq j \leq k} (1 - (N\mathfrak{p})^{-j-s})^{-1} .$$

Proof. - Set $\mathcal{P}_2 = \mathcal{P} - \mathcal{P}_1$. Then we clearly have $\mathfrak{a} = \mathfrak{a}_1 \mathfrak{a}_2$ with $\mathfrak{a}_i$ being the $\mathcal{P}_i$-part of $\mathfrak{a}$, and

$$w_k(f \circ \mathcal{P}_1 ; \mathfrak{a}) = \sum_{G(\mathfrak{a})} w_k(G)\, f(G_{\mathcal{P}_1}) = \sum_{G_1(\mathfrak{a}_1)} f(G_1)\, w_k(G_1) \sum_{G_2(\mathfrak{a}_2)} w_k(G_2)$$

and the first formula follows. The second one is a formal consequence of the definition of $\zeta_k(f ; s)$ and of corollary 3.7.

We can now give examples of u-probabilities and u-averages. For simplicity we assume $k = \infty$, but of course all the results can be obtained also for finite k. The proofs, being in general straightforward applications of the results of this section, will be omitted or only sketched.

It should be recalled at this point that all the constants like C_∞, $\eta_\infty(\mathfrak{p})$ etc... that have been introduced earlier, are relative to the domain A and should more properly be written C_∞^A, $\eta_\infty^A(\mathfrak{p})$, etc... .

Example 5.8. - Let $\alpha \in \mathbb{R}$. Then for $u > \alpha$ the u-average of $(\# G)^\alpha$ is equal to

$$M_u((\# G)^\alpha) = (C_u / C_\infty) \prod_{j \geq 1} \zeta_A(j + u - \alpha) .$$

In particular, if $u \geq 2$ the u-average of $\# G$ is $\zeta_A(u)$.

Example 5.9. - (i) Let L be a $\mathcal{P}_1$-group with $\# L = \ell$. Then the u-probability that the $\mathcal{P}_1$-part of an A-module be isomorphic to L is equal to :

$$\ell^{-u} (\# \operatorname{Aut} L)^{-1} \prod_{\mathfrak{p} \in \mathcal{P}_1} (\eta_\infty(\mathfrak{p}) / \eta_u(\mathfrak{p})) .$$

(ii) Assume that $\mathfrak{p} | \mathfrak{a} \Rightarrow \mathfrak{p} \in \mathcal{P}_1$. Then the u-probability that the $\mathcal{P}_1$-part of a group has A-cardinality equal to $\mathfrak{a}$ is equal to

$$(N\mathfrak{a})^{-u}\, w(\mathfrak{a}) \prod_{\mathfrak{p} \in \mathcal{P}_1} (\eta_\infty(\mathfrak{p}) / \eta_u(\mathfrak{p})) .$$

Example 5.10. - The u-probability that $G_{\mathfrak{p}} \neq 0$ is equal to

$$1 - \eta_\infty(\mathfrak{p}) / \eta_u(\mathfrak{p}) .$$

Example 5.11. - The u-probability that the $\mathcal{P}_1$-part of an A-module G is A-cyclic (i.e. $G_{\mathcal{P}_1} \simeq A/\mathfrak{a}$) is equal to

$$\prod_{\mathfrak{p} \in \mathcal{P}_1} \frac{1-(N\mathfrak{p})^{-1}+(N\mathfrak{p})^{-u-2}}{(1-(N\mathfrak{p})^{-u-1})(1-(N\mathfrak{p})^{-1})} \frac{\eta_\infty(\mathfrak{p})}{\eta_u(\mathfrak{p})} .$$

In particular for $u=0$ and $\mathcal{P}_1 = \mathcal{P}$ this is equal to

$$\varkappa_A \zeta_A(2) \zeta_A(3)/(\zeta_A(6) C_\infty^A)$$

(recall that $\varkappa_A$ is the residue at $s=1$ of $\zeta_A(s)$).

Example 5.12. - Let $\mathfrak{a}$ be an ideal. The u-average of the number of element x in a finite A-module whose annihilator is $\mathfrak{a}$ equals $(N\mathfrak{a})^{-u}$.

For example if $A = \mathbb{Z}$, the u-average of the number of elements of order $a \geq 1$ in an abelian group is a^{-u}.

Proof. - Simply note that the number of $x \in G$ such that $\mathrm{Ann}\, x = \mathfrak{a}$ is equal to

$$\varphi(\mathfrak{a}) \# \{ G_1 \leq G : G_1 \simeq A/\mathfrak{a} \}$$

where $\varphi(\mathfrak{a}) = \# (A/\mathfrak{a})^*$.

Example 5.13. - Call an A-module G elementary if for all $\mathfrak{p}$, $G_{\mathfrak{p}} \simeq (A/\mathfrak{p})^{k_{\mathfrak{p}}}$ for some $k_{\mathfrak{p}} \geq 0$, i.e. if no $A/\mathfrak{p}^\alpha$ occur in $G_{\mathfrak{p}}$ with $\alpha > 1$. Then :

(i) The 0-probability that a finite A-module is elementary equals

$$\Big(\prod_{\substack{k \not\equiv 1,4 \ (\mathrm{mod}\ 5) \\ k \geq 2}} \zeta_A(k) \Big)^{-1}$$

(ii) The 1-probability that a finite A-module is elementary equals

$$\Big(\prod_{\substack{k \not\equiv 2,3 \ (\mathrm{mod}\ 5) \\ k \geq 2}} \zeta_A(k) \Big)^{-1} .$$

(Example 5.13, (i) was suggested to us by D. Zagier.)

Proof. - Straightforward, using the easily proven fact that $\# \mathrm{Aut}\,(A/\mathfrak{p})^m = (N\mathfrak{p})^{m^2} \eta_m(\mathfrak{p})$ and the two identities of Rogers-Ramanujan.

§ 6. - u-probabilities and u-averages involving $\mathfrak{p}$-ranks

In this section we show how to obtain information on the distribution of $\mathfrak{p}$-ranks of finite A-modules, where $\mathfrak{p} \in \mathcal{P}$ is fixed. The first theorem is as follows :

THEOREM 6.1. - Let $\mathfrak{a}$ be an ideal of A , $\alpha = v_{\mathfrak{p}}(\mathfrak{a})$ and r a non negative integer such that $r \leq \alpha$ (otherwise the theorem is empty). Then :

$$\text{(i)} \quad \sum_{\substack{G(\mathfrak{a}) \\ r_{\mathfrak{p}}(G)=r}} w_k(G) = w_k(\mathfrak{a})(N\mathfrak{p})^{-r^2+r} \begin{bmatrix} k \\ r \end{bmatrix}_{\mathfrak{p}} \begin{bmatrix} \alpha-1 \\ r-1 \end{bmatrix}_{\mathfrak{p}} \Big/ \begin{bmatrix} \alpha+k-1 \\ \alpha \end{bmatrix}_{\mathfrak{p}}$$

and in particular

$$\sum_{\substack{G(\mathfrak{a}) \\ r_{\mathfrak{p}}(G)=r}} w(G) = w(\mathfrak{a})(N\mathfrak{p})^{-r^2+r} \begin{bmatrix} \alpha-1 \\ r-1 \end{bmatrix}_{\mathfrak{p}} \eta_\alpha(\mathfrak{p})/\eta_r(\mathfrak{p})$$

$$\text{(ii)} \quad \sum_{\substack{G \text{ up to isomorphism} \\ G\ \mathfrak{p}\text{-A-module} \\ r_{\mathfrak{p}}(G)=r}} w_k(G)(\# G)^{-s} = \begin{bmatrix} k \\ r \end{bmatrix}_{\mathfrak{p}} (N\mathfrak{p})^{-r(r+s)} \prod_{1\leq j\leq r} (1-(N\mathfrak{p})^{-j-s})^{-1} \quad \text{for } \operatorname{Re} s > -1$$

and in particular if $k \geq r$:

$$\sum_{\text{same}} w_k(G) = \frac{(N\mathfrak{p})^{-r^2}}{(\eta_r(\mathfrak{p}))^2} \, \frac{\eta_k(\mathfrak{p})}{\eta_{k-r}(\mathfrak{p})} \, .$$

Proof. - (i) Write $\mathfrak{a} = \mathfrak{p}^\alpha \mathfrak{b}$ with $\mathfrak{p} \nmid \mathfrak{b}$. Then

$$\sum_{\substack{G(\mathfrak{a}) \\ r_{\mathfrak{p}}(G)=r}} w_k(G) = \sum_{G_1(\mathfrak{b})} w_k(G_1) \sum_{\substack{G_2(\mathfrak{p}^\alpha) \\ r_{\mathfrak{p}}(G_2)=r}} w_k(G_2)$$

$$= w_k(\mathfrak{b}) \sum_{\substack{G(\mathfrak{p}^\alpha) \\ r_{\mathfrak{p}}(G)=r}} (\# \operatorname{Aut} G)^{-1} \eta_k(\mathfrak{p})/\eta_{k-r}(\mathfrak{p}) \quad \text{by proposition 3.1.}$$

Now it is easy to see that every $\mathfrak{p}$-A-module G of rank r is of the form $G = A^r/H$, and by proposition 3.1 the number of such H for a given G (with $\chi_A(G) = \mathfrak{p}^\alpha$) is equal to

$$((N\mathfrak{p})^{\alpha r}/\# \operatorname{Aut} G)\, \eta_r(\mathfrak{p}) \, .$$

Furthermore, given H, the conditions $\chi_A(A^r/H) = \mathfrak{p}^\alpha$ and $r_\mathfrak{p}(A^r/H) = r$ are equivalent to the conditions

$$H \subset \mathfrak{p}^r \quad \text{and} \quad \chi_A(\mathfrak{p}^r/H) = \mathfrak{p}^{\alpha - r} .$$

(Use the multiplicativity of χ_A on the exact sequence $0 \to \mathfrak{p}^r/H \to A^r/H \to (A/\mathfrak{p})^r \to 0$.) We obtain :

LEMMA 6.2. - With the above notations :

$$\sum_{\substack{G(\mathfrak{a}) \\ r_\mathfrak{p}(G) = r}} w_k(G) = \frac{w_k(\mathfrak{b})}{(N\mathfrak{p})^{\alpha r}} \left[{k \atop r}\right]_\mathfrak{p} \sum_{\substack{H \subset \mathfrak{p}^r \\ \chi_A(\mathfrak{p}^r/H) = \mathfrak{p}^{\alpha - r}}} 1 .$$

Now by proposition 6.1 applied to $J = \mathfrak{p}^r$ (hence "k" $= r$) this last sum is equal to

$$(N\mathfrak{p})^{r(\alpha - r)} w_r(\mathfrak{p}^{\alpha - r}) = (N\mathfrak{p})^{(r-1)(\alpha - 1)} \left[{\alpha - 1 \atop \alpha - r}\right]_\mathfrak{p} \qquad \text{(corollary 3.8)}$$

and theorem 6.1 (i) follows, using the fact that

$$w_k(\mathfrak{b}) = w_k(\mathfrak{a}) (N\mathfrak{p})^\alpha / \left[{\alpha + k - 1 \atop \alpha}\right]_\mathfrak{p} .$$

The rest of the assertions in the theorem follow easily from (i) and lemma 6.2.

Applying the techniques of section 5 we easily obtain :

THEOREM 6.3. - The u-probability that the $\mathfrak{p}$-rank of a finite A-module is r is equal to

$$(N\mathfrak{p})^{-r(r+u)} \eta_\infty(\mathfrak{p}) / (\eta_r(\mathfrak{p})\, \eta_{r+u}(\mathfrak{p})) .$$

The final result that we want about $\mathfrak{p}$-ranks is the one giving the u-average of $(N\mathfrak{p})^{r_\mathfrak{p}(G)}$ or more generally of $(N\mathfrak{p})^{\alpha r_\mathfrak{p}(G)}$ for $\alpha \geq 0$, α integral. This will follow from the following :

THEOREM 6.4. - Let $\alpha \geq 0$ be an integer, $\mathfrak{a}$ an ideal such that $\mathfrak{p}^\alpha | \mathfrak{a}$. Then :

$$(N\mathfrak{a})^{-u} \sum_{G(\mathfrak{a}), \varphi_u} w_k(G) \prod_{0 \leq i < \alpha} \left((N\mathfrak{p})^{r_\mathfrak{p}(G/\mathrm{Im}\,\varphi)} - (N\mathfrak{p})^i\right) =$$

$$= (N\mathfrak{p})^{-\alpha u} w_k(\mathfrak{a}\mathfrak{p}^{-\alpha})\, \eta_k(\mathfrak{p}) / \eta_{k-\alpha}(\mathfrak{p}) .$$

Proof. - We can deduce theorem 6.4 from theorem 6.1 using known q-identities. We shall use the converse approach, obtaining theorem 6.4 directly and deducing the q-identities.

We apply proposition 4.5 (i) to $K = (A/\mathfrak{p})^{\alpha}$. If G is a finite A-module, the number of submodules of G isomorphic to K is clearly equal to

$$(\# \mathrm{Aut}(A/\mathfrak{p})^{\alpha})^{-1} \# \{\varphi \in \mathrm{Hom}_A((A/\mathfrak{p})^{\alpha}, G),\ \varphi \text{ injective}\} .$$

Now if $G^{\mathfrak{p}}$ is the subgroup of G of elements annihilated by $\mathfrak{p}$, it is clear that

$$\# \{\varphi \in \mathrm{Hom}_A((A/\mathfrak{p})^{\alpha}, G),\ \varphi \text{ injective}\} = \# \{\varphi \in \mathrm{Hom}_A((A/\mathfrak{p})^{\alpha}, G^{\mathfrak{p}}),\ \varphi \text{ injective}\}$$
$$= \prod_{0 \leq i < \alpha} ((N\mathfrak{p})^{r_{\mathfrak{p}}(G)} - (N\mathfrak{p})^{i}) \quad \text{since } G^{\mathfrak{p}} \simeq (A/\mathfrak{p})^{r_{\mathfrak{p}}(G)} .$$

Hence proposition 4.5 (i) gives

$$\sum_{G(\alpha),\, \varphi_u} w_k(G) \prod_{0 \leq i < \alpha} ((N\mathfrak{p})^{r_{\mathfrak{p}}(G/\mathrm{Im}\,\varphi)} - (N\mathfrak{p})^{i}) =$$
$$= \# \mathrm{Aut}\,(A/\mathfrak{p})^{\alpha}\ N(\mathfrak{a}\mathfrak{p}^{-\alpha})^{u}\ w_k(\mathfrak{a}\mathfrak{p}^{-\alpha})\ w_k((A/\mathfrak{p})^{\alpha})$$

and the theorem follows from proposition 3.1.

From the definition of $M_{k,u}$ we obtain immediately :

COROLLARY 6.5. - We have :

$$M_{k,u}\Big(\prod_{0 \leq i < \alpha} (N\mathfrak{p}^{r_{\mathfrak{p}}(G)} - N\mathfrak{p}^{i})\Big) = (N\mathfrak{p})^{-\alpha u}\, \eta_k(\mathfrak{p}) / \eta_{k-\alpha}(\mathfrak{p}) .$$

Example 6.6. - The u-average of $(N\mathfrak{p})^{r_{\mathfrak{p}}(G)}$ is $1 + (N\mathfrak{p})^{-u}$; the u-average of $(N\mathfrak{p})^{2r_{\mathfrak{p}}(G)}$ is $1 + (N\mathfrak{p}+1)(N\mathfrak{p})^{-u} + (N\mathfrak{p})^{-2u}$.

As was mentioned earlier, one can easily obtain from the combination of preceding theorems some q-identities. We leave the proofs to the reader, noting that they can also be proved directly very simply :

COROLLARY 6.7. - Write $(q)_k = \prod_{1 \leq n \leq k} (1-q^n)$. Then for $k \geq \alpha$:

$$\sum_{\alpha \leq r \leq k} \frac{q^{(r+u)(r-\alpha)}}{(q)_{r-\alpha}\,(q)_{k-r}\,(q)_{r+u}} = \frac{1}{(q)_{k-\alpha}\,(q)_{k+u}} .$$

In particular for $k \to \infty$:

$$\sum_{r \geq \alpha} \frac{q^{(r+u)(r-\alpha)}}{(q)_{r-\alpha}\,(q)_{r+u}} = \frac{1}{(q)_\infty}$$

and with $\alpha = u = 0$:

$$\sum_{r \geq 0} q^{r^2} / (q)_r^2 = 1/(q)_\infty \quad .$$

§ 7. - An analytic digression : the function $\zeta_\infty(s)$

We study here the properties of the function $\zeta_\infty(s)$ as a meromorphic function. They will not be needed in the sequel, but may give some hints for the proofs of the conjectures that we will state in the next sections.

In what follows we assume that A is the ring of integers of a number field K of discriminant D, degree N over $\mathbb{Q}$, r_1 real places and $2r_2$ complex ones, with $r_1 + 2r_2 = N$. Then we recall that $\zeta_A(s)$ can be analytically continued to the whole complex plane with a single pole at $s = 1$, which is simple and with residue

$$\varkappa_A = 2^{r_1} (2\pi)^{r_2}\, h\, R / w\, |D|^{\frac{1}{2}} .$$

where as usual h is the class number, R the regulator and w the number of roots of unity in K. Furthermore if we set

$$\Lambda_A(s) = |D|^{s/2} (\pi^{-s/2}\, \Gamma(s/2))^{r_1} ((2\pi)^{-s}\, \Gamma(s))^{r_2}\, \zeta_A(s)$$

we have the functional equation

$$\Lambda_A(s) = \Lambda_A(1-s) \quad .$$

We want to study the function $\zeta_{\infty,A}(s) = \zeta_\infty(s)$ defined in corollary 3.7. We recall that

$$\zeta_\infty(s) = \prod_{j \geq 1} \zeta_A(s+j) \quad .$$

Note first that the Euler product is as follows :

$$\zeta_\infty(s) = \prod_{\mathfrak{p} \in \mathcal{P}} \prod_{j \geq 1} (1 - (N\mathfrak{p})^{-s-j})^{-1} \quad .$$

This is formally identical with the Euler product for the reciprocal of the Selberg zeta function $Z(s)$ (see e.g. [9]) where $\mathcal{P}$ denotes in that case the set of conjugacy classes of primitive hyperbolic matrices, and $N\mathfrak{p}$ is the norm of such a class. Helped by this analogy (which shouldn't be pushed too far since $Z(s)$ satisfies the

Riemann hypothesis while $\zeta_\infty(s)$ has its zeros spread out over the whole plane) we first note that $\zeta_\infty(s)$ is a meromorphic function of order 2 (more accurately, removing the poles, $(\sin \pi s)\,\zeta_\infty(s)$ is an entire function of order 2). This is easily proven and left to the reader.

Second, we can try to find a kind of functional equation for $\zeta_\infty(s)$, involving not only Γ-factors, but also Barnes' Γ_2-function, as for $Z(s)$ (see [16]). This is easily done as a consequence of the functional equation for $\zeta_A(s)$ itself. One such result is as follows :

THEOREM 7.1. - Set

$$W_\infty(s) = |D|^{-(s(s+1))/4} \left(\left(\Gamma\left(\tfrac{s}{2}\right)^{-1} \Gamma_2(s) \right)^{\frac{1}{2}} \pi^{s^2/4}\, 2^{(s-1)(s-2)/4} \right)^{r_1} \times$$
$$\times \left(\Gamma(s)^{-1} \Gamma_2(s) (2\pi)^{s(s+1)/2} \right)^{r_2} \zeta_\infty(s)$$

and

$$\Lambda_\infty(s) = W_\infty(s)\, W_\infty(-s)\, \frac{\sin^2 \pi s}{\pi^2\, C_\infty^2} \; .$$

Then Λ_∞ is an entire function of order 2 which is even and periodic of period 1, such that $\Lambda_\infty(n) = 1$ for $n \in \mathbb{Z}$.

Proof. - Simply note that $W_\infty(s)/W_\infty(s-1) = 1/\Lambda_A(s)$ and hence the periodicity of Λ_∞ is trivially equivalent to the functional equation of Λ_A.

Remark. - It is an easy exercise to check that $\left(\Gamma\left(\tfrac{s}{2}\right)^{-1} \Gamma_2(s)\right)^{\frac{1}{2}}$ is a (single valued) meromorphic function on $\mathbb{C}$. We choose the square root so that it is positive for s positive.

Having a natural periodic function at hand, it is natural to plot it for real values of s, and this is what the first author did on a computer in the case of $A = \mathbb{Z}$, hence $D = 1$, $r_1 = 1$, $r_2 = 0$. The astounding (and impossible) result was that $\Lambda_\infty(s)$ seemed to be constant equal to 1 for all real s. Of course this is absurd since it would then have to be equal to 1 for all complex s, which is impossible since $\Lambda_\infty(s)$ vanishes at all the complex zeroes (and their integer translates) of the Riemann zeta function $\zeta_{\mathbb{Z}}(s)$.

This apparent paradox was resolved a few days later by computing the Weierstrass product of the function $\Lambda_\infty(s)$, which turns out to be particulary simple. The result is as follows (for any domain A) :

THEOREM 7.2. -

$$\Lambda_\infty(s) = \prod_{\operatorname{Im}\rho>0} \left(1 - \frac{\sin^2 \pi s}{\sin^2 \pi\rho}\right)$$

<u>where the product is over the non trivial zeroes of</u> $\zeta_A(s)$ <u>with positive imaginary part</u>.

The proof is left to the reader.

In the case $A = \mathbb{Z}$, we have the first zero $\rho_1 \simeq \frac{1}{2} + 14.134\,i$ hence $\sin^2 \pi\rho_1 \simeq \operatorname{ch}^2(\pi \times 14.134) \simeq \frac{1}{4}\, e^{2\pi \times 14.134} \simeq 9 \times 10^{37}$ and it is easy to deduce from this and known estimates on the zeroes ρ, that for s real, $|\Lambda_\infty(s) - 1| < 2.10^{-38}$. Hence one would need multiprecision arithmetic to at least 40 decimals to be able to detect that $\Lambda_\infty(s) \neq 1$!

§ 8. - <u>The fundamental heuristic assumptions</u>

We begin here the second part of this paper. Except if explicitly stated otherwise, it must be considered that all the statements made in this part are <u>conjectural</u>. These conjectures all derive essentially from one heuristic principle which we now explain.

Let Γ be an abelian group of order N, and r_1, r_2 chosen such that $r_1 + 2r_2 = N$. Finally we let $A = A_\Gamma$ be the maximal order in the ring $\mathbb{Q}[\Gamma] / \sum_{g\in\Gamma} g$. It is well known that A_Γ is unique, and that it is a product of ring of integers of number fields.

Hence, as was mentioned at the beginning of section 2, the theory developed in part 1 is applicable to A.

<u>Examples</u>. - 1) If $\Gamma = \mathbb{Z}/N\mathbb{Z}$ with N <u>prime</u>, then $A_\Gamma = \mathbb{Z}[\sqrt[N]{1}]$, the ring of integers of the Nth cyclotomic field.

2) For $\Gamma = \mathbb{Z}/4\mathbb{Z}$ then $A_\Gamma = \mathbb{Z}[i] \times \mathbb{Z}$.

3) For $\Gamma = \mathbb{Z}/2\mathbb{Z} \times \mathbb{Z}/2\mathbb{Z}$ then $A_\Gamma = \mathbb{Z} \times \mathbb{Z} \times \mathbb{Z}$.

We will write $\mathfrak{F}_{\Gamma, r_1, r_2}$ (or simply $\mathfrak{F}$ when there is no ambiguity on Γ, r_1, r_2) for the set of isomorphism classes of abelian extensions of $\mathbb{Q}$ with Galois group Γ, r_1 real places and $2r_2$ complex ones. Note that $\mathfrak{F}_{\Gamma, r_1, r_2} = \emptyset$ unless $r_1 = N$, $r_2 = 0$ (the totally real case) or $r_1 = 0$, $r_2 = N/2$ (the totally complex case).

We assume the set of fields in $\mathfrak{F}$ ordered by the absolute value of the discriminant, and in the (rare) cases of equal discriminant, any ordering will do. If $K \in \mathfrak{F}$ and $\mathcal{K} = \mathcal{K}(K)$ is the <u>prime to</u> N <u>part</u> of the class group of K, it is easy to see that $\mathcal{K}$ is a finite A_Γ-module. Hence, if f is a function defined on isomorphism classes of finite A_Γ-modules (of order prime to N if necessary) we can define the <u>average</u> of f on the prime to N part of the class groups as the following limit, if it exists :

$$M(f) = \lim_{X \to \infty} \frac{\sum_{K \in \mathfrak{F},\, |D(K)| \leq X} f(\mathcal{K}(K))}{\sum_{K \in \mathfrak{F},\, |D(K)| \leq X} 1}$$

where D(K) is the discriminant of K and $\mathcal{K}(K)$ is the prime to N part of the class group.

FUNDAMENTAL ASSUMPTIONS 8.1. - <u>For all "reasonable" functions</u> f (<u>including probably non negative functions</u>) <u>we have</u> :

(1) (<u>Complex quadratic case</u>) <u>If</u> $r_1 = 0$, $r_2 = 1$ <u>then</u> M(f) <u>is the</u> 0-<u>average of</u> f <u>restricted to</u> A_Γ-<u>modules of order prime to</u> N . [<u>Here in fact</u> $A_\Gamma = \mathbb{Z}$, $N = 2$.]

(2) (<u>Totally real case</u>) <u>If</u> $r_1 = N$, $r_2 = 0$ <u>then</u> M(f) <u>is the</u> 1-<u>average of</u> f <u>restricted to</u> A_Γ - <u>modules of order prime to</u> N .

For lack of experimental evidence, we do not make any assumptions in the totally complex case, except when $N = 2$. We hope to come back to this in another paper (see also section 10). Note also that in both cases, we take the u-average, where u is the A_Γ-rank of the groups of units.

In the next section we will give some consequences of these fundamental assumptions. In the rest of this section we would like to try to justify them.

The first assumption, for the complex quadratic case, is exactly the assumption mentioned in section 1, i.e. weighting isomorphism classes of abelian groups G with weight proportional to $1/\#\,\mathrm{Aut}\;G$. Since the number of group structures on a set with n elements which are isomorphic to G is $n!/\#\,\mathrm{Aut}\;G$, the assumption above, <u>for a given</u> n , boils down to giving equal weight to each group <u>structure</u>. However for different n it is difficult to compare. Hence it would seem that one could define the 0-average of f as

$$M_o(f,\psi) = \lim_{x\to\infty} \frac{\sum_{n\le x} \psi(n) \sum_{G(n\mathbb{Z})} f(G)\, w(G)}{\sum_{n\le x} \psi(n) \sum_{G(n\mathbb{Z})} w(G)}$$

for some function ψ. Luckily, it turns out that for quite a wide class of functions ψ, including for instance the non zero polynomials, $M_o(f,\psi)$ is independent of ψ, whence the choice $\psi = 1$.

It is much more difficult to justify the second assumption. Let us assume $N = 2$ (i.e. the real quadratic case), the case of general N being a reasonable extrapolation from this case. Then it is well known that in terms of binary quadratic forms the class group can be obtained as follows : Consider the set of reduced binary quadratic forms having the right discriminant. This set is finite. In the imaginary quadratic case, composition of quadratic forms gives a group law on this set, and the group is exactly the class group. In the real quadratic case this is not true for several reasons which all boil down to the fact that the group of units is of rank 1 instead of 0. However in some sense which can be made precise, composition gives a group-like structure to this set, if we neglect a logarithmic number of reductions to be done. Furthermore this set breaks into cycles under the reduction operation, and in some sense one can interpret the principal cycle as being a "cyclic subgroup" ; finally the cycles do not have necessarily the same number of forms, but their length (in the sense of [11] or [13]) is the same, i.e. the regulator R. The number of these cycles being the class number, our heuristic assumption can be reformulated in the following way : the class group of a real quadratic field is of the form $G/\langle\sigma\rangle$, where G is a "random" group, weighted as usual with $1/\#\,\mathrm{Aut}\ G$, and σ is a random element in G (we denote by $\langle\sigma\rangle$ the cyclic subgroup of G generated by σ). The group G can then be thought of as the "group" of reduced quadratic forms, and $\langle\sigma\rangle$ as the principal cycle.

Another way of saying this is that we are trying to give a group theoretical interpretation of the trivial equality $h = hR/R$.

The "explanations" above have been put on more solid ground by the second author [11], and under this interpretation one should try to extend the techniques of the preceding sections to compact groups.

A very analogous situation was suggested to us by B. Gross. Let p be a fixed prime, and consider the set of <u>imaginary</u> quadratic fields K where p splits :

$p = \mathfrak{p}\bar{\mathfrak{p}}$. Then almost by definition $\mathcal{H}(O_K[1/\mathfrak{p}]) \simeq \mathcal{H}(O_K)/\langle\mathfrak{p}\rangle$, where $O_K[1/\mathfrak{p}] = \{x \in K / \mathfrak{p}x \subset O_K\}$. This is exactly of the type $G/\langle\sigma\rangle$, and G is weighted with $1/\#\,\mathrm{Aut}\,G$ if we assume assumption 1, and $\{\mathfrak{p}\}$ is random in $\mathcal{H}(O_K)$. Tables of such class groups reveal a striking similarity with tables of class groups of real quadratic fields.

§ 9. - Consequences of the heuristic assumptions

It must be again emphasized that all the results in this section are conjectural, except noted otherwise. No "proofs" are given since the conjectures are trivial consequences of the assumptions and the work done in the first part of the paper.

I. - Complex quadratic fields

Here $\mathcal{H}$ will denote the odd part of the class group, $h = \#\mathcal{H}$, $\mathcal{H}_p$ will be the p-part of $\mathcal{H}$, $r_p(\mathcal{H})$ will be the p-rank of $\mathcal{H}$, where p is always an odd prime.

All constants and zeta functions are relative to $A = \mathbb{Z}$.

(C 1) The probability that $\mathcal{H}$ is cyclic is equal to

$$\zeta(2)\,\zeta(3)/(3\,\zeta(6)\,C_\infty\,\eta_\infty(2)) \simeq 97.7575\ \% .$$

(C 2) The probability that p divides h is equal to

$$f(p) = 1-\eta_\infty(p) = p^{-1} + p^{-2} - p^{-5} - p^{-7} + \dots .$$

In particular

$$f(3) \simeq 43.987\ \% \ ; \quad f(5) \simeq 23.967\ \% \ ; \quad f(7) \simeq 16.320\ \% .$$

(C 3) The probability that	$\mathcal{H}_3 \simeq \mathbb{Z}/9\mathbb{Z}$	is close to	9.335 %
"	$\mathcal{H}_3 \simeq (\mathbb{Z}/3\mathbb{Z})^2$	"	1.167 %
"	$\mathcal{H}_3 \simeq (\mathbb{Z}/3\mathbb{Z})^3$	"	0.005 %
"	$\mathcal{H}_3 \simeq (\mathbb{Z}/3\mathbb{Z})^4$	"	2.3×10^{-8}
"	$\mathcal{H}_5 \simeq \mathbb{Z}/25\mathbb{Z}$	"	3.802 %
"	$\mathcal{H}_5 \simeq (\mathbb{Z}/5\mathbb{Z})^2$	"	0.158 %

(The exact formulas can easily be obtained from example 5.9 (i).).

(C 4) Let n be odd. The average number of elements of $\mathcal{H}$ of order exactly equal to n is 1.

(C 5) The probability that $r_p(\mathcal{H}) = r$ is equal to

$$p^{-r^2}\,\eta_\infty(p) \prod_{1\le k\le r} (1-p^{-k})^{-2} .$$

(C 6) The average of

$$\prod_{0\leq i<\alpha} (p^{r_p(\mathcal{K})} - p^i)$$

where α a fixed integer, is equal to 1. In particular the average of $p^{r_p(\mathcal{K})}$ is equal to 2 and that of $p^{2r_p(\mathcal{K})}$ is equal to $p+3$.

It is a consequence of a theorem of Heilbronn-Davenport (see [5]) that the average of $3^{r_3(\mathcal{K})}$ is equal to 2. Thus (C 6) is true for $\alpha = 1$, $p = 3$.

II. - <u>Real quadratic fields</u>

We keep the same notations as in the complex case. All the conjectural statements made in that case have an analog here. We give a few :

(C 7) The probability that p divides h is equal to

$$1 - \prod_{k\geq 2} (1 - p^{-k}) = p^{-2} + p^{-3} + p^{-4} - p^{-7} - \ldots \quad .$$

(C 8) Let n be odd. The average number of elements of $\mathcal{K}$ of order exactly equal to n is $1/n$.

(C 9) The probability that $r_p(\mathcal{K}) = r$ is equal to

$$p^{-r(r+1)} \eta_\infty(p) \prod_{1\leq k\leq r} (1 - p^{-k})^{-1} \prod_{1\leq k\leq r+1} (1 - p^{-k})^{-1} \quad .$$

(C 10) The average of

$$\prod_{0\leq i<\alpha} (p^{r_p(\mathcal{K})} - p^i)$$

where α is a fixed integer, is equal to $p^{-\alpha}$. In particular the average of $p^{r_p(\mathcal{K})}$ is equal to $1 + p^{-1}$ and that of $p^{2r_p(\mathcal{K})}$ is equal to $2 + p^{-1} + p^{-2}$.

It is again a consequence of a theorem of Heilbronn-Davenport (see [5]) that the average of $3^{r_3(\mathcal{K})}$ is equal to $4/3$. Thus (C 10) is true for $\alpha = 1$, $p = 3$.

A number of results are uninteresting in the complex case (for example the analogue of (C 11) would say that the probability that $\mathcal{K} \simeq L$ is equal to 0, which is true since the class number tends to infinity).

(C 11) If L is a group of odd order ℓ, the probability that $\mathcal{K}$ be isomorphic to L is equal to :

$$(2\ell\, C_\infty\, \eta_\infty(2)\ \#\,\mathrm{Aut}\, L)^{-1} .$$

In particular, if $p(\ell)$ is the probability that $\# \mathcal{K} = \ell$, we have :

$$p(1) \simeq 75.446\,\% \ ; \ p(3) \simeq 12.574\,\%; \ p(5) \simeq 3.772\,\% \ ;$$

$$p(7) \simeq 1.796\,\% \ ; \ p(9) \simeq 1.572\,\% \ .$$

If we make the extra assumption that fields with prime discriminants behave like the others with respect to the odd part of the class group, then $p(\ell)$ is the probability that $h = \ell$ when one restricts to prime discriminants.

(C 12) (Suggested by C. Hooley) Call $h(p)$ the class number of $\mathbb{Q}(\sqrt{p})$. Then, when p is restricted to the primes congruent to 1 mod. 4, and $x \to \infty$:

a) The probability that $h(p) > x$ is asymptotic to $\frac{1}{2x}$;

b) $\sum_{p \leq x} h(p) \sim x/8$.

III. - Higher degree fields

We give two examples :

(C 13) For cyclic cubic extensions (i.e. $\Gamma = \mathbb{Z}/3\mathbb{Z}$, $r_1 = 3$, $r_2 = 0$, $A_\Gamma = \mathbb{Z}[\sqrt[3]{1}]$) the probability that the class number is divisible by 2 (or by 4, which is the same) is equal to

$$1 - \prod_{k \geq 2} (1 - 4^{-k}) \simeq 8.195\,\% \ .$$

(C 14) For totally real extensions of prime degree p (including $p = 2$) (here $\Gamma = \mathbb{Z}/p\mathbb{Z}$, $r_1 = p$, $r_2 = 0$, $A_\Gamma = \mathbb{Z}[\sqrt[p]{1}]$) the probability that the prime to p part of the class number is 1, is equal to

$$(1 - p^{-1}) / (\eta_\infty(p) \prod_{k \geq 2} \zeta_{\mathbb{Q}(\sqrt[p]{1})}(k))$$

where $\zeta_{\mathbb{Q}(\sqrt[p]{1})}(s)$ is the Dedekind zeta function of the cyclotomic field $\mathbb{Q}(\sqrt[p]{1})$.

One can easily check that the above probability tends to 1 as $p \to \infty$. This would imply that, at least for $\mathbb{Z}/p\mathbb{Z}$-extensions, the non triviality of the class group comes only from the p-part. In fact, if we assume that we can restrict to prime conductors (as we did in the real quadratic case) the probability above is the probability that $h = 1$, when restricted to prime conductor. Hence, contrary to popular opinions, the proportion of class number 1 fields would seem to increase (and tend to 1) among fields of prime conductor. Apparently this had already been predicted by C. L. Siegel. (We thank D. Shanks for this information). Tables seem to agree with this : we have seen that the probability is 75.446 % in the real quadratic case, and it is close to 85.0 % in the cyclic cubic case, and both are close to the observed data ([14], [7]). We lack sufficient data in the cyclic quintic case.

§ 10. - Discussion of the conjectures. Further work

A) All the conjectures that we make are in close agreement with existing tables ([3], [4], [6], [7], [14]). Furthermore a conjecture like (C 5) helps to explain why class groups with high 3-rank (for instance) are difficult to find : to our knowledge, the record is 3-rank 5, and we have $3^{-25} \simeq 10^{-12}$ while $3^{-36} \simeq 7.10^{-18}$. This can help to give an indication of the difficulty of finding 3 - rank 6.

B) A very nice fact is that two particular cases of our conjectures ((C 6), $\alpha = 1$, $p = 3$ and (C 10), $\alpha = 1$, $p = 3$) are in fact theorems, due to Heilbronn-Davenport. Since all the conjectures are consequences of a single heuristic principle, this gives strong support for this principle, hence for the rest of the conjectures.

C) By a completely different heuristic method, C. Hooley has also conjectured (C 12). (Personal communication)

D) We can try to obtain statistical information on class groups of complex quadratic orders and not only on maximal orders. A priori, the only information available is the formula for the class number :

$$h(Df^2) = \Big[f \prod_{\substack{\ell \mid f \\ \ell \text{ prime}}} \big(1 - \big(\tfrac{D}{\ell}\big)/\ell\big) \Big] h(D)$$

where D is a fundamental discriminant.

With a naïve assumption of probabilistic independence, one can obtain from (C 2) the following conjecture :

(C'2) The probability that p divides the class number of a complex quadratic order (p an odd prime) is equal to

$$f'(p) = 1 - (1-p^{-3})\, \eta_\infty(p) \prod_{\substack{\ell \equiv \pm 1 \ (\mathrm{mod}\ p) \\ \ell \text{ prime},\ \ell > 2}} (1 - (\ell - 1)/2\ell^3) \times \begin{cases} 1 & \text{if } p > 3 \\ 11/12 & \text{if } p = 3 \end{cases} .$$

This gives for example

$$f'(3) \simeq 52.4664\ \% \quad ; \quad f'(5) = 25.1301\ \% \quad ; \quad f'(7) = 16.9271\ \% ,$$

in reasonable agreement with the tables.

E) It is interesting to notice that in many cases, the observed probabilities or averages do not oscillate around the predicted value, but seem to have a generally monotonic behavior (taken in a very wide sense) towards the predicted limit. For example, the probability that $3|h$ is around 42.5 or 43 % instead of 43.987 % for discriminants less than 10^9 (private communication of C. P. Schnorr) while in the real quadratic case with prime discriminant the proportion of class number 1 seems to decrease very slowly, and is still around 77 % for $D = 10^7$ ([14]).

F) In the totally complex case $r_1 = 0$, $r_2 = N/2$, for which we have not given any assumptions except for $N = 2$, J. Martinet (private communication) has suggested the following : if K is such a field let K_0 be its maximal real subfield. Then the 0-average of f should be the average of f taken on the <u>relative</u> class group, i.e. classes $c \in \mathcal{H}$ such that $c\bar{c} = 1$ in $\mathcal{H}$, where denotes complex conjugation.

G) It would be very interesting to extend the above conjectures to non abelian Γ, and in fact more generally to non Galois extensions of $\mathbb{Q}$. The first case to consider for which plenty of tables are available, is the case of non cyclic cubics, either with $r_1 = 1$, $r_2 = 1$, or totally real.

The behavior of the N-part, while certainly not random, should also be investigated.

H) In most of the conjectures, values of the function $\zeta_\infty(s)$ or of an Euler factor of that function occur (see typically (C 2), (C 11)). Since we believe these conjectures to be true at least in the complex quadratic case, we are led to believe that any <u>proof</u> of these conjectures <u>must</u> use analytic functions of order 2 like $\zeta_\infty(s)$ and in fact maybe $\zeta_\infty(s)$ itself.

A confirmation of this belief comes from the fact that the only cases where the conjectures have indeed been proved using existing mathematical tools (the Heilbronn Davenport theorems) are also the only cases in which the result does not contain Euler factors or values of functions of order 2 (with the exception of C 12, but here the difficulty lies probably in dealing with the regulator). It would in fact be very interesting to know if the Heilbronn-Davenport results can be extended to proving C 6 or C 10 with $p = 3$ and $\alpha = 2$ or with $p = 5$ and $\alpha = 1$.

Acknowledgements

It is a pleasure for us to thank our friends and colleagues B. Gross, J. Martinet D. Shanks, L. Washington, D. Zagier for valuable discussions during the preparation of this paper and D. Buell, C. P. Schnorr, D. Shanks and H. Williams for making available to us, or even computing for us, extensive tables which have not yet been published and which confirmed our conjectures.

-:-:-:-

REFERENCES

[1] N. BOURBAKI, Algèbre commutative, ch. 2.

[2] N. BOURBAKI, Algèbre commutative, ch. 7, § 4, ex. 10.

[3] D. A. BUELL, Class groups of quadratic fields, Math. Comp. 30 (1976), 610-623.

[4] D. A. BUELL, The expectation of good luck in factoring integers-some statistics on quadratic class numbers, technical report n° 83-006, Dep^{t} of Computer Science, Louisiana State University/Baton Rouge.

[5] H. DAVENPORT, H. HEILBRONN, On the density of discriminants of cubic fields II, Proc. Royal Soc., A 322 (1971), 405-420.

[6] M.-N. GRAS et G. GRAS, Nombre de classes des corps quadratiques réels $\mathbb{Q}(\sqrt{m})$, $m<10\,000$, Institut de Math. Pures, Grenoble (1971-72).

[7] M.-N. GRAS, Méthodes et algorithmes pour le calcul numérique du nombre de classes et des unités des extensions cubiques cycliques de $\mathbb{Q}$, J. reine und angew. Math. 277 (1975), 89-116.

[8] P. HALL, A partition formula connected with Abelian groups, Comment. Math. Helv. 11 (1938-39), 126-129.

[9] D. HEJHAL, The Selberg trace formula for $PSL(2, \mathbb{R})$ I, Springer Lecture notes 548 (1976) and II, Springer Lecture notes 1 001 (1983).

[10] I. KAPLANSKY, Commutative rings, Allyn and Bacon (1970), p. 146.

[11] H. W. LENSTRA, Jr., On the calculation of regulators and class numbers of quadratic fields, pp 123-150 in : J. V. Armitage (ed.), Journées Arithmétiques 1980, London Math. Soc. Lecture notes series 56, Cambridge University Press (1982).

[12] J.-P. SERRE, Corps locaux, Hermann (1966).

[13] D. SHANKS, The infrastructure of real quadratic fields and its applications, proc. 1972 number theory conference, Boulder (1972).

[14] D. SHANKS, H. WILLIAMS, in preparation.

[15] G. TENENBAUM, Cours de théorie analytique des nombres, Bordeaux (1980).

[16] M.-F. VIGNÉRAS, L'équation fonctionnelle de la fonction zêta de Selberg du groupe modulaire PSL(2, $\mathbb{Z}$), Astérisque 61 (1979), 235-249.

-:-:-:-

H. COHEN
L.A. au C.N.R.S. n° 226
Mathématiques et Informatique
Université de Bordeaux
351, cours de la Libération
33405 Talence Cedex (France)

H. W. LENSTRA, Jr.
Mathematisch Instituut
Universiteit van Amsterdam
Roetersstraat 15
1018 WB Amsterdam
(The Netherlands)

LOWER BOUNDS FOR REGULATORS

T. W. Cusick
Department of Mathematics
State University of New York at Buffalo
Buffalo, New York 14214

. Introduction.

The purpose of this paper is to give some lower bounds depending on he discriminant for the regulators of some algebraic number fields of egrees three and four. In the cubic case, our results are best possible.

Until the last section of the paper, we consider only totally real ields, so the regulator can be defined as follows: Let F be a totally eal field of degree n. If δ is an element of F, let $\delta, \delta', \delta'', \ldots, \delta^{(n-1)}$ enote its conjugates. If $\epsilon_1, \ldots, \epsilon_{n-1}$ is a fundamental system of units or F, then the regulator R of F equals $|\det[\log|\epsilon_i^{(j-1)}|]|$, $1 \leq i, j \leq n-1$.

Our result for cubic fields is

HEOREM 1. Let R denote the regulator of a totally real cubic field ith discriminant D; then

$$R \geq \frac{1}{16} \log^2(D/4).$$

or infinitely many fields, the constant 1/16 cannot be replaced by a arger number.

The situation for quartic fields is more complicated. If the field as no quadratic subfield, then we prove a lower bound for the regulator hich is best possible up to a constant factor, but the constant given by ur method may not be the best one.

HEOREM 2. Let R denote the regulator of a totally real quartic field ith discriminant D and having no quadratic subfield; then

$$R \geq (80\sqrt{10})^{-1} \log^3 (D/16).$$

If the quartic field has a quadratic subfield, then only the second ower of $\log D$ should appear in the lower bound. This is discussed later n.

Remak [7] was the first to give lower bounds for regulators which epend on the discriminant rather than merely on the degree of the field (as in Remak [6]). He showed that if F is a field of degree n and discriminant D, then the regulator R is bounded below by $c\log(|D|n^{-n})$ for some constant c depending on n, provided F is not a field "with unit defect". A field F has unit defect if it contains a proper subfield with the same number of multiplicatively independent units as F. Remak [7, p. 248] showed that F has unit defect if and only if F is a totally imaginary field of degree 2 over a totally real subfield. For such fields, Remak [7, pp. 249-250] gives an example to show that the regulator need not tend to infinity as $|D|$ does. Such fields are also called CM fields.

For the particular fields in our Theorems 1 and 2, the procedure of Remak [7, pp. 281-283] gives lower bounds with the correct power of $\log D$, but with worse constants.

There is a paper of Pohst [5] which gives improvements on the Remak estimates for various fields with degrees ≤ 11. However, Pohst only gives numerical lower bounds for the regulators of various specific fields, plus a method (based on solving certain extremal value problems) for computing such a numerical lower bound in any given case. He does not actually state any general lower bounds in terms of the field discriminant D. For cubic fields, the method of Pohst (see especially [5, Satz VIII, p. 478 and Tabelle I, p. 487]) would give the inequality in our Theorem 1. For small values of D, the Pohst method gives numerical lower bounds [5, Tabelle I for n = 3, p. 487] which are better than the ones given by the inequality in Theorem 1. However this will not be the case for large D, since the constant in Theorem 1 is best possible. Of course, for a cubic field with small D, it is possible to compute the regulator exactly by finding a fundamental pair of units (see for example [2]), so there is no need for any lower bounds.

For fields with only one fundamental unit, it is easy to obtain good lower bounds for the regulator. The case of real quadratic fields is trivial [5, p. 485]. For the other types of field with one fundamental unit, we prove the following results in Section 7 below:

THEOREM 3. Let R denote the regulator of a nontotally real cubic field with discriminant $-D$; then

$$R \geq \frac{1}{3}\log(D/27).$$

For infinitely many fields, the constant 1/3 cannot be replaced by a larger number.

THEOREM 4. Let R denote the regulator of a totally complex quartic field F with discriminant D; if F contains a fundamental unit which generates F, then

$$R \geq \frac{1}{4}\log(D/256) .$$

It is very likely that for infinitely many fields, the constant 1/4 cannot be replaced by a larger number.

2. The inequality in Theorem 1.

Let F be a totally real cubic field with discriminant D. Let ϵ be any unit $\neq \pm 1$ in F and let Δ denote the discriminant of ϵ, so

$$\Delta = (\epsilon - \epsilon')^2 (\epsilon - \epsilon'')^2 (\epsilon' - \epsilon'')^2 .$$

Let $r(1), r(2), r(3)$ denote the permutation of $0, 1, 2$ which gives

$$|\epsilon^{(r(1))}| \leq |\epsilon^{(r(2))}| \leq |\epsilon^{(r(3))}|$$

and let s denote the map defined by $s(r(j)+1) = j$ for $j = 1,2,3$. Since ϵ is a unit $\neq + 1$, the identity

$$\Delta = \prod_{1\leq i<j\leq 3} (1 - (\epsilon^{(r(i))}/\epsilon^{(r(j))}))^2 \prod_{i=1}^{3} (\epsilon^{(r(i))})^{2i}$$

implies

$$\log D \leq \log \Delta \leq \log 4 + 2\sum_{i=1}^{3} s(i) \log|\epsilon^{(i-1)}| ; \tag{1}$$

the term $\log 4$ in the last inequality comes from the fact that the function $f(x,y) = (1 - xy)(1 - x)(1 - y)$ satisfies $f(x,y) \leq 2$ when $|x| \leq 1$ and $|y| \leq 1$.

Define

$$y_i = \log|\epsilon^{(i-1)}| \qquad (i = 1,2,3) . \tag{2}$$

Let ϵ_1, ϵ_2 be a fundamental pair of units for F. Thus there exist integers x_1, x_2 such that

$$\log|\epsilon^{(i)}| = x_1 \log|\epsilon_1^{(i)}| + x_2 \log|\epsilon_2^{(i)}| \quad (i = 0,1,2) . \tag{3}$$

We define a quadratic form $Q(\vec{x})$ in x_1, x_2 by

$$Q(\vec{x}) = \sum_{i=1}^{3} y_i^2 = \sum_{i=1}^{2} \sum_{j=1}^{2} a_{ij}x_i x_j .$$

A calculation gives the determinant of the quadratic form Q :

(4) $$\det Q = \det [a_{ij}] = 3R^2 ,$$

where R is the regulator of F .

As in the work of Remak [7], our lower bound for R is derived from (1) and (4), as follows: Since $y_1 + y_2 + y_3 = 0$, if we introduce a parameter λ in (1) and use the Cauchy-Schwarz inequality, we obtain

(5) $$\log(D/4) \leq 2 \sum_{i=1}^{3} (s(i) - \lambda)y_i \leq 2 (Q(\vec{x}) \sum_{i=1}^{3} (s(i) - \lambda)^2)^{\frac{1}{2}} .$$

Choosing $\lambda = 2$ minimizes the sum on the right-hand side. An old result on definite binary quadratic forms (see Cassels [1, pp. 30-31]) states that there exist integers x_1, x_2 not both zero such that

(6) $$Q(\vec{x}) \leq (\tfrac{4}{3} \det Q)^{\frac{1}{2}} .$$

If we choose the unit ϵ so that (3) and (5) hold, then (4) and (5) give the inequality in Theorem 1.

3. Cubic fields with small regulators.

We shall prove that the constant 1/16 in Theorem 1 cannot be replaced by a smaller number by exhibiting a sequence of cubic fields such that the ratio $R/\log^2 D$ tends to 1/16 as the discriminant D of the fields tends to infinity.

Let K_a denote the cubic field defined by one of the roots of the polynomial

$$f_a(x) = x^3 + ax^2 - (a + 3)x + 1$$

with discriminant $D(f_a) = (a^2 + 3a + 9)^2$. The field K_a is cyclic (so it coincides with its conjugate fields) and the roots $\eta_a, \eta_a', \eta_a''$ of $f_a(x)$ satisfy

$$1 < \eta_a < 2, 0 < \eta_a' < 1, \eta_a'' < 0, \eta_a(\eta_a' - 1) = - 1 .$$

In fact, we have

$$\lim_{a \to \infty} \eta_a = \lim_{a \to \infty} a\eta_a' = \lim_{a \to \infty} -(\eta_a'' + a) = 1 . \tag{7}$$

Of course it may happen that $D(f_a)$ is larger than the discriminant of the field K_a. The following lemma shows that we can often be certain that this is not the case.

LEMMA 1. If $a^2 + 3a + 9$ is squarefree, then $D(f_a)$ is equal to the discriminant of the field K_a.

Proof. This follows from the fact that every prime p which divides a squarefree integer $a^2 + 3a + 9$ ramifies in K_a. We can see this, for example, by a Newton polygon argument like the one in an example of Weiss [9, p. 167].

Our next lemma says that the situation in Lemma 1 occurs infinitely often.

LEMMA 2. There are infinitely many positive integers a such that $a^2 + 3a + 9$ is squarefree.

Proof. A special case of an old result of Nagell [4, pp. 183-186] gives the lemma. Indeed, Nagell's result applies to any irreducible quadratic polynomial with integer coefficients.

Define

$$U(a) = \left| \det \begin{bmatrix} \log|\eta_a| & \log|\eta_a'| \\ \log|\eta_a'| & \log|\eta_a''| \end{bmatrix} \right|$$

Now a calculation using (7) gives

$$\lim_{a \to \infty} U(a)/\log^2 D(f_a) = \frac{1}{16} . \tag{8}$$

If a is large and $D(f_a)$ is the discriminant of the field K_a, then (8) and the inequality in Theorem 1 imply that $U(a)$ is equal to the regulator of K_a (because $U(a)$ must be an integer multiple of the regulator). Thus it follows from Lemmas 1 and 2 that a sequence of cubic fields with $R/\log^2 D$ tending to $1/16$ exists, as we wanted to show.

We remark in passing that the method of Cusick [2] can be used to prove directly that η_a, η'_a is a fundamental pair of units for K_a whenever $D(f_a)$ is the field discriminant; this gives another (but lengthier) proof that $U(a)$ equals the regulator of K_a.

We also remark that although the fields K_a which we used as examples above are all cyclic, there is no reason to believe that the inequality of Theorem 1 can be improved for noncyclic fields. For instance, let L_a denote the cubic field defined by the root $\theta > 0$ of of the polynomial

$$g_a(x) = x^3 + ax^2 - 2x - a$$

with discriminant $D(g_a) = 4a^4 + 13a^2 + 32$. It can be shown that $\theta - 1$, $\theta + 1$ is a fundamental pair of units for L_a, and that if $V(a)$ denotes the regulator of L_a, then

$$\lim_{a \to \infty} V(a)/\log^2 D(g_a) = \frac{1}{16} .$$

The fields L_a are never cyclic if $a > 1$, for then $D(g_a)$ lies between $(2a^2 + 3)^2$ and $(2a^2 + 4)^2$. Thus if there exist infinitely many a such that $D(g_a)$ is the discriminant of L_a (which seems very likely although the results of Nagell [4] do not suffice to prove it), then we have a sequence of noncyclic fields with the desired small regulators.

4. The inequality in Theorem 2.

Let F be a totally real quartic field with discriminant D and having no quadratic subfield. We shall use an argument similar to that in Section 2 in order to deduce the inequality in Theorem 2.

Let ϵ be any unit $\neq \pm 1$ in F and let Δ denote the discriminant of ϵ. Let $r(1), r(2), r(3), r(4)$ denote that permutation of $0,1,2,3$ which gives

$$|\epsilon^{(r(i))}| \leq |\epsilon^{(r(i+1))}| \qquad (i = 1,2,3) \tag{9}$$

and let s denote the map defined by $s(r(j) + 1) = j$ for $j = 1,2,3,4$ Since ϵ is a unit $\neq \pm 1$ and F has no quadratic subfield, ϵ generat the field F. Therefore $\log D \leq \log \Delta$ and the identity

(10) $$\Delta = \prod_{1\le i<j\le 4} (1-(\epsilon^{(r(i))}/\epsilon^{(r(j))}))^2 \prod_{i=1}^{4} (\epsilon^{(r(i))})^{2i}$$

implies

(11) $$\log D \le \log \Delta \le \log 16 + 2 \sum_{i=1}^{4} s(i) \log|\epsilon^{(i-1)}| .$$

The term $\log 16$ in (11) comes from the fact that the first product in (10) is bounded above by 16 whenever (9) holds. The proof of this is a tedious calculation; Pohst [5, pp. 467-470] gives the details for one way of doing it.

Define y_i for $i = 1,2,3,4$ by (2). Let $\epsilon_1,\epsilon_2,\epsilon_3$ be a fundamental triple of units for F. Thus there exist integers x_1,x_2,x_3 such that

$$\log|\epsilon^{(j)}| = \sum_{i=1}^{3} x_i \log|\epsilon_i^{(j)}| \quad (j = 0,1,2,3) .$$

We define a quadratic form $Q(\vec{x})$ in x_1,x_2,x_3 by

$$Q(\vec{x}) = \sum_{i=1}^{4} y_i^2 = \sum_{i=1}^{3}\sum_{j=1}^{3} a_{ij}x_ix_j .$$

A calculation gives

(12) $$\det Q = \det[a_{ij}] = 4R^2 ,$$

where R is the regulator of F.

Since $y_1 + \ldots + y_4 = 0$, if we introduce a parameter λ in (11) and use the Cauchy-Schwarz inequality, we obtain

(13) $$\log(D/16) \le 2 \sum_{i=1}^{4} (s(i) - \lambda)y_i \le 2(Q(\vec{x}) \sum_{i=1}^{4} (s(i) - \lambda)^2)^{\frac{1}{2}} .$$

Choosing $\lambda = 5/2$ minimizes the sum on the right-hand side. There exist integers x_1,x_2,x_3 not all zero such that

(14) $$|Q(\vec{x})| \le (2\det Q)^{1/3}$$

(for this result on definite ternary quadratic forms, which goes back to Gauss, see Cassels [1, pp. 33-35]). If we choose the unit ϵ so that (14) holds, then (12) and (13) give the inequality in Theorem 2.

5. Quartic fields with small regulators.

We give some examples which make it very likely that, for infinitel[y] many fields, the constant $(80\sqrt{10})^{-1} = (252.98)^{-1}$ cannot be replaced by a number larger than $(216)^{-1}$. Only the lack of suitable analogues for Lemmas 1 and 2 prevents us from actually proving that our examples work.

Let H_a denote the quartic field defined by the smallest positive root of the polynomial

$$h_a(x) = x^4 + ax^3 - (3a+6)x^2 + (2a+3)x + 1 .$$

The roots $\mu_a^{(i)}$ $(i = 0,1,2,3)$ of $h_a(x)$ satisfy

$$0 < \mu_a < 1,\ 2 < \mu_a' < 3,\ -a-3 < \mu_a'' < -a-2,\ -1 < \mu_a''' < 0$$

and in fact

$$(15) \quad \lim_{a\to\infty} a(1-\mu_a) = \lim_{a\to\infty} 2a(\mu_a' - 2) = \lim_{a\to\infty} (\mu_a'' + a + 4) = \lim_{a\to\infty} -2a\mu''' = 1 .$$

The discriminant of $x^4 + px^3 + qx^2 + rx + s$ is given by

$$\frac{1}{27}(4(q^2 - 3pr + 12s)^3 - (2q^3 - 72qs + 27p^2s - 9pqr + 27r^2)^2) ,$$

so the polynomial discriminant $D(h_a)$ satisfies

$$(16) \qquad \lim_{a\to\infty} D(h_a)(4a^6)^{-1} = 1 .$$

Let D_a denote the discriminant of the field H_a. It is likely that $D_a = D(h_a)$ for infinitely many a; indeed, many would conjecture that $D(h_a)$ is equal to a prime for infinitely many a, and this would suffice. For the rest of this section we consider only values of a fo[r] which $D_a = D(h_a)$.

We can see that H_a has no quadratic subfield for $a \geq 0$ as follows: We have $D(h_a) \equiv 5 \bmod 8$, so D_a is not a square. A resolven[t] cubic polynomial for $h_a(x)$ is

$$r(x) = x^3 + (3a+6)x^2 + (2a^2+3a-4)x - (5a^2+24a+33) ,$$ and $r(x)$ is irreducible for $a \geq 0$. These facts together imply that the Galois group of H_a is the symmetric group S_4, and so H_a has no quadrati[c] subfield. (Incidentally, $r(x)$ is reducible for $a = -1$, and H_{-1}

is the totally real quartic field with smallest possible discriminant, viz. 725.)

It is easy to see that $\mu_a - 1$ and $\mu_a - 2$ are units of norm -1 in H_a. If we define $\epsilon_1 = \mu_a$, $\epsilon_2 = \mu_a - 1$, $\epsilon_3 = \mu_a - 2$ and let

$$W(a) = |\det[\log|\epsilon_i^{(j-1)}|]| \qquad (i,j = 1,2,3),$$

then a computation using (15), (16) and $D_a = D(h_a)$ gives

$$\lim_{a\to\infty} W(a)/\log^3 D_a = \frac{1}{216}. \tag{17}$$

For a large a, it follows from (17) and the inequality in Theorem 2 that $W(a)$ is equal to the regulator of H_a. Thus we have the examples referred to at the beginning of the section, subject to our assumption that there exist infinitely many a with $D_a = D(h_a)$.

6. Quartic fields with quadratic subfields.

The method of Section 4 does not give a satisfactory result if the totally real quartic field F being considered has a quadratic subfield. The difficulty comes from the fact that the unit ϵ corresponding to a solution of (14) may lie in a quadratic subfield; then it does not generate F and its discriminant Δ is 0, so we cannot make use of (11). It is possible to modify the argument so that a unit ϵ which does generate F is obtained, but this gives weaker results. The method of Remak [7] gives a lower bound for the regulator R of the form $c\log D$, where D is the discriminant of F. The method of Pohst [5, Satz XII, p. 484] would give $R \geq c\log^2 D$. This is the correct form for the lower bound, as the following lemma shows.

LEMMA 3. Let R denote the regulator and D the discriminant for the totally real quartic field J_a generated by $\sqrt{2}$ and $\sqrt{a^2+1}$, where $a \geq 2$ is one of the values of y for which $x^2 - 2y^2 = 1$ is solvable; then $R \leq c\log^2 D$ for some absolute constant c.

Proof. Let $b = a^2+1$, so that $D = 64b^2$ (note that a is always even under our hypotheses). The units in J_a defined by

$$\epsilon_1 = 1 + \sqrt{2},\ \epsilon_2 = a + \sqrt{b},\ \epsilon_3 = x + \sqrt{2b},$$

where x is the positive integer such that $x^2 - 2a^2 = 1$, each lie in one of the three quadratic subfields of J_a. Indeed, ϵ_1 is a fundamental unit in its subfield, and if $b > 5$ is squarefree then ϵ_2 and ϵ_3 are fundamental units in their respective subfields. If we let

$$T(a) = |\det[\log|\epsilon_i^{(j-1)}|]| \qquad (i,j = 1,2,3),$$

then a calculation gives

$$T(a) \leq c_1 \log^2 b \leq c_2 \log^2 D$$

for some constants c_1 and c_2. Since $T(a)$ is an integral multiple of R, this proves the lemma.

Let F be an algebraic number field of degree d with regulator R and absolute value of discriminant D. Let $r = r(F)$ be the rank of the unit group in F and let s be the maximum of $r(k)$ as k ranges over all proper subfields of F. Joseph Silverman [8] has proved that

$$R > c_d(\log a_d D)^{r-s}$$

where the positive constants a_d and c_d depend only on d. He gives heuristic reasoning which suggests that the exponent $r - s$ is best possible; Lemma 3 above establishes this for the case of totally real quartic fields with quadratic subfields.

7. Proof of Theorems 3 and 4.

We first let F be a nontotally real cubic field with discriminant $-D$. Following the argument in Section 2 above gives

$$\log D \leq \log \Delta \leq \log 27 + 2 \sum_{i=1}^{3} s(i) \log|\epsilon^{(i-1)}| \tag{18}$$

in place of (1); the constant $\log 27$ comes from an old inequality of I. Schur, quoted by Remak in [7, p. 253].

We can choose the conjugate fields of F so that ϵ is real, and we can suppose $\epsilon < 1$. Thus

$$\epsilon < |\epsilon'| = |\epsilon''| = \epsilon^{-1/2},$$

so that $s(i) = i$ for $i = 1,2,3$. Hence the sum in (18) is $3\log|\epsilon'| = -(3/2)\log\epsilon$. If ϵ is a fundamental unit, then the regulator R is $-\log\epsilon$, so (18) gives

$$\log D \leq \log 27 + 3R.$$

This proves the inequality in Theorem 3.

To prove that the constant 1/3 in Theorem 3 is best possible, we consider the polynomials

$$f_a(x) = x^3 - ax^2 - 1$$

for $a \geq -1$. The discriminant $D(f_a)$ is $-4a^3 - 27$. It follows from a result of Erdös [3] that $D(f_a)$ is squarefree for infinitely many positive integers a. Using this result, we can argue as in Section 3 to show that for infinitely many cubic fields, the constant 1/3 in Theorem 3 cannot be replaced by a larger number.

Now let F be a totally complex quartic field. This means F may be a CM field, but we can still prove the analogue of (11) if F contains a fundamental unit ϵ which generates F; we need only replace the constant log 16 by log 256, using the inequality of I. Schur again. Now a simple argument as in the nontotally real cubic case gives the inequality in Theorem 4. We remark that Remak [7, p. 285] also obtained the inequality in Theorem 4 by a more involved argument.

We could show that the constant 1/4 in Theorem 4 is best possible by considering the polynomial

$$g_a(x) = x^4 + ax^2 + 1$$

with discriminant $D(g_a)$, provided that the corresponding field has this same discriminant for infinitely many a. As in Section 5, this seems to require a very probable but still unproved result about values of polynomials. This completes the proof of Theorem 4.

REFERENCES

1. J. W. S. Cassels, An Introduction to the Geometry of Numbers (Springer, 1959).

2. T. W. Cusick, Finding fundamental units in cubic fields, Math. Proc. Cambridge Phil. Soc. 92 (1982), 385-389.

3. P. Erdös, Arithmetical properties of polynomials, J. London Math. Soc. 28 (1953), 416-425.

4. T. Nagell, Zur Arithmetik der Polynome, Abh. Math. Sem. Hamburg 1 (1922), 178-193.

5. M. Pohst, Regulatorabschätzungen für total reelle algebraische Zahlkörper, J. No. Theory 9 (1977), 459-492.

6. R. Remak, Über die Abschätzung des absoluten Betrages des Regulators eines algebraischen Zahlkörpers nach unten, J. Reine Angew. Math 167 (1932), 360-378.

7. R. Remak, Über Grössenbeziehungen zwischen Diskriminante und Regulator eines algebraischen Zahlkörpers, Compos. Math. 10 (1952), 245-285.

8. J. H. Silverman, An inequality relating the regulator and the discriminant of a number field, preprint.

9. E. Weiss, Algebraic Number Theory (McGraw-Hill, 1963).

SUR QUELQUES MOYENNES DES COEFFICIENTS DE FOURIER DE FORMES MODULAIRES

par

Jean-Marc DESHOUILLERS

-:-:-:-

Cette rédaction diffère sensiblement de l'exposé oral en ceci qu'elle ne présente qu'un des points abordés dans la conférence ; nous espérons que le lecteur intéressé par une introduction à la "Kloostermanie", selon la terminologie introduite par Huxley, trouvera une réponse à sa quête dans les premières pages de l'article [Deshouillers-Iwaniec 1] .

Dans les dernières années, différents auteurs se sont intéressés à l'évaluation asymptotique (lorsque X tend vers l'infini) de sommes du type

$$(1) \qquad \sum_{1\leq n\leq X} a_n \overline{b_{n+1}} \ , \quad \sum_{n\geq 1} w\left(\frac{n}{X}\right) a_n \overline{b_{n+1}}$$

où a_m, b_m sont des coefficients de Fourier de formes modulaires, et w un poids lisse à décroissance exponentielle de moyenne 1 ; en voici quelques exemples :

. $a_m = b_m = \tau(m)$, où τ désigne la fonction de Ramanujan ($\Delta(z) = e^{2\pi iz} \prod_{n\geq 1} (1-e^{2\pi inz})^{24} = \sum_{m\geq 1} \tau(m)\, e^{2\pi imz}$, définie pour $\mathrm{Im}\, z > 0$, est une forme modulaire holomorphe parabolique de poids 12 (cf. [Serre])). On a les évaluations

$$\sum_{n\geq 1} e^{-(n/X)} \tau(n)\,\tau(n+1) = O_\varepsilon(X^{12-\frac{1}{2}+\varepsilon}) \qquad \text{[Goldfeld]}$$

$$\sum_{1\leq n\leq X} \tau(n)\,\tau(n+1) = O_\varepsilon(X^{12-\frac{1}{3}+\varepsilon}) \qquad \text{[Good]}$$

. $a_m = b_m = d(m) = \sum_{d|m} 1$; on notera que la fonction

$$g(z) = \frac{1}{2} y^{\frac12} (\mathrm{Log}\, y + \gamma - \mathrm{Log}(4\pi)) + y^{\frac12} \sum_{n\neq 0} d(n) K_o(2\pi |n| y) e^{2\pi i n x}$$

définie pour $\mathrm{Im}\, z > 0$ est une forme modulaire réelle analytique (non parabolique) de poids nul [Maass] . On a les évaluations

$$\sum_{1\leq n\leq X} d(n)\, d(n+1) = X\, P_2 (\mathrm{Log}\ X) + O_\varepsilon (X^{2/3+\varepsilon})\ ^{(*)}$$ [Deshouillers-Iwaniec 2]

ce qui améliore les résultats précédents [Estermann], [Heath-Brown] et un inédit de Wirsing).

$$\sum w(\tfrac{n}{X})\, d(n)\, d(n+1) = X\, P_2 (\mathrm{Log}\ X) + O_\varepsilon (X^{\frac12+\varepsilon})$$

e résultat se déduit très facilement de [Deshouillers - Iwaniec 2] .

. $a_m = \square(m)$, fonction caractéristique des carrés ; on notera que $(z) = \frac12 + \sum_{n\geq 1} \square(n)\, e^{2\pi i n z}$ est une forme modulaire holomorphe (non parabolique) de poids $\frac12$ relativement au groupe $\Gamma_o(4)$ [Rankin] . Dans la majoration

$$\sum_{1\leq n\leq X} \square(n)\, d(n+1) = \sum_{1\leq n\leq X} d(n^2+1) = X\, P_1 (\mathrm{Log}\ X) + O_\varepsilon (X^{8/9+\varepsilon})$$ [Hooley]

l'exposant $\frac{8}{9}$ du terme d erreur peut être remplacé par $\frac{2}{3}$, et, dans le cas où un poids lisse est introduit, il peut être ramené à $\frac{1}{2}$ (Deshouillers - Iwaniec, non publié) ; voir aussi [Sarnak] et [Bikovski] .

On notera dès maintenant que tous ces résultats ont la même structure, l'absence du terme principal n'étant dû qu'à la parabolicité d'une des deux formes considérées.

Ainsi que le note Selberg, deux abords sont possibles :

. la "voie Rankin" qui passe par l'étude de la série de Dirichlet $\sum a_n b_{n+1} (n+\frac12)^{-s}$

. la "voie Kloosterman", alias méthode du cercle, et ses variantes.

Le premier chemin est celui qui a été suivi jusqu'à présent ; il a été présenté par Goldfeld et Good dans le cadre des Journées Arithmétiques et ne sera pas plus développé ici ; mentionnons simplement pour mémoire les résultats de Hejhal concernant des expressions similaires à celles considérées en (1) et relatives à des sous-groupes co-compacts de $PSL_2(\mathbb{Z})$.

(*) La notation P_k désigne un polynôme de degré k , pas nécessairement le même d'une ligne à l'autre.

Fouvry et moi avons exploré la seconde route ; nous nous bornerons ici à la présenter dans le cas où $a_m = b_m = \tau(m)$. En considérant le développement de Fourier de Δ indiqué ci-dessus, on obtient aisément

$$e^{-(1/2X)} \sum_{n\geq 1} e^{-n/X} \tau(n)\, \tau(n+1) = \int_0^1 |\Delta(x+\frac{i}{4\pi X})|^2\, e(x)\, dx$$

avec la notation usuelle $e(u) = e^{2\pi i u}$. On remarquera dès à présent que le traitement trivial de l'intégrale (l'équivalent de la méthode de Hecke) conduit à la majoration X^{12}, que nous appellerons ultérieurement majoration triviale. Pour évaluer l'intégrale du second membre, il est possible de recourir à la méthode du cercle (c'est d'ailleurs ce qu'il faudrait faire pour des formes de faible poids, cf. [Lehner]), mais ici on obtient un résultat plus net (i. e. sans terme d'erreur) en passant par la méthode de Petersson : soit

$$\mathfrak{P}_1(z) = \sum_{\left(\begin{smallmatrix} a & b \\ c & d\end{smallmatrix}\right)\in \mathcal{R}} (cz+d)^{-12}\, e(\frac{az+b}{cz+d}) = \sum_{n\geq 1} p_1(n)\, e(nz)$$

la première série de Poincaré de poids 12, où $\mathcal{R}$ désigne un système de représentants de $\Gamma_\infty \backslash PSL_2(\mathbb{Z})$.

Notons

$$I = \int_0^1 \mathfrak{P}_1(x+\frac{i}{4\pi X})\, \overline{\Delta(x+\frac{i}{4\pi X})}\, e(x)\, dx = p_1(1) \int_0^1 |\Delta(x+\frac{i}{4\pi X})|^2\, e(x)\, dx \;;$$

d'après la définiton de $\mathfrak{P}_1$, on a

$$I = \int_0^1 e(z)\, \Delta(z)\, e(x)\, dx + \int_0^1 \sum^{*}_{\left(\begin{smallmatrix} a & b \\ c & d\end{smallmatrix}\right)\in \mathcal{R}} (cz+d)^{-12}\, e(\frac{az+b}{cz+d})\, \overline{\Delta(z)}\, e(x)\, dx$$

où $z = x + \frac{i}{4\pi X}$, et où l'étoile indique que l'on omet la classe de l'identité dans la sommation. On utilise la décomposition de Bruhat

$$PSL_2(\mathbb{Z}) = \Gamma_\infty \cup \bigcup_{c>0} \bigcup_{\substack{d \bmod. c \\ (d,c)=1}} \Gamma_\infty\, \sigma_{c,d}\, \Gamma_\infty ,$$

où $\sigma_{c,d}$ est représentée par une matrice $\left(\begin{smallmatrix} * & * \\ c & d\end{smallmatrix}\right)$ de $SL_2(\mathbb{Z})$, disons $\left(\begin{smallmatrix} a & b \\ c & d\end{smallmatrix}\right)$. D'après la modularité de Δ, on a

$$I = \int_{-\infty}^{\infty} \sum_{c>0} \sum_{\substack{d \bmod. c \\ (d,c)=1}} |cz+d|^{-24}\, \overline{\Delta(\frac{az+b}{cz+d})}\, e(\frac{az+b}{cz+d})\, e(x)\, dx .$$

On développe alors Δ en série de Fourier ; en notant que $\overline{e(z)} = e(-\bar{z})$, il vient

$$I = \sum_{m\geq 1} \tau(m) \sum_{c>0} \sum_{\substack{d \bmod. c \\ (d,c)=1}} \int_{-\infty}^{\infty} |cz+d|^{-24} e(-m\frac{a\bar{z}+b}{c\bar{z}+d} + \frac{az+b}{cz+d} + x)\, dx$$

on décompose $\frac{az+b}{cz+d} = \frac{a}{c} - \frac{1}{c(cz+d)} \equiv \frac{\bar{d}}{c} - \frac{1}{c(cz+d)}$ (mod. $\mathbb{Z}$), où $d\bar{d} \equiv 1$ (mod. c) et on effectue le changement de variable $cu = c.z + d$, il vient

$$\int_{-\infty}^{\infty} = c^{-24} e(\frac{(1-m)\bar{d}-d}{c}) \int_{-\infty}^{\infty} (u^2 + \frac{1}{16\pi^2 X^2})^{-12} e(\frac{m}{c^2(u-i/(4\pi X))} - \frac{1}{c^2(u+i/(4\pi X))} + u)\, du$$

En notant $S(m, n ; c)$ la somme de Kloosterman $\sum_{\substack{d \bmod. c \\ (d,c)=1}} e(\frac{m\bar{d}+nd}{c})$, nous avons donc obtenu

THÉORÈME (Deshouillers - Fouvry). - <u>On a</u>

$$(2) \qquad \sum_{n\geq 1} e^{-n/X} \tau(n)\tau(n+1) = e^{1/2X}(p_1(1))^{-1} \sum_{m\geq 1} \tau(m) \sum_{c>0} c^{-24} S(1-m, -1; c)\, g_X(m,c) ,$$

<u>où</u> $p_1(1)$ <u>est le premier coefficient de Fourier de la première série de Poincaré, et</u>

$$g_X(m,c) = \int_{-\infty}^{\infty} (u^2 + \frac{1}{16\pi^2 X^2})^{-12} \exp(-\frac{(m+1)}{2Xc^2(u^2+1/(16\pi^2X^2))})\, e(\frac{(m-1)u}{c^2(u^2+1/(16\pi^2X^2))} + u)\, du .$$

La fonction g n'est pas très agréable à regarder, mais ce n'est qu'un exercice d'en expliciter ses propriétés principales (ordre de grandeur en fonction de m et c, dépendance lisse en c). Des majorations (2) et

$$(3) \qquad S(1-m, -1; c) \ll_\varepsilon c^{1-\delta+\varepsilon}$$

on déduit facilement

$$\sum_{n\geq 1} e^{-n/X} \tau(n)\tau(n+1) \ll_\varepsilon X^{12+\varepsilon-\delta/2} ,$$

c'est-à-dire que toute majoration non triviale des sommes de Kloosterman conduit à une majoration non triviale de la somme que nous considérons, ce résultat étant obtenu sans utiliser la théorie de Maass-Selberg (cf. l'introduction de [Good]).

Cette théorie permet en revanche de majorer les sommes de Kloosterman en moyenne [Kuznetsov] et de montrer que lorsqu'elles sont pondérées par une fonction lisse, on a, en moyenne, $\delta = 1$ dans (3) (e.g. [Deshouillers-Iwaniec 1], th. 9). La dépendance lisse de $g_X(m,c)$ en c permet alors de retrouver le Théorème (3) de [Goldfeld].

On notera à ce propos que la somme $\sum_{n\geq 1} d^2(n)\, K_o^2(\frac{2\pi n}{X})$ considérée par J. Unterberger peut être traitée de façon similaire, et que le terme d'erreur $O(X)$ obtenu par l'auteur peut être réduit à $cX + O(X^{\frac{1}{2}+\varepsilon})$; en effet, dans ce cas les sommes de Kloosterman se réduisent à des sommes de Ramanujan, triviales pour $m = 1$, ce qui fournit le terme principal. Bien entendu, dans ce problème, la méthode de Rankin est la voie royale, comme l'a remarqué Godement.

Pour conclure, signalons l'existence d'une méthode alternative pour traiter les sommes (1) dans le cas particulier $a_m = b_m = d(m)$ qui utilise simplement le fait que d est le carré de convolution de Dirichlet de la fonction $n \longmapsto 1$. Utilisant cette méthode, Iwaniec et moi (cf. [Deshouillers]) avons également pu traiter des sommes du type $\sum_n w(n)\, d(n)\, d_3(n+1)$, où d_3 est le nombre de représentations d'un entier comme produit de trois facteurs. Cela laisse quelqu'espoir de pouvoir traiter la somme $\sum_n w(n)\, d_3(n)\, d_3(n+1)$ considérée par Vinogradov et Takhtadzhyan, cette somme d'écrivant $\sum_n w(n)\, r'(n)\, d_3(n+1)$ où $r'(n)$ est coefficient de Fourier d'une série θ.

-:-:-:-

BIBLIOGRAPHIE

[BIKOVSKI], Propriétés asymptotiques des points entiers (a_1, a_2) satisfaisant la congruence $a_1 a_2 \equiv \ell$ (mod. q). Zapiski Nauč. Sem. LOMI 112 (1981), 5-25.

[DESHOUILLERS], Majorations en moyenne de sommes de Kloosterman, Sém. Th. Nb. Bordeaux (1980-1981), exposé n° 3, 5p.

[DESHOUILLERS - IWANIEC 1], Kloosterman sums and Fourier coefficients of cusp forms, Invent. Math. 70 (1982), 219-288.

[DESHOUILLERS - IWANIEC 2], An additive divisor problem, J. London math. soc. (2) 26 (1982), 1-14.

[ESTERMANN], Über die Darstellung einer Zahl als Differenz von zwei Produkten, J. reine ang. math. 164 (1931), 173-182.

[GODEMENT], Evaluation d'une somme arithmétique. Remarques, Bull. S.M.F. 10 (1973), 125-127.

[GOLDFELD], Analytic and arithmetic theory of Poincaré series, Astérisque 61 (1979), 95-107.

[GOOD], On various means involving the Fourier coefficients of cusp forms, Prepri du Forschungsinstitut für Mathematik ETH Zürich (1982).

[GUNNING], Lectures on modular forms, Annals of Mathematics Studies n° 48, Princeton University Press (1962).

[HEATH-BROWN], The fourth power moment of the Riemann zeta function, Proc. London Math. Soc. (3) 38 (1979), 385-422.

[HEJHAL], Sur certaines séries de Dirichlet dont les pôles sont sur des lignes critiques, C.R.A.S. 287 (1978), 383-385.
(voir également C.R.A.S. 294 (1982), 273-276, 509-512 et 637-640).

[HOOLEY], An asymptotic formula in the theory of numbers, J. London Math. Soc. (3) 7 (1957), 396-413.

[KUZNETSOV], Petersson hypothesis for parabolic forms of weight zero and Linnik hypothesis. Sums of Kloosterman sums, Mat. Sb. 111 (153) (1980), 334-383.

[MAASS], Über eine neue Art von nichtanalytischen automorphen Funktionen, Math. Ann. 121 (1949), 141-183.

[RANKIN], Modular forms and functions, Cambridge University Press, Cambridge (1977).

[SARNAK], Additive number theory and Maass forms, Manuscrit (1982).

[SELBERG], On the estimation of Fourier coefficients of modular forms, in Proc. Symposia on pure math. VIII, A.M.S., Providence (1965).

[VINOGRADOV - TAKHTADZHYAN], Theory of Eisenstein series for the group SL(3, $\mathbb{R}$) and its application to a binary problem, J. Soviet. Math. 18 (1982), 293-324.

[UNTERBERGER], Evaluation d'une somme arithmétique associée à une forme modulaire non analytique, Bull. S.M.F. 101 (1973), 113-124.

-:-:-:-

Jean-Marc DESHOUILLERS
L. A. au C.N.R.S. n° 226
U.E.R. de Math. et d'Informatique
Université de Bordeaux I
351, cours de la Libération
F - 33405 TALENCE CEDEX

ZUR ALGEBRAISCHEN UNABHÄNGIGKEIT GEWISSER WERTE DER EXPONENTIALFUNKTION

R. Endell
Theodor-Heuß-Ring 15
D-3000 Hannover

1. Einleitung

A.O. Gelfond ([6]) bewies 1949 mit einer neuen Methode die algebraische Unabhängigkeit von wenigstens zwei Zahlen unter gewissen Werten der Expo nentialfunktion. Eines seiner Ergebnisse war die algebraische Unabhängig keit von α^{β} und α^{β^2} für algebraische Zahlen $\alpha \neq 0,1$ und β mit Grad $\beta=3$. Se ne Methode wurde von verschiedenen Autoren weiterentwickelt; siehe z.B. D.W. Brownawell [1], G.V. Chudnovsky [3], A.A. Shmelev [10] und M. Waldschmidt [12].

1974 veröffentlichte G.V. Chudnovsky [4] eine Erweiterung von Gelfon Methode, mit der er die algebraische Unabhängigkeit von r Zahlen unter g wissen Werten der Exponentialfunktion zeigte. In seiner Arbeit führt er mehrere wichtige neuen Techniken ein, jedoch enthält der Beweis einige w sentliche Lücken und einige Punkte seiner Beweisführung bleiben unklar.

Bevor wir die Ergebnisse angeben, haben wir einige Bezeichnungen ein zuführen. Für ganze Zahlen $n \geq 1$, $m \geq 1$ seien $v_1, \ldots, v_n$ und $w_1, \ldots, w_m$ komple Zahlen, die jeweils linear unabhängig über $\mathbb{Q}$ sind und darüberhinaus ein Maß für lineare Unabhängigkeit der folgenden Gestalt besitzen:

Für $\underline{x}=(x_1,\ldots,x_n) \in \mathbb{Z}^n$, $\underline{y}=(y_1,\ldots,y_m) \in \mathbb{Z}^m$, $\underline{x} \neq 0$ und $\underline{y} \neq 0$ gelte mit $\underline{v}=(v_1,\ldots,v_n)$ und $\underline{w}=(w_1,\ldots,w_m)$ stets die folgenden Ungleichungen:

$$|\langle \underline{x}, \underline{v} \rangle| > \exp(-\tau |x| \ln |x|) \qquad ($$
$$|\langle \underline{y}, \underline{w} \rangle| > \exp(-\tau |y| \ln |y|) \qquad ($$

Hierbei sind $\tau > 0$ eine Konstante, die nur von $\underline{v}$ und $\underline{w}$ abhängig ist, $|x| = \max_{1 \le i \le n} |x_i|$, $|y| = \max_{1 \le j \le m} |y_j|$ und $\langle .,. \rangle$ bedeutet das Skalarprodukt zweie Vektoren.

Schließlich bezeichnen wir mit $r > 0$ eine ganze Zahl. Dann gilt das folgende Resultat:

<u>THEOREM 1:</u> *Ist* $2^r \leq \frac{mn}{m+n}$, *dann sind r+1 Zahlen aus der Menge*

$$S_1=\{e^{v_i w_j},\ 1\leq i\leq n,\ 1\leq j\leq m\}$$

algebraisch unabhängig über $\mathbb{Q}$.

<u>Beispiel zu Theorem 1:</u> Sei $\alpha>0$ algebraisch und t transzendent. Dann sind 2 der sieben Zahlen

$$e^t,\ e^{te^{\alpha}},\ldots,\ e^{te^{6\alpha}}$$

algebraisch unabhängig. ($v_i=e^{(i-1)\alpha}$, $w_i=te^{(i-1)\alpha}$, $i=1,2,3,4$)

<u>THEOREM 2:</u> *Ist* $2^r \leq \frac{m(n+1)}{m+n}$, *dann sind r+1 Zahlen aus*

$$S_2=\{w_j, e^{v_i w_j},\ 1\leq i\leq n,\ 1\leq j\leq m\}$$

algebraisch unabhängig über $\mathbb{Q}$.

Bemerkung: Läßt man $r=0$ zu, dann sind die Theoreme 1 und 2 gültig, falls $1<\frac{mn}{m+n}$ bzw. $1<\frac{m(n+1)}{m+n}$ ist.

<u>Beispiel zu Theorem 2:</u> Sei β irrational algebraisch und $\alpha\neq 0,1$ ebenfalls algebraisch. Dann ist α^{β} transzendent (Gelfond-Schneider). ($n=1, m=2$, $v_1=\log\alpha$, $w_j=\beta^{j-1}$, $j=1,2$). Weiter folgt hieraus das schon eingangs erwähnte Resultat von Gelfond. ($n=m=3, v_i=\beta^{i-1}\log\alpha, w_i=\beta^{i-1}, i=1,2,3$)

<u>THEOREM 3:</u> *Ist* $2^r < \frac{mn+m+n}{m+n}$, *dann sind r+1 Zahlen aus der Menge*

$$S_3=\{v_i, w_j, e^{v_i w_j},\ 1\leq i\leq n,\ 1\leq j\leq m\}$$

algebraisch unabhängig über $\mathbb{Q}$.

<u>Beispiel zu Theorem 3:</u> Sei n eine positive ganze Zahl und sei die ganze Zahl r maximal gewählt mit $2^{r+1}<n+3$. Dann sind mindestens r+1 der Zahlen $e, e^e, e^{e^2},\ldots, e^{e^{2n}}$ algebraisch unabhängig.

In dieser Note wollen wir kurz die Beweise skizzieren; ausführliche Beweise geben wir in einer nachfolgenden Arbeit. Unsere Beweise unterscheiden sich in einigen Punkten erheblich von denen Chudnovskys. Sie führen - im Gegensatz zu denen von Chudnovsky - zu einem anwendbaren Kriterium für algebraische Unabhängigkeit (siehe hierzu Waldschmidt [14]).

2. Abschätzungen für die Norm gewisser Teilmengen aus $\mathbb{Z}[X_1,\ldots,X_q] \times \prod_{i=1}^{q} \mathbb{C}[X_i,\ldots,X_q]$

Bevor wir die Ergebnisse angeben, wollen wir einige Bezeichnungen einführen. Mit $X_1,\ldots,X_q$ bezeichnen wir unabhängige Variablen. Für reelle Zahl $\psi>0$ und Vektoren $\underline{a}=(a_1,\ldots,a_q)\in\mathbb{C}^q$ definieren wir Mengen $\Gamma(\underline{a},\psi)$ durch

$$\Gamma(\underline{a},\psi)=\{\underline{z}'=(z_1',\ldots,z_q')\in\mathbb{C}^q,\ \max_{1\le i\le q}|z_i'-a_i| < e^{-\psi}\}$$

$\underline{z}=(z_1,\ldots,z_q)$ sei ein beliebig vorgegebener Vektor aus $\mathbb{C}^q$. Mit K_1 bezeichnen wir stets den Körper der rationalen Zahlen und mit K_i, $2\le i\le q$, die Quotientenkörper von $\mathbb{Z}[z_1,\ldots,z_{i-1}]$. $R_i\in K_i[X_i]$ sei das normierte Minimalpolynom von z_i über K_i, falls z_i über K_i algebraisch ist, bzw. 1 sonst. Wir betrachten die Teilmenge

$$S(\underline{z}) \subseteq \mathbb{Z}[X_1,\ldots,X_q] \times \prod_{i=1}^{q}\mathbb{C}[X_i,\ldots,X_q]$$

die definiert ist durch

$$S(\underline{z})= \{(P_0,\ldots,P_q),\ P_0=R_1^{s_1}P_1,\ P_i(z_i,X_{i+1},\ldots,X_q)=R_{i+1}^{s_{i+1}}P_{i+1},$$
$$P_q(z_q)\neq 0,\ 1\le i<q,\ s_1,\ldots,s_q\ge 0,\ \text{ganz}\}$$

Ist $t(P)$ das Maß des Polynoms P (siehe [12]), so definieren wir für reel Zahlen $N>0$ die Mengen

$$S(\underline{z},N)= \{(P_0,\ldots,P_q)\in S(\underline{z}),\ t(P_0) < N\}.$$

Die Norm von $S(\underline{z},N)$ sei definiert durch

$$||S(\underline{z},N)||=\inf_{(P_0,\ldots,P_q)\in S(\underline{z},N)} |P_q(z_q)|$$

__Bemerkung:__ Der Beweis der Theoreme und der nachfolgenden Proposition 1 macht von der folgenden Konstruktion Gebrauch:
Sind natürliche Zahlen r und l, $r>2l>0$ und Polynome F_{hj}, $0\le h<r$, $0\le j<l$, a $\mathbb{Z}[X_1,\ldots,X_q]$ gegeben, so kann man mit Hilfe des Lemmas von Siegel zeigen daß es eine Konstante $c_1>0$ gibt und daß Polynome $b_h \in K_q[X_q]$ existieren, die das Gleichungssystem

$$\sum_{h=0}^{r-1} b_h(X_q)F_{hj}(z_1,\ldots,z_{q-1},X_q)=0 \quad (0\le j<l)$$

lösen und die folgenden Eigenschaften besitzen:

1) $\max_{0\le h<r} |b_h(z_q)|\neq 0$,
2) Ist $b_h(z_q)\neq 0$, dann ist b_h aus $pr_{q+1}S(\underline{z},c_1(\max_{h,j} t(F_{hj})+\ln r))$,

wobei pr_{q+1} die Projektion auf den q+1-ten Faktor bezeichne.
Zunächst kann man mit dem Lemma von Siegel zeigen, daß Polynome b_{oh} aus $[X_1,\ldots,X_q]$ existieren, die das folgende Gleichungssystem

$$\sum_{h=0}^{r-1} b_{oh}(X_1,\ldots,X_q)F_{hj}(X_1,\ldots,X_q)=0 \quad (0\le j<l)$$

nichttrivial lösen und deren Maß durch

$$t(b_{oh})\le c_1(\max_{h,j} t(F_{hj})+\ln r) \text{ für } 0\le h<r$$

abgeschätzt werden kann.
Im allgemeinen gilt nicht, daß $\max_{0\le h<r} |b_{oh}(z_1,\ldots,z_q)|\neq 0$ ist. Durch simultanes "Herausziehen" geeigneter Potenzen der Polynome R_i, kann man zeigen, daß Polynome $b_h, 0\le h<r$ aus $K_q[X_q]$ mit obigen Eigenschaften existieren.

Schließlich definieren wir für i=1,2,3 Funktionen $f_i:\mathbb{R}^+\longrightarrow\mathbb{R}$ durch
$f_1(M)=M^{\frac{mn}{m+n}-2^r}\ln M$, $f_2(M)=M^{\frac{m(n+1)}{m+n}-2^r}\ln^{\frac{n}{m+n}}M$ und $f_3(M)=M^{\frac{mn+m+n}{n+m}-2^r}$ für $M\in\mathbb{R}^+$.
Dann gilt die folgende

PROPOSITION 1: *Seien die Voraussetzungen von Theorem i erfüllt.* $\underline{\theta}_i=(\theta_{1i},\ldots,\theta_{qi})$ *sei eine Tranzendenzbasis von* $\mathbb{Q}(S_i)$. *(i=1,2,3)*
Dann gibt es positive Konstanten $c_2\ge 1$, $c_3\ge 1$, $c_4\le 1$ *und eine reelle Zahl* $M'>0$, *so daß für jede reelle Zahl* $M\ge M'$ *und jeden Vektor* $\underline{z}\in\Gamma(\underline{\theta}_i,c_2M^{2^r}f_i(M))$ *gilt:*

$$||S(\underline{z},c_3M)||\le \exp(-c_4M^{2^r}f_i(M)).$$

Der Beweis der Proposition 1 benutzt klassische Hilfsmittel wie das Lemma von Siegel, die Hermitesche Interpolationsformel und das "Small Value Theorem" von Tijdeman [11].

Seien $\underline{a}=(a_1,\ldots,a_q)$ aus $\mathbb{C}^q$ und eine reelle Zahl b mit $0<b\le 1$ beliebig vorgegeben. Setzt man $c_5=14q^2(q+3+\max_{1\le i\le q}|a_i|)^2$ und $c_6=2c_5^{2^{q+1}}$, dann gilt das folgende Resultat:

PROPOSITION 2: *Zu jeder reellen Zahl* $N>0$ *gibt es eine Zahl* $\psi=\psi(N)>0$ *und einen Vektor* $\underline{z}'$ *aus* $\mathbb{C}^q$ *mit den folgenden Eigenschaften:*
) $c_5b^{-1}N\le\psi\le c_6b^{-2^q}N^{2^q}$
) $\underline{z}'\in\Gamma(\underline{a},\psi)$
) $||S(\underline{z}',N)||>\exp(-2qb\psi)$

Der Beweis der Proposition 2 wird durch Induktion geführt. Wichtigste Hilfsmittel sind die Semi-Resultante und ihre Eigenschaften. Sie gestatt es ein sukzessives Eliminationsverfahren durchzuführen.

3. Beweis der Theoreme und weitere Bemerkungen

Beweis von Theorem 1: Die Voraussetzungen von Theorem 1 seien erfüllt. D Transzendenzgrad von $\mathbb{Q}(S_1)$ sei q. Aufgrund der Ergebnisse in [12], kann man ohne Einschränkung der Allgemeinheit annehmen, daß q größer als Null ist. Dann existieren q algebraisch unabhängige Zahlen $\theta_1,\ldots,\theta_q$ und eine über $\mathbb{Z}[\theta_1,\ldots,\theta_q]$ ganzalgebraische Zahl θ_{q+1}, die $\mathbb{Q}(S_1)$ erzeugen. Wir ze gen, daß $q\geq r+1$ ist.
Dazu betrachten wir die Norm von $S(\underline{z},N)$ für gewisse Vektoren $\underline{z}$ aus $\mathbb{C}^q$ un zeigen, daß eine reelle Zahl $N_0>0$ und Konstanten $c_7,c_8>0$ existieren, so d für jedes $N>N_0$ die beiden folgenden Aussagen gültig sind:

(1) Für jede reelle Zahl ζ mit $c_7N\leq \zeta \leq c_8N^{2^r}$ und für jedes $\underline{z}\in\Gamma((\theta_1,.,\theta_q),$
gilt: $\quad ||S(\underline{z},N)||\leq \exp(-c_4c_2^{-1}\zeta)$ (c_2,c_4 wie in Proposition

(2) Es gibt eine reelle Zahl ψ mit $c_7N\leq \psi \leq c_8N^{2^q}$ und ein $\underline{z}'\in\Gamma((\theta_1,.,\theta_q),$
so daß: $\quad ||S(\underline{z}',N)||> \exp(-c_4c_2^{-1}\psi)$

Aus (1) und (2) gewinnt man für ψ folgende Abschätzungen:

$$c_8N^{2^r} < \psi \leq c_8N^{2^q}$$

Daraus folgt $q>r$. Da q ganz ist, gilt $q\geq r+1$ wie in Theorem 1 verlangt.

Es genügt daher die Aussagen (1) und (2) zu beweisen. Wir wenden Proposition 2 mit $\underline{a}=(\theta_1,\ldots,\theta_q)$, $b=c_4c_2^{-1}(2q)^{-1}$, $c_7=c_5b^{-1}$ und $c_8=c_6b^{-2^q}$ an. Dann gilt (2) für jede reelle Zahl $N>0$.
Die Aussage (1) ist eine Folgerung aus Proposition 1.
Sind zunächst positive reelle Zahlen ζ und N mit $c_7N\leq \zeta \leq c_8N^{2^r}$ beliebig vorgegeben, so ist es aus Monotoniegründen stets möglich, $M=M(\zeta)$ durch

$$c_2M^{2^r}f_1(M)=\zeta$$

eindeutig zu bestimmen. Aus denselben Gründen ist es möglich, die Zahl M_o durch die Bedingung $f_1(M_0)=c_8c_2^{-1}c_3^{2^r}$ eindeutig festzulegen. Ist M' die Z aus Proposition 1, so wird N_0 definiert durch

$$M(c_7N_0)=1+\max(M_0,M').$$

Für $N \geq N_0$ gilt dann stets $M(\zeta) \geq 1+\max(M_0, M')$. Diese Ungleichung und die obere Schranke für ζ benutzt man, um für $N \geq N_0$ die Ungleichung $c_3 M \geq N$ herzuleiten. Dies zieht nach Definition sofort $||S(\underline{z},N)|| \leq ||S(\underline{z},c_3M)||$ nach sich. Für jedes $\underline{z} \in \Gamma((\theta_1,..,\theta_q),\zeta)$ kann die rechte Seite mit Proposition 1 durch $||S(\underline{z},c_3M)|| \leq \exp(-c_4 c_2^{-1} \zeta)$ abgeschätzt werden.

Für $N \geq N_0$ ist hiermit die Aussage (1) gezeigt.

Die Beweise der Theoreme 2 und 3 verlaufen analog. Wir ersetzen f_1 durch f_2 bzw. durch f_3.

Mit der hier vorgestellten Methode kann man die Transzendenzmaße in [5] beweisen. Für allgemeine algebraische Gruppen scheint die Methode in Ermanglung eines "Small Value Theorems" nicht anwendbar zu sein.

<u>Literatur</u>

1. Brownawell, W.D.: On the Development of Gelfond's Method, Kapitel 4 in Number Theory:Carbondale 1979(Proceedings), Berlin-Heidelberg-New York, 1979
2. Brownawell, W.D.: Some Remarks on Semi-Resultants, Kapitel 14 in Trancendence Theory, Advances and Applications, London 1977
3. Chudnovsky, G.V.: Algebraic Independence of Some Values of the Exponentialfunction, Math. Notes 15 (1974),391-398 (Übersetzung)
4. Chudnovsky, G.V.: Some Analytic Methods in the Theory of Transcendental Numbers und Analytic Methods in Diophantine Approximations, Preprint IM-74-8 u. 74-9, Kiev 1974, Ukrainian SSR Academy of Sciences
5. Chudnovsky, G.V.: A Mutual Transcendence Measure for some Classes of Numbers, Soviet Math. Dokl. 15(1974), 1424-1428 (Übersetzung)
6. Gelfond, A.O.: Transcendental and Algebraic Numbers, New York,1960
7. Nesterenko, J.V.: Approximations Diophantiennes et Nombres Transcendants, Luminy 1982, Progress in Math. 31, Birkhäuser(1983),199-220
8. Philippon, P.: Indépendance algebrique de valeurs de fonctions exponentielles p-adiques, J. reine angew. Math, 329(1981), 42-51
9. Reyssat, E.: Un critère d'indépendance algébrique, J. reine angew. Math., 329(1981),66-81
10. Shmelev, A.A.: On the Problem of the Algebraic Independence of algebraic Numbers, Math. Notes 11(1972),387-392
11. Tijdeman, R.: An Auxiliary Result in the Theory of Transcendental Numbers II, Duke Math. J. 42 (1975), 239-247
12. Waldschmidt, M.: Nombres Transcendants, Berlin-Heidelberg-New York 1974
13. Waldschmidt, M.,Zhu Yao Chen: Une généralisation en plusieurs variables d'un critère de transcendance de Gel'fond, (wird erscheinen)
14. Waldschmidt, M.: Algebraic Independence of Transcendental Numbers, (wird erscheinen)
15. Warkentin, P.: Algebraische Unabhängigkeit gewisser p-adischer Zahlen, Diplomarbeit, Freiburg, 1978

ANALYTICAL AND ARITHMETICAL METHODS IN THE THEORY OF FUCHSIAN GROUPS

A. Good, Forschungsinstitut für Mathematik
ETH-Zentrum, CH-8092 Zürich, Switzerland

1. Introduction

Ever since Riemann, people have looked for functions with properties similar to those of the Riemann zeta-function. This led on one side to the zeta-functions of quadratic forms and on the other to the zeta-functions of algebraic number fields. Essentially the same functions arise from binary quadratic forms and from quadratic number fields respectively. Nowadays all these functions are subsumed under the zeta-functions attached to automorphic forms via a kind of Mellin transformation. In particular, their functional equations are implied by the condition of automorphy and under certain assumptions the converse also holds.

There is yet another general principle for obtaining functional equations. This principle is based on the fact that selfadjoint operators have only real eigenvalues. Here this principle will be applied to Poincaré series which are automorphic eigenfunctions with singularities. As a result we obtain functional equations for all Fourier coefficients attached to these Poincaré series. 'Generically', these Fourier coefficients are meromorphic functions ressembling the logarithmic derivative of Selberg's zeta-functions. In a semi-degenerate situation the coefficients coincide essentially with Hecke's zeta-functions of quadratic fields. Thus there is a third context in which these functions turn up. Finally, a degenerate case is equivalent with the theory of the constant term matrix for Eisenstein series.

The methods we employ have links with different branches of mathematics. We mention just one example from algebra (structure of discrete groups; see later on), analysis (spectral theory of Laplacians; see later on), geometry (distribution of closed geodesics in Riemann surfaces; cf. [6]), number theory (mean-values for zeta-functions; cf. [2], [5], [7]) and automorphic forms (Rankin-Selberg convolution method for non-arithmetic groups; cf. [4]).

2. Preliminaries

We start recalling familar things (cf. [8], [13]). Let $G = SL_2(\mathbb{R})$ act on the upper half-plane $\mathfrak{H} = \{z = x+iy \mid x \text{ real}, y > 0\}$ by

$$(M,z) \longmapsto M(z) = \frac{az+b}{cz+d} \quad , \quad M = \begin{pmatrix} a & b \\ c & d \end{pmatrix} \quad . \tag{1}$$

We often write $z_M = x_M + iy_M$ instead of $M(z)$. Then the (hyperbolic) Laplacian

$$\Delta = y^2\left(\frac{\partial^2}{\partial x^2} + \frac{\partial^2}{\partial y^2}\right)$$

and the (hyperbolic) measure

$$d\omega(z) = \frac{dxdy}{y^2}$$

are invariant under this action of G on $\mathfrak{H}$. Thus Δ and ω have natural projections to the orbit space $\Gamma\backslash\mathfrak{H}$ if Γ is a discrete subgroup of G . We assume that the volume $\omega(\Gamma\backslash\mathfrak{H})$ is finite. This includes in particular the groups of finite index in the modular group $SL_2(\mathbb{Z})$ as well as the universal covering transformation groups of compact Riemann surfaces of genus greater than one. Let $L_2(\Gamma\backslash\mathfrak{H})$ denote the Hilbert space of Γ-invariant functions f on $\mathfrak{H}$ with $\langle f,f\rangle < \infty$, where the inner product is given by

$$\langle f,g\rangle = \int_{\Gamma\backslash\mathfrak{H}} f(z)\bar{g}(z)\,d\omega(z) \quad .$$

The Laplacian then defines a non-positive selfadjoint operator in $L_2(\Gamma\backslash\mathfrak{H})$ which for simplicity we still denote by Δ . Moreover, a Γ-invariant, bounded and smooth function f admits an absolutely and locally uniformly convergent expansion into eigenfunctions of Δ :

$$f(z) = \sum_{j\geq 0} \langle f,e_j\rangle e_j(z) + \frac{1}{4\pi i}\sum_{\iota=1}^{\kappa}\int_{(1/2)} \langle f,E_\iota(\cdot,s)\rangle E_\iota(z,s)\,ds \quad . \tag{2}$$

Here $(e_j)_{j\geq 0}$ denotes a maximal system of orthonormal functions in $L_2(\Gamma\backslash\mathfrak{H})$ such that

$$\Delta e_j(z) + s_j(1-s_j)e_j(z) = 0 \quad , \quad z \text{ in } \mathfrak{H} \quad .$$

We may assume that $\frac{1}{2} < s_j \leq 1$ or $s_j = \frac{1}{2} + it_j$, $t_j \geq 0$. These are the eigenfunctions tied up with the discrete spectrum of Δ . If Γ contains parabolic elements there is also a continuous spectrum. It gives rise to the κ Eisenstein series $E_\iota(z,s)$, one for each Γ-equivalence class of cusps. If $s = \sigma+it$ lies in $\sigma > 1$, the Eisenstein series are given by absolutely convergent series and satisfy

$$\Delta E_\iota(z,s) + s(1-s)E_\iota(z,s) = 0 \ , \quad z \text{ in } \mathfrak{H} \, .$$

By Selberg's theory they extend to meromorphic functions in s with no poles on $\sigma = 1/2$, the line over which one integrates in (2) . The expansion (2) is of a rather abstract nature since we still know very little about the e_j or $E_\iota(z,s)$ on $\sigma = 1/2$.

Automorphic functions have many Fourier series expansions. These always arise in connection with stabilizers Γ_ξ in Γ , where one distinguishes three types:

$$\xi = \begin{cases} \zeta = \text{any point in } \mathfrak{H} \text{ (elliptic type)}, \\ \vartheta = \text{cusp of } \Gamma \text{ (parabolic type)}, \\ \eta = (\eta_1, \eta_2) = \text{pair of fixpoints for a hyperbolic matrix in} \\ \Gamma \text{ (hyperbolic type)}. \end{cases}$$

Every Γ_ξ acts on $\mathfrak{H}$ in a cyclic manner. This allows to attach Fourier coefficients to a Γ-automorphic eigenfunction of Δ at every ξ . It also suggests to study Γ by reducing Γ modulo its stabilizers. In this way we are led to two basic problems:

(I) Analyze e_j and $E_\iota(z,s)$ through their Fourier coefficients arising from stabilizers in Γ.

(II) Determine the structure of the double coset space $\Gamma_\xi \backslash \Gamma / \Gamma_\chi$ for an arbitrary pair $(\Gamma_\xi, \Gamma_\chi)$ of stabilizers in Γ .

The analytical problem (I) and the algebraic problem (II) are most effectively studied together since there is a kind of duality between them. It results from intimate relations between the spectral data of Δ and the algebraic data of Γ . First examples of such relations appeared in [14], where Selberg linked eigenvalues with conjugacy classes in his celebrated trace formula. Here we shall sketch a unified approach to the two problems above. As a result we obtain a <u>complete</u> list of the identities relating (I) and (II). Special cases of these problems have been considered before. For instance, Selberg's theory of point pair invariants [14] is contained in one of the infinitely many identities arising if both ξ and χ are of the elliptic type. Moreover, the sum formulae of Bruggeman [1], Kuznietsov [9] and Proskurin [12] are among the identities obtained in case both ξ and

χ are of the parabolic type.

At least on a formal level the unified approach allows us to treat the nine different pairs of types in (II) simultaneously. Some differences of mostly analytical nature do remain but they merge to a large extent if one of the parameters gets large. Thus the unified approach puts on equal terms so different things as the hyperbolic lattice point problem, the estimation of sums of Kloosterman sums and certain distribution problems for closed geodesics in Riemann surfaces. In special cases, it also yields new insights into old number-theoretical problems as the following example shows.

The 2-dimensional Poisson summation formula provides a good attack on the classical lattice point problem in Euclidean circles. In particular, it proves the uniform distribution modulo 1 of the numbers $\frac{1}{2\pi}\arg w$ when w runs through the Gaussian integers according to any ordering compatible with their norms. By the Moebius inversion formula this continues to hold if one admits only primitive w, i.e. w with coprime real and imaginary parts. On the other hand, the restriction to primitive lattice points allows to refine the distribution problem. For then there exists a w^* such that w and w^* form an integral basis for the Gaussian integers. Since the numbers w^* are not uniquely determined it is not sensible to study the distribution of their arguments. One notice, however, that $\frac{w^*}{w}$ is uniquely determined modulo 1 by w and $\operatorname{Im}\frac{w^*}{w} > 0$. Under the latter condition the pairs $(\frac{1}{2\pi}\arg w, \operatorname{Re}\frac{w^*}{w})$ prove to be uniformly distributed in the unit square, when w runs through the primitive Gaussian integers ordered as before. This result can no longer be deduced from the Poisson summation formula but requires the use of harmonic analysis on the modular curve. Indeed it is a consequence of our identities in case ξ is of the elliptic and χ of the parabolic type. We infer it from the analytic properties of the generating functions

$$Z(s,m,n) = \sum_{w \text{ primitive}} \left(\frac{w}{|w|}\right)^m e\left(n \operatorname{Re}\frac{w^*}{w}\right) |w|^{-2s}, \quad \sigma > 1 \ ; \ m,n \in \mathbb{Z} ,$$

where $e(x) = \exp(2\pi i x)$. Note that for $n = 0$ these functions are closely related to Hecke's zeta-functions with Grössencharacters for $\mathbb{Q}(\sqrt{-1})$.

Our approach differs in many ways from known proofs in special cases. We split our arguments into an algebraic and an analytical part.

On the algebraic side we require a number of double coset decompositions of G among which one finds e.g. the Iwasawa or Bruhat decompositions. We give these decompositions in terms of maps from G to its Lie algebra while they are usually defined by maps in the opposite direction. There are two reasons for it. First of all we start with a discrete subgroup Γ whose elements we want to parametrize. Our decompositions then allow us to define generalized Kloosterman sums in Lie algebraic terms. These sums describe the discrete content of the group-theoretical side in our identies. Secondly, our maps enable us to give integral representations for the functions which govern the effect of the regular representation on periodic eigenfunctions. These functions describe the continuous content of the group-theoretical side in our identities. This splitting strongly reminds one of the Hardy-Littlewood method with which the whole subject is indeed connected.

In the analytical part we make systematic use of eigenfunctions, integral representations and functional equations. We follow Selberg [14] by deducing everything from the basic eigenfunctions

$$z \longmapsto (\mathrm{Im}\, z)^s .$$

We do no longer discuss the periodic eigenfunctions in terms of solutions to ordinary differential equations by separating variables. Instead we introduce them by integrals involving those basic eigenfunctions. This directly leads to the integral representations mentioned in the preceding paragraph. The periodic eigenfunctions which decay near the boundary of $\mathfrak{H}$ can be used to define Poincaré series. The functions we are really after then appear as Fourier coefficients of the Poincaré series. On the analytical side of our identities, the contribution of the discrete spectrum comes from the residues and that of the continuous spectrum comes from the functional equations of those functions.

In our approach it proves to be irrelevant to identify the occurring integrals with the corresponding special functions. Thus no books on special functions have to be consulted. On the other hand, if such an identification is made proofs for numerous results on special functions (especially but not only on Bessel and Legendre functions) are obtained as a by-product. Moreover, large parts of [3], [10] and [11] are then seen to be subsumed in our statements on Poincaré series

Detailed proofs of our results are given in [6].

3. Decompositions of G

The elements of G are in 1-1 correspondence with the points on the hypersurface ad-bc = 1 in R^4. However, this does not reveal very much about G as a group. We first discuss various parametrizations of G which are especially tailored to its group structure.

All non-trivial 1-parameter subgroups of G are conjugate to

$$H_\xi = \{e^{\tau X_\xi} | \tau \text{ in } R\} ,$$

where X_ξ equals

$$\begin{pmatrix} 0 & 1/2 \\ -1/2 & 0 \end{pmatrix}, \begin{pmatrix} 0 & 1 \\ 0 & 0 \end{pmatrix} \quad \text{or} \quad \begin{pmatrix} 1/2 & 0 \\ 0 & -1/2 \end{pmatrix}$$

for $\xi = \zeta$, ϑ or η respectively and e^X denotes the image of X under the exponential map, i.e. in our case under the exponential function for a matrix argument. We denote by f_ξ the set of fixpoints of e^{X_ξ} on the Riemann sphere and put

$$\mathcal{F}_\xi = \begin{cases} f_\xi \ , & \text{if } \xi = \zeta \text{ or } \vartheta \ , \\ f_\xi \cup \{iy | y > 0\}, & \text{if } \xi = \eta \ . \end{cases}$$

For every pair (ξ, χ) of types we split G into a disjoint union of three left H_ξ- and right H_χ-invariant subsets

$$_\xi\mathcal{S}_\chi \ , \quad _\xi g_\chi = {_\xi\mathcal{f}_\chi} - {_\xi\mathcal{S}_\chi} \quad \text{and} \quad _\xi G_\chi = G - {_\xi\mathcal{f}_\chi} \ ,$$

where

$$_\xi\mathcal{S}_\chi = \{M \text{ in } G \,|\, M(z) \text{ in } f_\xi \text{ for some } z \text{ in } f_\chi\}$$

and

$$_\xi\mathcal{f}_\chi = \{M \text{ in } G \,|\, M(z) \text{ in } \mathcal{F}_\xi \text{ for some } z \text{ in } \mathcal{F}_\chi\}$$

Here $_\xi\mathcal{S}_\chi$ is of dimension < 3 and plays the rôle of a singular set. On $_\xi g_\chi \cup {_\xi G_\chi}$ we define real functions $_\xi\Lambda^\ell_\chi$ and $_\xi\Lambda^r_\chi$ such that

$$M = \pm\, e^{_\xi\Lambda^\ell_\chi(M) X_\xi} \, {_\xi N_\chi}(\nu) \, e^{_\xi\Lambda^r_\chi(M) X_\chi}$$

and the dependence of the matrix $_\xi N_\chi$ on the real parameter $\nu = {_\xi\nu_\chi}(M)$ is explicitly determined. These real functions are related by simple but important identities, namely

$$_\xi\Lambda^r_\chi(M) = -{_\chi\Lambda^\ell_\xi}(M^{-1}) \ , \quad _\xi\nu_\chi(M) = {_\chi\nu_\xi}(M^{-1}) \ .$$

Moreover an invariant measure on G is then given by

$$dM = \nu d\Lambda^{\ell} d\Lambda^{r} d\nu ,$$

where for simplicity we suppressed all indices on the right hand side. While from an analytical point of view the coordinates ${}_{\xi}\Lambda^{\ell}_{\chi}$, ${}_{\xi}\Lambda^{r}_{\chi}$ and ${}_{\xi}\nu_{\chi}$ can be chosen in many ways, our decompositions are unique from a geometrical point of view.

<u>Example</u> (Bruhat decomposition): If both ξ and χ are parabolic then ${}_{\xi}\mathfrak{s}_{\chi} = \{M|c=0\}$ and ${}_{\xi}g_{\chi}$ is empty. In the notation of (1) the coordinates are then given by

$${}_{\xi}\Lambda^{\ell}_{\chi}(M) = M(\infty) = \frac{a}{c} , \quad {}_{\xi}\Lambda^{r}_{\chi}(M) = -M^{-1}(\infty) = \frac{d}{c} \quad \text{and} \quad {}_{\xi}\nu_{\chi}(M) = |c| .$$

<u>Remark</u>: (i) The set ${}_{\xi}g_{\chi}$ is empty unless ξ or χ are hyperbolic. (ii) Although we are primarily interested in G we also require similar decompositions for $SL_2(\mathbb{C})$. The reason for this need will become clearer in the next section.

4. <u>Periodic eigenfunctions</u>

Since Δ is G-invariant and $\Delta y^s = s(s-1)y^s$ every well-defined assignment

$$z \longmapsto \int_G y_M^s d\mu(M)$$

yields an eigenfunction of Δ with eigenvalue $s(s-1)$. The class of eigenfunctions thus produced is at the same time too general and too restrictive for our purposes. For the integration over a 1-parameter subgroup already yields periodic eigenfunctions. As long as we integrate inside G , these eigenfunctions are smooth on $\mathfrak{H}$ but do no decay near the boundary of $\mathfrak{H}$. In order to obtain decay we have to sacrifice smoothness by moving contours into the complexification of G . This partly explains the subsequent definitions. A deeper comprehension can be obtained from a closer look at our decompositions of G

Let $r_{\zeta} = 0$, $r_{\vartheta} = \pi$, $r_{\eta} = \frac{\pi}{2}$ and put

$$R_{\xi} = e^{r_{\xi} X_{\xi}} .$$

For M in $SL_2(\mathbb{C})$ and $z = x+iy$ in $\mathfrak{H}$ we set

$$Y(M,z) = \frac{y}{(cz+d)(c\bar{z}+d)} ,$$

where the entries of M are denoted as in (1) and $\bar{z}$ is the complex conjugate of z . Then we define

$$U_\xi(z,s,\lambda) = \int_{\mathcal{C}_\xi} e(-\lambda\varrho)\, Y^s(R_\xi e^{\varrho X_\xi}, z)\, d\varrho$$

and

$$V_\xi(z,s,\lambda) = \frac{1}{i} \int_{\mathcal{D}_\xi(z)} e(-\lambda\varrho)\, Y^{1-s}(R_\xi e^{\varrho X_\xi}, z)\, d\varrho \ ,$$

where $\mathcal{C}_\xi$ is a path on the real line and $\mathcal{D}_\xi(z)$ is a vertical line segment in the ϱ-plane connecting two poles of $\varrho \longmapsto Y(R_\xi e^{\varrho X_\xi}, z)$.

The U_ξ and V_ξ are periodic eigenfunctions of Δ . The U_ξ are smooth on $\mathfrak{H}$ and the V_ξ behave nicely near the boundary of $\mathfrak{H}$.

Example: If ξ is parabolic then

$$U_\xi(z,s,\lambda) = \begin{cases} B(1/2, s-1/2)\, y^{1-s} , & \text{if } \lambda = 0 , \\ e(\lambda x)\, y^{1/2} K_{s-1/2}(2\pi|\lambda|y)\, \dfrac{2\pi^s |\lambda|^{s-1/2}}{\Gamma(s)} , & \text{if } \lambda \neq 0 , \end{cases}$$

and

$$V_\xi(z,s,\lambda) = \begin{cases} B(1/2, s)\, y^{s} , & \text{if } \lambda = 0 , \\ e(\lambda x)\, y^{1/2} I_{s-1/2}(2\pi|\lambda|y)\, \pi^{1-s}\Gamma(s)\, |\lambda|^{1/2-s} , & \text{if } \lambda \neq 0 , \end{cases}$$

where $\Gamma(s)$ denotes the gamma- , $B(\alpha,\beta)$ the beta- and $I_\nu(z)$, $K_\nu(z)$ the usual modified Bessel functions.

The behaviour of U_ξ and V_ξ near their singularities can directly be determined from their integral representations. Luckily we always find that this 'just' requires the computation of a beta- or gamma-integral. We use the asymptotic behaviour of U_ξ and V_ξ to characterize them and to derive their functional equations.

Characterization: If $\Delta f(z) = s(s-1) f(z)$, $s \neq 1/2$, and $f(e^{\tau X_\xi} z) = e(\lambda\tau) f(z)$ for z in an open set $\mathfrak{U}$ then there are numbers C,D such that

$$f(z) = C\, U_\xi(z,s,\lambda) + D\, V_\xi(z,s,\lambda) \ , \quad z \text{ in } \mathfrak{U} \ .$$

Functional equations: There are simple expressions $\gamma_\xi(s,\lambda)$ in exponentials and gamma-functions such that

$$U_\xi(z,1-s,\lambda) = \gamma_\xi(s,\lambda)U_\xi(z,s,\lambda) \quad , \text{ unless } \quad \xi = \vartheta \quad \text{and} \quad \lambda = 0 \ ,$$

and

$$V_\xi(z,1-s,\lambda) = -\mathrm{ctg}\pi s \; U_\xi(z,s,\lambda) + \gamma_\xi(1-s,\lambda)V_\xi(z,s,\lambda) \ .$$

Example:

$$\gamma_\vartheta(s,\lambda) = \begin{cases} 0 \quad , \text{ if } \quad \lambda = 0 \ , \\ \frac{\Gamma(s)}{\Gamma(1-s)} \; (\pi|\lambda|)^{1-2s} \quad , \text{ if } \lambda \neq 0 \ . \end{cases}$$

5. Eigenfunctions and the regular representation

If M is in G then $U_\xi(z_M,s,\lambda)$ is also an eigenfunction though in general not a constant multiple of $U_\xi(z,s,\lambda)$. Nevertheless the former is expressible by functions of the latter kind thanks to our decompositions of G . More specifically, there are functions ${}_\xi I_\chi$ and ${}_\xi J_\chi$ such that for M in ${}_\xi G_\chi$

$$U_\xi(z_M,1-s,\lambda) = e(\lambda\Lambda^\ell) \int_{\hat{\mathscr{C}}_\chi} e(\lambda'\Lambda^r) \, {}_\xi I_\chi(\nu,s,\lambda,\lambda') U_\chi(z,s,\lambda') d\lambda'$$

and

$$V_\xi(z_M,s,\lambda) = e(\lambda\Lambda^\ell) \int_{\hat{\mathscr{C}}_\chi} e(\lambda'\Lambda^r) \, {}_\xi J_\chi(\nu,s,\lambda,\lambda') U_\chi(z,s,\lambda') d\lambda' \ ,$$

where $\Lambda^\ell = {}_\xi\Lambda^\ell_\chi(M)$, $\Lambda^r = {}_\xi\Lambda^r_\chi(M)$, $\nu = {}_\xi\nu_\chi(M)$ and $\hat{\mathscr{C}}_\chi$ denotes a dual path to $\mathscr{C}_\chi$. Due to the singularities of V_ξ the second equation may hold only on parts of $\mathfrak{H}$. There is also a function ${}_\xi i_\chi(\nu,s,\lambda,\lambda')$ which similarly relates $U_\xi(z_M,1-s,\lambda)$ with $U_\chi(z,s,\lambda')$ for M in ${}_\xi g_\chi$.

This close connection with the action of the regular representation on periodic eigenfunctions is reason enough to study ${}_\xi i_\chi$, ${}_\xi I_\chi$ and ${}_\xi J_\chi$ thoroughly. We base our investigations on integral representations which we deduce from the material in the previous two sections. Formally we only have to interchange two integrations. Rigorous proofs, however, require some efforts since technical complications arise. We find that ${}_\xi I_\chi(\nu,s,\lambda,\lambda')$ and ${}_\xi J_\chi(\nu,s,\lambda,\lambda')$ are essentially given by integrals of the form

$$\int e(-\lambda\tau+\lambda' \, {}_\xi\Phi_\chi(\tau,\nu)) \, {}_\xi Q_\chi^{s-1}(\tau,\nu) d\tau$$

with explicit computable functions ${}_\xi\Phi_\chi$ and ${}_\xi Q_\chi$. For instance ${}_\vartheta\Phi_\vartheta(\tau,\nu) = \frac{-1}{\nu^2\tau}$ and ${}_\vartheta Q_\vartheta(\tau,\nu) = (\nu\tau)^2$. The contour that has to be taken in the above integral depends on ξ and χ. It lies on the real line in the τ-plane for ${}_\xi I_\chi$ but not for ${}_\xi J_\chi$.

We say that $(\xi,\chi,\lambda,\lambda')$ is <u>degenerate</u> if ξ is parabolic and $\lambda = 0$ or if χ is parabolic and $\lambda' = 0$. Otherwise we call it <u>generic</u>. The following results can then be deduced from the integral representations:

<u>Functional equations</u>: In the generic cases we have

$$ {}_\xi i_\chi(\nu,s,\lambda,\lambda') = \gamma_\xi(s,\lambda)\gamma_\chi(s,\lambda')\,{}_\xi i_\chi(\nu,1-s,\lambda,\lambda') , $$

$$ {}_\xi I_\chi(\nu,s,\lambda,\lambda') = \gamma_\xi(s,\lambda)\gamma_\chi(s,\lambda')\,{}_\xi I_\chi(\nu,1-s,\lambda,\lambda') $$

and

$$ {}_\xi J_\chi(\nu,s,\lambda,\lambda') = \gamma_\xi(s,\lambda)\gamma_\chi(s,\lambda')\,{}_\xi J_\chi(\nu,1-s,\lambda,\lambda') + \operatorname{ctg}\pi s\,{}_\xi I_\chi(\nu,s,\lambda,\lambda') . $$

<u>Explicit evaluations</u>: In the degenerate cases

$$ {}_\xi J_\chi(\nu,s,\lambda,\lambda') = \beta_\xi B(1/2,s)\,\nu^{-2s} , $$

where $\beta_\xi = 1$ for $\xi = \vartheta$ and $\beta_\xi = 2$ otherwise.

<u>Asymptotic behaviour</u>: It is always true that

$$ {}_\xi J_\chi(\nu,s,\lambda,\lambda') \sim \beta_\xi\, B(1/2,s)\,\nu^{-2s} , \quad \text{if} \quad \nu \to \infty . $$

6. Fourier coefficients, Poincaré series and Kloosterman sums

In the previous three sections we dealt with G as a continuous group. We now return to the discrete subgroup Γ and its stabilizers Γ_ξ. For every ξ we choose an M_ξ in G and a positive number λ_ξ such that the conjugate stabilizer $\Gamma'_\xi = M_\xi \Gamma_\xi M_\xi^{-1}$ is generated by $\pm e^{\lambda_\xi X_\xi}$.

Then there are Fourier coefficients $\alpha_{j\xi}(n)$, $\alpha_{\iota\xi}(s,n)$ such that the eigenfunctions in (2) can be written in the form

$$ e_j(M_\xi^{-1}(z)) = \sum_{n\in Z} \alpha_{j\xi}(n) U_\xi(z,s_j,\frac{n}{\lambda_\xi}) , \quad j \geq 0 , $$

and

$$ E_\iota(M_\xi^{-1}(z),s) = \delta(\xi,\vartheta_\iota)y^s + \sum_{n\in Z} \alpha_{\iota\xi}(s,n) U_\xi(z,s,\frac{n}{\lambda_\xi}) , \quad \iota = 1,\dots,\kappa , $$

where E_ι corresponds to the cusp ϑ_ι and

$$\delta(\xi,\chi) = \begin{cases} 1 \ , & \text{if } \xi = M(\chi) \text{ for an } M \text{ in } \Gamma \ , \\ 0 \ , & \text{otherwise.} \end{cases}$$

By the characterization in section 4 a Fourier series expansion in terms of U_ξ and V_ξ clearly exists. It remains to verify that the coefficient of V_ξ has to vanish in almost all cases.

On the other hand, the V_ξ are needed to form the Poincaré series

$$P_\xi(z,s,m) = \sum_{M \in \Gamma_\xi \setminus \Gamma} V_\xi(z_{M_\xi M}, s, \frac{m}{\lambda_\xi}) \ , \ \sigma > 1 \ \text{ and } \ m \ \text{ in } \ \mathbb{Z} \ . \quad (3)$$

We also consider a truncated version P_ξ^ψ differing from P_ξ only by a finite number of terms. It arises from replacing V_ξ in (3) by ψV_ξ where ψ is a cut-off function which kills the singularity of V_ξ and equals 1 near the boundary of $\mathfrak{H}$. By standard methods one establishes absolute and locally uniform convergence of $P_\xi^\psi(z,s,m)$ for s in $\sigma > 1$. Moreover P_ξ^ψ then is a Γ-invariant, bounded and smooth function of its first variable.

Up to constant multiples $P_\vartheta(z,s,o)$ is an Eisenstein series and $(\zeta,z) \longmapsto P_\zeta(z,s,o)$ is the resolvent kernel of Δ to the eigenvalue $s(s-1)$.

As a Γ-invariant eigenfunction P_ξ has a Fourier series expansion at every χ . In order to compute the associated Fourier coefficients we sum in (3) first over double cosets $\Gamma_\xi N \Gamma_\chi$ and then over $\Gamma_\xi \setminus \Gamma / \Gamma_\chi$. On grouping together those $\Gamma_\xi N \Gamma_\chi$ which determine the same double coset in $H_\xi \setminus G / H_\chi$ we are led to define <u>generalized Kloosterman sums</u> by

$${}_\xi S_\chi(m,n,\nu) = \sum e\left(\frac{m}{\lambda_\xi}\, {}_\xi\Lambda_\chi^\ell(M) + \frac{n}{\lambda_\chi}\, {}_\xi\Lambda_\chi^r(M)\right)$$

with summation over all $\Gamma_\xi' M \Gamma_\chi'$ in $\Gamma_\xi' \setminus M_\xi \Gamma M_\chi^{-1} \cap {}_\xi G_\chi / \Gamma_\chi'$ satisfying ${}_\xi\nu_\chi(M) = \nu$. On replacing ${}_\xi G_\chi$ by ${}_\xi g_\chi$ we similarly introduce exponential sums ${}_\xi s_\chi(m,n,\nu)$. Empty sums are considered to be zero. If $\Gamma = SL_2(\mathbb{Z})$, $\xi = \chi = \infty$ (i.e. parabolic) and M_ξ = identity, then ${}_\xi S_\chi(m,n,\nu) = 0$ unless ν is equal to a positive integer c . By the example in section 3 we then have

$${}_\xi S_\chi(m,n,\nu) = \sum_{\substack{a \bmod c \\ ad \equiv 1 (c)}} e\left(\frac{ma+nd}{c}\right) \ ,$$

i.e. it is a classical Kloosterman sum.

The Fourier coefficients of P_ξ at χ are essentially given by

$${}_{\xi}P_{\chi}(s,m,n) = \frac{1}{\lambda_{\chi}} \sum_{\nu} {}_{\xi}S_{\chi}(m,n,\nu)\, {}_{\xi}J_{\chi}\left(\nu, s, \frac{m}{\lambda_{\xi}}, \frac{n}{\lambda_{\chi}}\right)$$

and

$${}_{\xi}p_{\chi}(s,m,n) = \frac{1}{\lambda_{\chi}} \sum_{\nu} {}_{\xi}s_{\chi}(m,n,\nu)\, {}_{\xi}i_{\chi}\left(\nu, s, \frac{m}{\lambda_{\xi}}, \frac{n}{\lambda_{\chi}}\right),$$

where the first sum is an infinite sum converging absolutely in $\sigma > 1$ and the second is a finite sum. For the results in section 3 and 5 yield

$$P_{\xi}(M_{\chi}^{-1}(z), s, m) = \sum_{n \in Z} {}_{\xi}P_{\chi}(s,m,n)\, U_{\chi}\left(z, s, \frac{n}{\lambda_{\chi}}\right) + {}_{\xi}\wp_{\chi}(z,s,m) + {}_{\xi}\mathcal{U}_{\chi}(z,s,m)$$

and $${}_{\xi}\wp_{\chi}(z,s,m) - \gamma_{\xi}\left(s, \frac{m}{\lambda_{\xi}}\right) {}_{\xi}\wp_{\chi}(z,1-s,m) = \operatorname{ctg} \pi s \sum_{n \in Z} {}_{\xi}p_{\chi}(s,m,n)\, U_{\chi}\left(z, s, \frac{n}{\lambda_{\chi}}\right),$$

if z is close enough to χ. The term ${}_{\xi}\mathcal{U}_{\chi}$ is due to ${}_{\xi}\mathcal{S}_{\chi}$ and vanishes in most cases. At any rate it is made up of not more than two summands occurring in (3).

7. Analytic continuations and functional equations

The Poincaré series P_{ξ} and their Fourier coefficients would be of little use if they were not analytically continuable in the variable s. To prove this we first continue P_{ξ}^{ψ} by using the eigenfunction expansion (2) with $f(z) = P_{\xi}^{\psi}(z,s,m)$. We show in particular that

$$\langle P_{\xi}^{\psi}(\cdot,s,m), e_j \rangle = \bar{\alpha}_{j\xi}(m) \frac{\psi_{\xi}(s,s_j,m)}{(s-s_j)(s-1+s_j)},$$

where $\psi_{\xi}(s,s_j,m)$ is an analytic function of s in $\sigma > 0$ and

$$\psi_{\xi}(s,\bar{s},m) = 2\pi\beta_{\xi}\lambda_{\xi}, \quad \psi_{\xi}(1-s,\bar{s},m) = \gamma_{\xi}\left(1-s, \frac{m}{\lambda_{\xi}}\right)\psi_{\xi}(s,\bar{s},m).$$

A similar expression for $\langle P_{\xi}^{\psi}(\cdot,s,m), E_{\iota}(\cdot,s') \rangle$ and crude bounds for ψ_{ξ} then yield the analytic continuation in a form from which the principal parts of $P_{\xi}^{\psi}(z,s,m)$ in $\sigma \geq 1/2$ can readily be read off. Since the difference $P_{\xi}(z,s,m) - P_{\xi}^{\psi}(z,s,m)$ is analytic in $\sigma > 0$ the analytic continuation of P_{ξ} into this half-plane now follows. Moreover, the Poincaré series satisfy the functional equation

$$P_{\xi}(z,1-s,m) = \gamma_{\xi}\left(1-s, \frac{m}{\lambda_{\xi}}\right) P_{\xi}(z,s,m) + \frac{2\pi\beta_{\xi}\lambda_{\xi}}{1-2s} \sum_{\iota=1}^{\kappa} \alpha_{\iota\xi}(1-s,-m)\, E_{\iota}(z,s). \quad (4)$$

For the difference of the two sides in (4) has to vanish since it is an eigenfunction of the selfadjoint Δ in $L_2(\Gamma\backslash\mathfrak{H})$ with the generally non-real eigenvalue $s(s-1)$. Its square integrability follows from the functional equations in sect. 4, the explicit evaluations in sect. 5, the Fourier series expansions in sect. 6 and the fact that

$$ {}_{\xi}S_{\chi}(m,n,\nu) = {}_{\chi}S_{\xi}(-n,-m,\nu) . $$

The analytic continuations and functional equations of the Poincaré series have the following effects on their Fourier coefficients:

<u>Singularities</u>: $(2s-1)\,{}_{\xi}P_{\chi}(s,m,n)$ is analytic in $\sigma \geq 1/2$ except for simple poles at $s = s_j$ or $s = 1-s_j$. There, up to a simple factor, e_j contributes $\bar{\alpha}_{j\xi}(m)\alpha_{j\chi}(n)$ to its residue.

<u>Functional equations</u>: (i) The ${}_{\xi}p_{\chi}(s,m,n)$ are analytic in $0<\sigma<1$ and satisfy

$$ \gamma_{\chi}\left(s,\frac{n}{\lambda_{\chi}}\right){}_{\xi}p_{\chi}(1-s,m,n) = \gamma_{\xi}\left(1-s,\frac{m}{\lambda_{\xi}}\right){}_{\xi}p_{\chi}(s,m,n) . $$

(ii) In all cases

$$ \gamma_{\chi}\left(s,\frac{n}{\lambda_{\chi}}\right)\{{}_{\xi}P_{\chi}(1-s,m,n) + \frac{1}{2}\operatorname{ctg}\pi(1-s)\,{}_{\xi}p_{\chi}(1-s,m,n)\} = \gamma_{\xi}\left(1-s,\frac{m}{\lambda_{\xi}}\right)\{{}_{\xi}P_{\chi}(s,m,n $$

$$ + \frac{1}{2}\operatorname{ctg}\pi s\,{}_{\xi}p_{\chi}(s,m,n)\} + {}_{\xi}A_{\chi}(m,n)\operatorname{ctg}\pi s + \frac{2\pi\beta_{\xi}\lambda_{\xi}}{1-2s}\sum_{\iota=1}^{\kappa}\alpha_{\iota\xi}(1-s,-m)\,\alpha_{\iota\chi}(s,n) , $$

where ${}_{\xi}A_{\chi}(m,n) = 0$ unless $|m| = |n|$ and ξ is Γ-equivalent to χ .

<u>Remark</u>: The above results establish a strong analytic resemblance between ${}_{\xi}P_{\chi}$ and the logarithmic derivative of <u>Selberg zeta-functions</u>

8. <u>The identities</u>

The identities relating the basic problems (I) and (II) have the form of sum formulae, i.e. they contain a rather general weight function on one side and integral transforms of it on the other side. We state these formulae in two ways. The first is more useful to study averages over the spectrum and the second deals better with asymptotic results on the double coset space $\Gamma_{\xi}\backslash\Gamma/\Gamma_{\chi}$.

<u>Sum formulae (first form)</u>: In the generic cases we have

$$ \frac{1}{4\pi i}\sum_{\iota=1}^{\kappa}\int_{(1/2)}\bar{\alpha}_{\iota\xi}(s,m)\,\alpha_{\iota\chi}(s,n)\,h(s)\,ds + \sum_{t_j\geq 0}\bar{\alpha}_{j\xi}(m)\,\alpha_{j\chi}(n)\,h(s_j) + $$

$$\sum_{1/2<s_j\leq 1} \bar{\alpha}_{j\xi}(m)\alpha_{j\chi}(n)h(s_j)\gamma_\xi(1-s_j,\frac{m}{\lambda_\xi}) = \frac{{}_\xi A_\chi(m,n)}{4\pi^2\beta_\xi\lambda_\xi i}\int_{(1/2)} h(s)(s-1/2)\operatorname{ctg}\pi s\,ds +$$

$$\frac{1}{\lambda_\xi\lambda_\chi}\sum_\nu {}_\xi s_\chi(m,n,\nu)\,{}_\xi h_\chi\left(\frac{m}{\lambda_\xi},\frac{n}{\lambda_\chi},\nu\right) + \frac{1}{\lambda_\xi\lambda_\chi}\sum_\nu {}_\xi S_\chi(m,n,\nu)\,{}_\xi H_\chi\left(\frac{m}{\lambda_\xi};\frac{n}{\lambda_\chi},\nu\right) .$$

Here h satisfies

$$\gamma_\xi(s,\frac{m}{\lambda_\xi})h(1-s) = \gamma_\chi(s,\frac{n}{\lambda_\chi})h(s)$$

besides some growth and regularity conditions. The ${}_\xi h_\chi$ and ${}_\xi H_\chi$ denote integral transforms of h defined in terms of ${}_\xi i_\chi$ and ${}_\xi I_\chi$ respectively.

Examples: (i) (Selberg [14]) If both ξ and χ are of the elliptic type and $m = n = 0$, one obtains a sum formula for the spectral kernel function of Δ . Then, up to normalizations, ${}_\xi H_\chi$ is the so-called Selberg transform.

(ii) If $\Gamma = SL_2(\mathbb{Z})$, $\xi = \chi = \infty$ and $mn \neq 0$ we recover the identities in [1].

In the degenerate cases the sum formulae are of a slightly simpler nature. Our integral representations enable us to express the above integral transforms by a double integral of which the inner always is an ordinary Fourier transformation. This is useful to prove the following

spectral average: For every ξ and m

$$\sum_{t_j\leq T}|\alpha_{j\xi}(m)|^2 + \frac{1}{4\pi}\sum_{\iota=1}^{\kappa}\int_{-T}^{T}|\alpha_{\iota\xi}(\tfrac{1}{2}+it,m)|^2 dt = \frac{{}_\xi A_\xi(m,m)}{4\pi^2\beta_\xi\lambda_\xi}T^2 + O(T) ,$$

as $T\to\infty$.

If we were able to invert the above integral transforms we could also study the asymptotic behaviour of the generalized Kloosterman sums. In the exact inversion for the parabolic types the Fourier coefficients of holomorphic cusp forms showed up. Here we develop an alternative approach without bringing them into the picture. We make only an approximate inversion which quickly follows from our functional equations and an asymptotic expansion of ${}_\xi J_\chi$. The approximate inversion suffices to prove asymptotic results.

Sum formulae (second form): Let $\mathcal{W}$ denote a C^{∞}-function of compact support in $\mathbb{R}$. If

$$M_{\mathcal{W}}(s) = 2\int_0^{\infty} \mathcal{W}(\varrho)\,\varrho^{2s-1}\,d\varrho \quad , \quad \sigma > 0 ,$$

and

$${}_{\xi}\mathcal{W}_{\chi}(s,\lambda,\lambda') = M_{\mathcal{W}}(s)\,B(1/2,s-1/2)\,\gamma_{\xi}(s,\lambda)$$

$$+ M_{\mathcal{W}}(1-s)\,B(1/2,1/2-s)\,\gamma_{\chi}(1-s,\lambda') ,$$

then

$$\frac{1}{\lambda_{\xi}\lambda_{\chi}} \sum_{\nu} {}_{\xi}S_{\chi}(m,n,\nu)\mathcal{W}(\nu) = \sum_{1/2<s_j\leq 1} \bar{\alpha}_{j\xi}(m)\,\alpha_{j\chi}(n)\,M_{\mathcal{W}}(s_j)\,B(1/2,s_j-1/2)$$

$$+ \sum_{t_j\geq 0} \bar{\alpha}_{j\xi}(m)\,\alpha_{j\chi}(n)\,{}_{\xi}\mathcal{W}_{\chi}\left(s_j,\frac{m}{\lambda_{\xi}},\frac{n}{\lambda_{\chi}}\right) +$$

$$+ \frac{1}{4\pi i} \sum_{\iota=1}^{\kappa} \int_{(1/2)} \bar{\alpha}_{\iota\xi}(s,m)\,\alpha_{\iota\chi}(s,n)\,{}_{\xi}\mathcal{W}_{\chi}\left(s,\frac{m}{\lambda_{\xi}},\frac{n}{\lambda_{\chi}}\right)ds +$$

+ 'something'.

In asymptotic results the 'something' is much smaller than the bounds we can prove for the two terms preceding it. Thus the inversion of the integral transforms is essentially given by simple Mellin transformations. It is now trivial to obtain the

Group average: If δ_{mn} denotes the Kronecker symbol, then

$$\sum_{\nu<X} {}_{\xi}S_{\chi}(m,n,\nu) = \frac{\delta_{om}\delta_{on}\lambda_{\xi}\lambda_{\chi}}{\pi\beta_{\xi}\beta_{\chi}\omega(\Gamma\backslash\mathfrak{H})} X^2 + \sum_{1/2<s_j<1} {}_{\xi}C_{\chi}(m,n,s_j)X^{2s_j} + O(X^{4/3}) ,$$

as $X \to \infty$, for suitable numbers ${}_{\xi}C_{\chi}(m,n,s_j)$.

Remark: The above remainder term is as strong as the usual bound for the hyperbolic lattice point problem which arises if both ξ and χ are of the elliptic type and $m = n = 0$. It falls short of Kuznietsov bound [9] $O(X^{7/6+\varepsilon})$, $\varepsilon > 0$, for the classical Kloosterman sums onl because no A. Weil's estimate is available in general. This estimate leads to further savings in the summation of Kloosterman sums over sho intervals. If corresponding savings took place in other cases a simila improvement of the error term would then result as well.

The above averages and Weyl's criterion now yield the following

Uniform distribution in $\Gamma_\xi \backslash \Gamma / \Gamma_\chi$: If $0 \leq a_1 < a_2 \leq \lambda_\xi$ and $0 \leq b_1 < b_2 \leq \lambda_\chi$, the number of double cosets $\Gamma_\xi M_\xi^{-1} M M_\chi \Gamma_\chi$ in Γ such that

$$a_1 \leq {}_\xi\Lambda_\chi^\ell(M) \leq a_2 \pmod{\lambda_\xi} \ , \ b_1 \leq {}_\xi\Lambda_\chi^r(M) \leq b_2 \pmod{\lambda_\chi} \ \text{and} \ {}_\xi\nu_\chi(M) \leq X$$

is asymptotically equal to

$$\frac{(a_2-a_1)(b_2-b_1)}{\pi\beta_\xi\beta_\chi\omega(\Gamma\backslash\mathfrak{H})} X^2 \quad , \quad \text{as} \quad X \to \infty \ .$$

9. Applications

The asymptotic results in sect. 8 follow in a rather straightforward way from a single identity and trivial bounds for Fourier coefficients or Kloosterman sums. More complex applications appear in [2], [4], [5] and [7], where further cancellations are obtained from the simultaneous use of many identities and where the dependence on certain parameters is important. At this conference J.M. Deshouillers reports on such applications in more detail. In an impressive series of joint papers, he and H. Iwaniec always work with congruence subgroups of $SL_2(\mathbb{Z})$ and consider only parabolic types. Many of their results presumably carry over to the wider context taken here although their proofs do not. Applications in other directions include:

Relations of the Maass-Selberg type: The Maass-Selberg relations are inner product formulae for truncated Eisenstein series. Similar formulae exist for inner products of our truncated Poincaré series P_ξ^ψ . These formulae can be used, for instance, to set up simple relations between ${}_\xi P_\chi(s,m,n)$ and ${}_\chi P_\xi(s,-n,-m)$. These relations for parabolic ξ,χ and $m = n = 0$ are equivalent with the symmetry of the constant term matrix of the Eisenstein series. While this symmetry is a rather obvious fact, the same can no longer be said of those relations in general.

Limit formulae: These are concerned with the values of ${}_\xi P_\chi$ at special points. We mention as an example that

$$2\pi^2 m \, {}_\xi P_\chi(1,m,n) = \begin{cases} \delta_{mn} \ , & n > 0 \ , \\ \text{integer} \ , & n < 0 \ , \end{cases}$$

if $\Gamma = SL_2(\mathbb{Z})$, $\xi = \chi = \infty$ and m positive.

<u>Caution</u>: We must confess that we oversimplified the exposition in case ξ or χ are of the hyperbolic type. For then the technical details are actually more involved. One of the reasons for this is the fact that geodesics can be run through forwards and backwards. Thus one should consult [6] for accurate statements in these cases.

References

[1] Bruggeman, R.W.: Fourier coefficients of cusp forms. Invent. Math 45, 1-18(1978).

[2] Deshouillers, J.M., Iwaniec, H.: Power mean-values for the Riemann zeta-function. Mathematika 29, 202-212(1982).

[3] Fay, J.D.: Fourier coefficients of the resolvent for a Fuchsian group. J. Reine Angew. Math. 293/294, 143-203 (1977).

[4] Good, A.: Cusp forms and eigenfunctions of the Laplacian. Math. Ann. 255, 523-548 (1981).

[5] Good, A.: The square mean of Dirichlet series associated to cusp forms. Mathematika 29, 278-295 (1982).

[6] Good, A.: Local analysis of Selberg's trace formula. To appear as Lecture Notes in Math.

[7] Iwaniec, H.: Fourier coefficients of cusp forms and the Riemann zeta-function. Sém. Th.Nb. Bordeaux (1979/80), exposé 18, 36 p.

[8] Kubota, T.: Elementary theory of Eisenstein series. New York: Wiley 1973.

[9] Kuznietsov, N.V.:Petersson hypothesis for parabolic forms of weight zero and Linnik hypothesis. Sums of Kloosterman sums. Math. Sborn. 111(153), 334-383 (1980).

[10] Neunhöffer, H.: Ueber die analytische Fortsetzung von Poincaré-reihen. Sitzb. Heidelberg Akad. Wiss. (Mat.-Nat,Kl.) 2 Abh., 33-90 (1973).

[11] Niebur, D.: A class of nonanalytic automorphic functions. Nagoya Math. J. 52, 133-145 (1973).

[12] Proskurin, N.V.: Summation formulas for generalized Kloosterman sums. Zap. Naucn. Sem. Leningrad. Otdel.Math.Inst. Steklov 82, 103-135 (1979).

[13] Roelcke, W.: Ueber die Wellengleichung bei Grenzkreisgruppen erster Art. Sitzb. Heidelberg Akad. Wiss. (Mat.-Nat.Kl.) 4 Abh., 159-267 (1956).

[14] Selberg, A.: Harmonic analysis and discontinuous groups in weakly symmetric Riemannian spaces with applications to Dirichlet series. J. Indian Math. Soc. 20, 47-87 (1956).

CUBIC FORMS IN 10 VARIABLES

D.R. Heath-Brown
Magdalen College
Oxford OX1 4AU, England

Let $Q(\underline{x}) = Q(x_1,\ldots,x_n) \in \mathbb{Z}[x_1,\ldots,x_n]$ be a quadratic form. Necessary and sufficient conditions for the solvability of $Q(\underline{x}) = 0$, with $\underline{x} \in \mathbb{Z}^n - \{\underline{0}\}$ are given by the well-known Hasse-Minkowski local-to-global principle. The analogous situation for a cubic form $F(\underline{x})$ is much more complicated. In particular examples are known with $n = 3$ and 4 for whic $F(\underline{x}) = 0$ is everywhere locally solvable, but has no global solutions. I $n \geqslant 10$ then $F(\underline{x}) = 0$ is automatically everywhere locally solvable, so one is lead to ask: <u>Does there exist</u> n_0 $(n_0 = 10\ ?)$ <u>such that</u> $F(\underline{x}) = 0$ <u>is solvable with</u> $\underline{x} \in \mathbb{Z}^n - \{\underline{0}\}$, <u>whenever</u> $n \geqslant n_0$? This was answered affirmatively by Lewis [5], who gave the value $n_0 = 1000$. Soon afterwards Davenport produced, by a quite different method, the value $n_0 = 32$. Later, see [2], he established the result for $n_0 = 16$, which is still the best known.

Davenport's proof used the Hardy-Littlewood circle method, and showed that "usually" one has

$$\#\{\underline{x} \in \mathbb{Z}^n;\ |x_i| \leqslant P,\ F(\underline{x}) = 0\} \sim C_F P^{n-3} \quad \text{as } P \to \infty. \tag{1}$$

However (1) is not true for all forms, as is shown by the example

$$F = x_1^{\,3} + x_2(x_3^{\,2} + \ldots + x_n^{\,2}),$$

which is zero whenever $x_1 = x_2 = 0$. Thus one needs some geometric condition on F before (1) can hold. Recently [4] the following has been established.

<u>THEOREM</u> <u>Let</u> $F(\underline{x}) \in \mathbb{Z}[x_1,\ldots,x_n]$ <u>be a cubic form with</u> $n \geqslant 10$. <u>Then</u> $F(\underline{x}) = 0$, $\underline{x} \in \mathbb{Z}^n - \{\underline{0}\}$, <u>is solvable if</u> F <u>is non-singular</u>.

Thus one has the right bound $(n_0 = 10)$ under a condition that is presumably unnecessary.

<u>COROLLARY</u> <u>Let</u> F <u>be as above, with</u> $n \geqslant 9$. <u>Then if</u> F <u>is non-singul</u> <u>one can solve</u> $F(\underline{x}) = a$, $\underline{x} \in \mathbb{Q}^n$, <u>for any</u> $a \in \mathbb{Q}$.

Here $n \geqslant 9$ is probably not best possible, since the corresponding local problem is solvable whenever $n \geqslant 8$.

The proof is based on the circle method, and uses the weighted exponential sum

$$S(\alpha) = \sum_{\underline{x}\in\mathbb{Z}^n} w(\underline{x})e(\alpha F(\underline{x})),$$

where

$$w(\underline{x}) = \exp\left(-\frac{\|\underline{x} - P\underline{x}_0\|^2}{P_0^{\,2}}\right), \qquad e(t) = \exp(2\pi i t).$$

Here $\|.\|$ is the Euclidean norm, $P_0 = P(\log P)^{-2}$, where P is a parameter thought of as tending to infinity, and $\underline{x}_0$ is a "good" real point on $F = 0$. The effect of the weight w is that $S(\alpha)$ essentially counts those $\underline{x}$ which are "close" to the "good" point $P\underline{x}_0$.

The idea of the circle method is to consider

$$\int_0^1 S(\alpha)\, d\alpha = \sum_{\underline{x}\in\mathbb{Z}^n,\ F(\underline{x})=0} w(\underline{x}),$$

and to show, by estimating $S(\alpha)$, that this expression tends to infinity with P. We estimate $S(\alpha)$ using an n-dimensional version of a technique introduced by Vaughan [6]. (This is the first major point where our method differs from Davenport's, since he used Weyl's inequality to bound $S(\alpha)$.) We have

$$S(\tfrac{a}{q} + z) = \sum_{\underline{c}\,(\mathrm{mod}\ q)} e_q(aF(\underline{c})) \sum_{\underline{d}\in\mathbb{Z}^n} f(\underline{d}), \tag{2}$$

where

$$e_q(t) = e(t/q), \qquad f(\underline{d}) = w(\underline{c} + q\underline{d})e(zF(\underline{c} + q\underline{d})).$$

Thus if we apply the Poisson summation formula to the inner sum of (2) we obtain

$$S(\tfrac{a}{q} + z) = q^{-n} \sum_{\underline{b}\in\mathbb{Z}^n} S(q;\underline{b},a)I(z,\tfrac{1}{q}\underline{b}), \tag{3}$$

where

$$S(q;\underline{b},a) = \sum_{\underline{c}\,(\mathrm{mod}\ q)} e_q(aF(\underline{c}) - \underline{b}.\underline{c}),\quad I(z,\underline{\beta}) = \int_{\mathbb{R}^n} w(\underline{x})e(zF(\underline{x}) + \underline{\beta}.\underline{x})dV.$$

One has now to estimate $I(z,\underline{\beta})$, and this turns out to be quite tricky. It is here that conditions are imposed on the real point $\underline{x}_0$. The outcome is essentially that $I(z,\underline{\beta})$ is negligible for

$$\|\underline{\beta}\| \geq P^{\varepsilon}(P^{-1} + |z| P^2),$$

and otherwise that

$$I(z,\underline{\beta}) \ll P^{\varepsilon}\text{Min}(P^n, (P|z|)^{-n/2}). \tag{4}$$

One has also to estimate $S(q;\underline{b},a)$. These sums factorize according to the prime power decomposition of q, so that each term of (3) already has the familiar structure of the singular series and integral. If q is prime one can use an estimate of Deligne [3; Theorem 8.4].

<u>LEMMA</u> <u>Let</u> $H(x_1,\dots,x_n)$ <u>be a form of degree</u> d, <u>such that</u> $H=0$ <u>defines a non-singular irreducible projective variety over the field of</u> p <u>elements. Let</u> $E(x_1,\dots,x_n)$ <u>be a polynomial of degree</u> $\leq d-1$. <u>Then</u>

$$\Big|\sum_{\underline{x}(\text{mod } p)} e_p(H(\underline{x}) + E(\underline{x}))\Big| \leq (d-1)^n p^{n/2},$$

<u>providing that</u> $p \nmid d$.

This is the most important point at which the non-singularity of F is used. Of course F is non-singular to all but finitely many prime moduli. We now have

$$S(p;\underline{b},a) \ll p^{n/2}.$$

The corresponding estimate for prime powers is false in general, but can be obtained on average, using the summation over $\underline{b}$ which occurs in (3). The derivation of this average bound is one of the most difficult parts of the analysis.

If we suppose, for the purpose of this discussion, that

$$S(q;\underline{b},a) \ll q^{n/2+\varepsilon},$$

then (3) and (4) lead to

$$S(\tfrac{a}{q} + z) = q^{-n}S(q;\underline{0},a)I(z,\underline{0}) + O(P^{\varepsilon}q^{n/2}(1+|z|P^3)^{n/2}).$$

By Dirichlet's theorem, for any α we can solve

$$|\alpha - \tfrac{a}{q}| \leq \frac{1}{qQ}, \quad q \leq Q.$$

On taking $Q = P^{3/2}$ this yields

$$S(\frac{a}{q} + z) = \text{Main Term} + O(P^{3n/4+\varepsilon}).$$

Consequently one obtains

$$\int_0^1 S(\alpha)\, d\alpha = \text{Main Term} + O(P^{3n/4+\varepsilon}). \qquad (5)$$

Since the main term here is of order P^{n-3} this is satisfactory if $n > 12$.

A further $q^{-1/2}$ can be saved by the use of the Kloosterman refinement, sometimes known as "Littlewood's fuzzy ends". This is the device that Kloosterman introduced in his work on representations by definite quaternary quadratic forms, in which he encountered Kloosterman sums for the first time. The method involves keeping track of the precise location of the endpoints of the Farey dissection. The ultimate effect is to replace the error term in (5) by $O(P^{(3n-3)/4+\varepsilon})$. In place of $S(q;\underline{b},a)$ one now encounters

$$S_t(q;\underline{b}) = \sum_{\substack{s=1\\(s,q)=1}}^{q} \sum_{\underline{c} \pmod q} e_q(\bar{s}F(\underline{c}) - \underline{b}.\underline{c} + st),$$

where $\bar{s}$ is the inverse of $s \pmod q$. A new problem arises in the estimation of $S_t(p,\underline{b})$ when $p|t$. It turns out that sums of the form

$$\sum_{\substack{\underline{c} \pmod p\\ p|\underline{b}.\underline{c}}} e_p(F(\underline{c}))$$

occur. If we denote by $F_{\underline{b}}$ the (n-1)-ary form obtained from F by setting $\underline{b}.\underline{x} = 0$, we need to know that $F_{\underline{b}}$ is non-singular (mod p), so that we can apply Deligne's result. A sufficient condition for this is that $p \nmid G(\underline{b})$, where G is a certain form depending only on F. Thus vectors $\underline{b}$ for which $G(\underline{b}) = 0$ play a special rôle, and we are lead to ask: <u>For how many $\underline{b} \in \mathbb{Z}^n$, $\|\underline{b}\| \leq B$, is $G(\underline{b}) = 0$?</u> This question seems to be of interest in its own right. From Cohen [1; Theorems 2.1 and 2.2] one can deduce that the number of vectors $\underline{b}$ is $O(B^{n-3/2+\varepsilon})$ if G has no linear factor over $\mathbb{Q}$.

Finally one might ask whether the method can be used for forms of higher degree. Here one hopes for an asymptotic formula of the shape (5) in which the main term is of order P^{n-d}, where d is the degree of the form. However the method used here leads essentially to an error term $O(P^{dn/4+\varepsilon})$, and the condition $dn/4 < n-d$ requires $d < 4$.

REFERENCES

1 S.D. Cohen, The distribution of galois groups and Hilbert's irreducibility theorem, Proc. London Math. Soc. (3), 43 (1981), 227-250.

2 H. Davenport, Cubic forms in sixteen variables, Proc. Roy. Soc. London Ser. A, 272 (1963), 285-303.

3 P. Deligne, La conjecture de Weil. I, Publications Mathématiques 43 (Institut des Hautes Etudes Scientifiques, Paris, 1974), 273-307.

4 D.R. Heath-Brown, Cubic forms in ten variables, Proc. London Math. Soc. (3), to appear.

5 D.J. Lewis, Cubic polynomials over algebraic number fields, Mathematika, 4 (1957), 97-101.

6 R.C. Vaughan, Some remarks on Weyl sums, Coll. Math. Soc. János Bolyai 34, Topics in classical number theory (Elsevier North Holland, Amsterdam, to appear).

ON THE STRUCTURE OF GALOIS GROUPS AS GALOIS MODULES

Uwe Jannsen

Fakultät für Mathematik
Universitätsstr. 31, 8400 Regensburg
Bundesrepublik Deutschland

Classical class field theory tells us about the structure of the Galois groups of the abelian extensions of a global or local field. One obvious next step is to take a Galois extension K/k with Galois group G (to be thought of as given and known) and then to investigate the structure of the Galois groups of abelian extensions of K as G-modules. This has been done by several authors, mainly for tame extensions or p-extensions of local fields (see [10],[12],[3] and [13] for example and further literature) and for some infinite extensions of global fields, where the group algebra has some nice structure (Iwasawa theory). The aim of these notes is to show that one can get some results for arbitrary Galois groups by using the purely algebraic concept of class formations introduced by Tate.

1. Relation modules.

Given a presentation

$$1 \to R_m \to F_m \to G \to 1$$

of a finite group G by a (discrete) free group F_m on m free generators, the factor commutator group $R_m^{ab} = R_m/[R_m,R_m]$ becomes a finitely generated $\mathbb{Z}[G]$-module via the conjugation in F_m. By Lyndon [19] and Gruenberg [8]§2 we have

1.1. PROPOSITION. a) *There is an exact sequence of* $\mathbb{Z}[G]$*-modules*

(1) $$0 \to R_m^{ab} \to \mathbb{Z}[G]^m \to I(G) \to 0 \quad ,$$

where I(G) *is the augmentation ideal, defined by the exact sequence*

(2) $$0 \to I(G) \to \mathbb{Z}[G] \xrightarrow{aug} \mathbb{Z} \to 0, \qquad aug(\sum_{\sigma\in G} a_\sigma \sigma) = \sum_{\sigma\in G} a_\sigma .$$

b) $\mathbb{Q} \otimes_{\mathbb{Z}} R_m^{ab} \cong \mathbb{Q}[G]^{m-1} \oplus \mathbb{Q}$ *as* $\mathbb{Q}[G]$*-module.*

c) *For a second presentation* $1 \to R_n \to F_n \to G \to 1$ *one has*

$$R_n^{ab} \oplus \mathbb{Z}[G]^m \cong R_m^{ab} \oplus \mathbb{Z}[G]^n$$

and therefore

(3) $$R_{n,p}^{ab} \cong R_{m,p}^{ab} \oplus \mathbb{Z}_p[G]^{n-m} \qquad (n \geq m),$$

for every prime p, *if we set* $R_{m,p}^{ab} = \mathbb{Z}_p \otimes_{\mathbb{Z}} R_m^{ab}$ ($\mathbb{Z}_p$ *the ring of integers in the field* $\mathbb{Q}_p$ *of* p*-adic numbers).*

In particular the G-structure of $R_{m,p}^{ab}$ only depends on m; for R_m^{ab} itself and minimal m this is still an open problem, see [8]. One has $R_m^{ab} \cong I(G) \otimes_{\mathbb{Z}} I(G)$ for $m = (G:1)-1$, and $R_m^{ab} \cong \mathbb{Z}[G]^{m-1} \oplus \mathbb{Z}$ for cyclic G ($\mathbb{Z}, \mathbb{Q}, \mathbb{Z}_p$ and $\mathbb{Q}_p$ are always equipped with the trivial G-action). If G is a p-group, $R_{m,p}^{ab}$ is just the factor commutator group of $\hat{R}_m$ in any presentation $1 \to \hat{R}_m \to \hat{F}_m \to G \to 1$ of G by a free pro-p-group on m free generators. If the order of G is prime to p, one has $R_{m,p}^{ab} \cong \mathbb{Z}_p[G]^{m-1} \oplus \mathbb{Z}_p$.

Tate has shown (see [16]) that R_m^{ab} is a class formation module for G, i.e.,

(4) $$H^i(U, R_m^{ab}) \cong H^{i-2}(U, \mathbb{Z})$$

for all subgroups U of G and all $i \in \mathbb{Z}$ (here and in the following we take the modified (Tate) cohomology groups), where the isomorphism is obtained by taking cupproduct with the restriction of a generating element of $H^2(G, R_m^{ab}) \cong \mathbb{Z}/(G:1)\mathbb{Z}$. It turns out that R_m^{ab} has to be regarded as a standard object with this property - all other class formation modules only differing by "projective kernels":

<u>1.2. THEOREM</u>. *Let* G *be a finite group,* G_p *a* p*-Sylow subgroup and* M *a finitely generated* $\mathbb{Z}_p[G]$*-module with the property*

(*) $$\begin{aligned} H^1(G_p, M) &= 0 \\ H^2(G_p, M) &\cong \mathbb{Z}_p/(G_p: 1)\mathbb{Z}_p \\ H^2(G, M) &\cong \mathbb{Z}_p/(G: 1)\mathbb{Z}_p. \end{aligned}$$

a) *There is an exact sequence*

(5) $$0 \to X \to R_{m,p}^{ab} \to M \to 0$$

for some $m \in \mathbb{N}$ *and some projective* $\mathbb{Z}_p[G]$*-module* X.

b) *If* M *is torsion free (as* $\mathbb{Z}_p$*-module), the sequence (5) splits, so*

$$M \oplus X \cong R^{ab}_{m,p}$$

c) *There is an exact sequence*

(6) $$0 \to M \to M' \to I_p(G) \to 0$$

with a cohomologically trivial $\mathbb{Z}_p[G]$*-module* M' *and* $I_p(G) = \mathbb{Z}_p \otimes I(G)$.

Proof. The proof of a) is nearly as in [13]I: Let

$$0 \to M \to E \to G \to 1$$

be the group extension corresponding to a generating element of $H^2(G,M)$, and choose a homomorphism $\varphi: F_m \to E$ with dense image (m suitable). This induces a surjection $\bar{\varphi}: R^{ab}_{m,p} \to M$, let $X = \ker \bar{\varphi}$. From the long exact sequence of cohomology under G_p we get $H^2(G_p,X) = 0 = H^3(G_p,X)$, so X is cohomologically trivial, i.e., projective, as X is torsion free.

b) is clear (compare [13] 1.5), and M' is defined by the exact commutative diagram

$$\begin{array}{ccccccccc}
 & & & & I_p(G) & = & I_p(G) & & \\
 & & & & \uparrow & & \uparrow & & \\
0 & \longrightarrow & X & \longrightarrow & \mathbb{Z}_p[G]^m & \longrightarrow & M' & \longrightarrow & 0 \\
 & & \| & & \uparrow & & \uparrow & & \\
0 & \longrightarrow & X & \longrightarrow & R^{ab}_{m,p} & \longrightarrow & M & \longrightarrow & 0,
\end{array}$$

where the middle column is given by 1.1.a).

2. Cohomologically trivial $\mathbb{Z}_p[G]$-modules.

We fix the following notations. For a finitely generated $\mathbb{Z}_p[G]$-module M, Tor(M) will denote the $\mathbb{Z}_p$-torsion submodule of M, $M^* = \mathrm{Hom}(M,\mathbb{Q}_p/\mathbb{Z}_p)$ is the Pontrjagin dual of M (with the operation $(\sigma f)(m) = f(\sigma^{-1}m)$), and $d_G(M)$ is the minimal number of $\mathbb{Z}_p[G]$-generators for M. Tensor products now are taken over $\mathbb{Z}_p$, if not denoted otherwise. For a pro-finite group A (abelian or not) A(p) is the maximal pro-p-quotient.

A $\mathbb{Z}_p[G]$-module P is projective iff it is torsion free and cohomologically trivial, and then determined by the structure of $\mathbb{Q}_p \otimes P$ as $\mathbb{Q}_p[G]$-module by a theorem of Swan [23] 6.4. This generalizes to

2.1. THEOREM ([12] 1.2.). *For cohomologically trivial, finitely generated* $\mathbb{Z}_p[G]$*-modules* M *and* M' *the following statements are equivalent:*

i) $M \cong M'$.
ii) $\mathrm{Tor}(M) \cong \mathrm{Tor}(M')$ *and* $\mathbb{Q}_p \otimes M \cong \mathbb{Q}_p \otimes M'$.

Furthermore we have the following construction, which follows from [12] 1.8-1.10.

2.2. LEMMA. *Let* N *be a finite* $\mathbb{Z}_p[G]$*-module.*
a) *There is a presentation (exact sequence)*

$$\mathbb{Z}_p[G]^{\ell} \xrightarrow{f} \mathbb{Z}_p[G]^m \longrightarrow N^* \longrightarrow 0$$

if and only if there is an exact sequence

$$0 \to \mathbb{Z}_p[G]^m \xrightarrow{f^+} \mathbb{Z}_p[G]^{\ell} \to M \to 0$$

for the cohomologically trivial $\mathbb{Z}_p[G]$*-module* M *with* $\mathrm{Tor}(M) \cong N$ *and* $\mathbb{Q}_p \otimes M \cong \mathbb{Q}_p[G]^{\ell-m}$.
b) *In the above statement,* f^+ *can be chosen to be the transpose of* f *in the following sense: If* f *is given by the matrix* (α_{ij}) *with* $\alpha_{ij} \in \mathbb{Z}_p[G]$, f^+ *is then given by the matrix* (α_{ji}^+), *where* $^+$ *is the anti-involution of* $\mathbb{Z}_p[G]$ *given by*

$$\Big(\sum_{\sigma\in G} a_\sigma \sigma\Big)^+ = \sum_{\sigma\in G} a_\sigma \sigma^{-1} .$$

3. Applications to number fields.

Let K/k be a finite Galois extension of local or global number fields with Galois group G (function fields can be treated similarly). Fix a prime p and let $\bar{K}$ be
i) the maximal p-extension of K, if k is local,
ii) the maximal p-extension of K unramified outside S, if k is global; here S is a finite set of non-archimedean primes of K closed under the action of G and containing all primes above p and all primes ramified in K/k.

For every field $k \subseteq L \subseteq K$ we set $G_L = \mathrm{Gal}(\bar{K}/L)$, and we want to consider the finitely generated $\mathbb{Z}_p[G]$-module G_K^{ab}.
Notations: for any field L, μ_L is the group of roots of unity in L, and $\mu_n = \{\zeta \in \mu_\Omega \mid \zeta^n = 1\}$ for an algebraic closure Ω of L.

i) For local fields the only interesting case is where p equals the residue characteristic. The following theorem generalizes the results for p-groups due to Borevič, K. Wingberg and the author (see [2], [3], [13] and [25]):

3.1. THEOREM. *Let* k *be of degree* n *over* $\mathbb{Q}_p$.

a) G *is generated by* n+2 *elements, and there is an exact sequence*

(7) $$0 \to \mathbb{Z}_p[G] \to R^{ab}_{n+2,p} \to G^{ab}_K \to 0 .$$

b) *If* K *is regular* (i.e., $\mu_p \not\subseteq K$), G *is generated by* n+1 *elements, and there is an isomorphism*

(8) $$G^{ab}_K \cong R^{ab}_{n+1,p} .$$

Proof. We only show a), because b) is similar, using the splitting of (7). As the reciprocity map induces an isomorphism between G^{ab}_K and the projective limit over the groups $K^{\times}/K^{\times p^n}$ for all n, G^{ab}_K has the property (*), and using the p-adic logarithm we get an isomorphism

(9) $$\mathbb{Q}_p \otimes G^{ab}_K \cong \mathbb{Q}_p[G]^n \oplus \mathbb{Q}_p .$$

Let R be defined by the exact commutative diagram

$$\begin{array}{ccccccccc}
 & & & & I_p(G) & = & I_p(G) & & \\
 & & & & \uparrow & & \uparrow & & \\
0 & \to & \mathbb{Z}_p[G] & \to & \mathbb{Z}_p[G]^{n+2} & \to & M' & \to & 0 \\
 & & \| & & \uparrow & & \uparrow & & \\
0 & \to & \mathbb{Z}_p[G] & \to & R & \to & G^{ab}_K & \to & 0 ,
\end{array}$$

where the right column is given by 1.2.c) and the middle row exists by 2.2., because M' is cohomologically trivial, $\mathrm{Tor}(M') \cong \mu_K(p)$ is cyclic and $\mathbb{Q}_p \otimes M' \cong \mathbb{Q}_p[G]^{n+1}$ by (9). If we can show that G is generated by n+2 elements, we are done, because then $R \cong R^{ab}_{n+2,p}$ by applying Schanuel's lemma to the middle column and 1.1.a).

For this we may assume that the ramification group of G is abelian, by Burnside's theorem on p-groups. If L is the fixed field of the ramification group, G is then a quotient of the middle group in the extension

(10) $$1 \to G^{ab}_L \to G_k/[G_L,G_L] \to \overline{G} \to 1,$$

where $\overline{G} = Gal(L/k)$ is generated by 2 elements. Applying the above to L instead of K we get a surjection

$$\overline{\varphi}: \quad R^{ab}_{n+2,p}(\overline{G}) \to G^{ab}_L ,$$

which induces an isomorphism in cohomology. As $H^2(\overline{G}, G^{ab}_L)$ is generated by the element x_1 belonging to (10) (proposition of Weil-Safarevic) and $H^2(\overline{G}, R^{ab}_{n+2})$ by x_0 belonging to

$$(11) \qquad 1 \to R^{ab}_{n+2} \to F_{n+2}/[R_{n+2}, R_{n+2}] \to \overline{G} \to 1$$

(Tate, see [16]13.), after possibly multiplying $\overline{\varphi}$ by a unit in $\mathbb{Z}_p$, we may assume that the image of x_0 in $H^2(\overline{G}, R^{ab}_{n+2,p})$ is mapped to x_1 under the map induced by $\overline{\varphi}$. This means ([1]p. 179) there exists a lifting

$$\varphi: \quad F_{n+2}/[R_{n+2}/R_{n+2}] \to G_k/[G_L, G_L]$$

that induces $\overline{\varphi}$ and therefore has dense image. So $G_k/[G_L, G_L]$ is (topologically) generated by n+2 elements.

3.2. COROLLARY. *The absolute Galois group of a* $\mathfrak{p}$*-adic number field* k *is generated by* n+2 *elements,* $n = [k:\mathbb{Q}_p]$, *and this number is minimal.*

Indeed, if $K = k(\mu_p)$ and G_k was generated by n+1 elements, G^{ab}_K would be generated by $[K:k]+1$ elements as $\mathbb{Z}_p$-module (using (7) and the rank of R^{ab}_{n+1}), which is not true. 3.2. was shown in [15]1.4.d) for $\mu_p \subseteq k$.

ii) In the case of a global number field we assume that k is totally imaginary for p = 2 and fix the following notations.

r_1 and r_2 are the numbers of the real and complex places of k, respectively, and $r_1' \leq r_1$ is the cardinality of the set S'_∞ of the real places of k which ramify in K/k. S_p is the set of primes above p in k, and for any set T of primes in k and any extension L/k we let T(L) be the set of primes in L lying above primes in T. Finally, $L_\mathfrak{p}$ denotes the completion of L with respect to the prime $\mathfrak{p}$ of L, and d(H) is the minimal number of generators for a finitely generated profinite group H.

If $S'_\infty = \phi$ and hence $r_1' = 0$ (e.g., for K/k a p-extension), all statements are remarkably simplified, and we have a complete analogy with the local case.

3.3. THEOREM. *Let* k *be a finite extension of* $\mathbb{Q}$. *If Leopoldt's conjecture with respect to* p *is true for* K ([6] p. 274), G_K^{ab} *has the property (*), and the following holds.*

a) *If* $d(G_k) \leq d$, *there is an exact sequence*

$$0 \to X \to R_{d,p}^{ab} \to G_K^{ab} \to 0 \tag{12}$$

with a projective $\mathbb{Z}_p[G]$*-module* X, *whose structure is defined by the isomorphism*

$$X \oplus \mathbb{Z}_p[G]^{r_1'} \cong Y_{S_\infty'} \oplus \mathbb{Z}_p[G]^{d-r_1-1} \quad , \tag{13}$$

where $Y_{S_\infty'}$ *is the free* $\mathbb{Z}_p$*-module with basis* $S_\infty'(K)$ *and the natural (left) action of* G.

b) *One has* $d_G(\mathrm{Tor}(G_K^{ab})^*) \leq d(G_k)-r_2-1-r_1'+d_G(Y_{S_\infty'})$ *and conversely* $d(G_k) \leq \max(d(G), d_G(\mathrm{Tor}(G_K^{ab})^*)+r_2+1+r_1'-r_1''),$ *if* $Y_{S_\infty'}$ *has* r_1'' *free* $\mathbb{Z}_p[G]$*-summands.*

c) *If* G_K^{ab} *is torsion free, the sequence (12) is splitting, and there is an isomorphism*

$$G_K^{ab} \oplus \mathbb{Z}_p[G]^{d-r_2-1} \cong Z_{S_\infty'} \oplus R_{d,p}^{ab} \quad , \tag{14}$$

where $Z_{S_\infty'}$ *is defined by the property*

$$Y_{S_\infty'} \oplus Z_{S_\infty'} \cong \mathbb{Z}_p[G]^{r_1'} \quad . \tag{15}$$

d) *If* G_K^{ab} *is torsion free and* $\mu_p \subseteq K$, G *is generated by* $r_2+1+r_1'-r_1''$ *elements (and so is* G_k *by* c)). *So for* $S_\infty' = \phi$ *we then get an isomorphism*

$$G_K^{ab} \cong R_{r_2+1,p}^{ab} \quad . \tag{16}$$

3.4. Remarks. a) By the existence of r_2+1 linear independent $\mathbb{Z}_p$-extensions over k one has always $d \geq r_2+1$ in (13). X is well defined by the Krull-Schmidt theorem for $\mathbb{Z}_p[G]$-modules, in particular

$$X \cong Y_{S_\infty'} \oplus \mathbb{Z}_p[G]^{d-r_2-1-r_1'} \tag{17}$$

for $d - r_2 - 1 - r_1' \geq 0$ and

$$X \cong \mathbb{Z}_p[G]^{d-r_2-1}$$

for $S'_\infty = \phi$.

b) Choosing one decomposition group $G_{\mathfrak{p}} \subseteq G$ for every prime $\mathfrak{p} \in S'_\infty$, the module $Y_{S'_\infty}$ can be described as

(18) $$Y_{S'_\infty} = \bigoplus_{\mathfrak{p}\in S'_\infty} \mathrm{Ind}_{G_{\mathfrak{p}}}^{G} (\mathbb{Z}_p),$$

where $\mathrm{Ind}_{G_{\mathfrak{p}}}^{G}$ means induction from $G_{\mathfrak{p}}$ to G. As well

(19) $$Z_{S'_\infty} = \bigoplus_{\mathfrak{p}\in S'_\infty} \mathrm{Ind}_{G_{\mathfrak{p}}}^{G} (\mathbb{Z}_p(-1)),$$

where $\mathbb{Z}_p(-1)$ is the module $\mathbb{Z}_p$, on which the non-trivial element of $G_{\mathfrak{p}}$ acts by multiplication with -1. For $p \neq 2$ one has $\mathbb{Z}_p[G_{\mathfrak{p}}] = \mathbb{Z}_p \oplus \mathbb{Z}_p(-1)$, which shows (15) (recall that $S'_\infty = \phi$ for $p = 2$ by assumption).

Proof of 3.3. As S contains S_p and k is totally imaginary for $p = 2$, we have $cd_p(G_k) \leq 2$ (see [4] 2.11), and

(20) $$H^2(G_K, \mathbb{Q}_p/\mathbb{Z}_p) = 0,$$

if and only if the Leopoldt conjecture is true for K and p (using the same arguments as in [9]4.4). In this case, $cd_p(G_k) \leq 2$ implies

(21) $$H^i(G_L, \mathbb{Q}_p/\mathbb{Z}_p) = 0 \qquad \text{for all} \quad k \subseteq L \subseteq K$$

not only for $i \geq 3$ but also for $i = 2$, using (20) and the surjectivity of the corestriction, see [22]I 3.3. Using the spectral sequence

(22) $$H^i(G(K/L), H^j(G_K, \mathbb{Q}_p/\mathbb{Z}_p)) \Longrightarrow H^{i+j}(G_L, \mathbb{Q}_p/\mathbb{Z}_p)$$

one shows as in [16], App., or [9] 2.3, that G_K^{ab} has the property (*) and $H^2(G, G_K^{ab})$ is generated by the element belonging to

$$0 \to G_K^{ab} \to G_k/[G_K, G_K] \to G \to 0 .$$

Proceeding as in the proof of 1.2.a), we get an exact sequence

(23) $$0 \to X \to R_{d,p}^{ab} \to G_K^{ab} \to 0$$

with projective X, if $G_K/[G_K,G_K]$ is generated by d elements.

On the other hand, class field theory gives us an exact sequence

(24) $$U_K\otimes_{\mathbb{Z}}\mathbb{Z}_p \to \prod_{\mathfrak{p}\in S(K)} U_{K_{\mathfrak{p}}}(p) \to G_K^{ab} \to Cl_K(p) \to 0,$$

where U_K (resp. $U_{K_{\mathfrak{p}}}$) denotes the group of units in K (resp. $K_{\mathfrak{p}}$) and Cl_K is the class group of K.

If the Leopoldt conjecture is true for K and p, the first map in (24) is injective and we may compute $\mathbb{Q}_p \otimes G_K^{ab}$. By Dirichlet's theorem $\mathbb{Q} \oplus \mathbb{Q}\otimes U_K$ is isomorphic to the $\mathbb{Q}$-vector space with all archimedean places of K as a basis and the natural permutation action of G on this basis. Therefore

(25) $$\mathbb{Q}_p \oplus \mathbb{Q}_p\otimes_{\mathbb{Z}}U_K \cong \mathbb{Q}_p\otimes Y_{S'_\infty} \oplus \mathbb{Q}_p[G]^{r_2+r_1-r_1'} .$$

On the other hand, by the local theory one gets

(26) $$\mathbb{Q}_p \otimes \Big(\prod_{\mathfrak{p}\in S(K)} U_{K_{\mathfrak{p}}}(p)\Big) \cong \mathbb{Q}_p[G]^n = \mathbb{Q}_p[G]^{r_1+2r_2} ,$$

with $n = [k:\mathbb{Q}] = \sum_{\mathfrak{p}\in S_p} [k_{\mathfrak{p}}: \mathbb{Q}_p]$. By (24) we calculate

(27) $$\mathbb{Q}_p\otimes G_K^{ab} \oplus \mathbb{Q}_p\otimes Y_{S'_\infty} \oplus \mathbb{Q}_p[G]^{r_2+r_1-r_1'} \cong \mathbb{Q}_p[G]^{r_1+2r_2}\oplus \mathbb{Q}_p,$$

while (23) and 1.1.b) imply

(28) $$\mathbb{Q}_p\otimes G_K^{ab} \oplus \mathbb{Q}_p\otimes X \cong \mathbb{Q}_p\otimes R_{d,p}^{ab} \cong \mathbb{Q}_p[G]^{d-1} \oplus \mathbb{Q}_p.$$

Combining (27) and (28) we get

$$\mathbb{Q}_p \otimes (X \oplus \mathbb{Z}_p[G]^{r_1'}) \cong \mathbb{Q}_p \otimes (Y_{S'_\infty} \oplus \mathbb{Z}_p[G]^{d-r_2-1}),$$

which implies (13) by Swan's theorem.

To show the first part of b), we apply the functor $M \rightsquigarrow M^+ = \mathrm{Hom}(M,\mathbb{Z}_p)$ to (12) and get the exact sequence

$$0 \to (G_K^{ab})^+ \to (R_{d,p}^{ab})^+ \to X^+ \to \mathrm{Tor}(G_K^{ab})^* \to 0,$$

because of the canonical isomorphism $\mathrm{Ext}^1_{\mathbb{Z}_p}(M,\mathbb{Z}_p) \cong \mathrm{Tor}(M)^*$. As $d_G(M\oplus\mathbb{Z}_p[G]) = d_G(M)+1$ for a finitely generated $\mathbb{Z}_p[G]$-module, see [8] 5.8, we get

$$d_G(\mathrm{Tor}(G_K^{ab})^*) \le d_G(X^+) = d_G(Y_{S'_\infty}{}^+) + d - r_2 - 1 - r_1$$

by (13) and the isomorphism $\mathbb{Z}_p[G]^+ \cong \mathbb{Z}_p[G]$, which also implies $d_G(P^+) = d_G(P)$ for projective P.

For the second part of b) one proceeds as in the proof of 3.1. (where we had $d_G(\mathrm{Tor}(G_K^{ab})^*) = 1$ and $d(G) \le n+2$), by considering $G_K^{ab} \oplus \tilde{Y}_{S'_\infty}$ for $Y_{S'_\infty} \cong \tilde{Y}_{S'_\infty} \oplus \mathbb{Z}_p[G]^{r''_1}$, and c) is clear.

For d) we use the fact that $H^2(G_K,\mathbb{Z}/p\mathbb{Z}) = 0$ for torsion free G_K^{ab} (which follows from (20) and the cohomology sequence for $0 \to \mathbb{Z}/p\mathbb{Z} \to \mathbb{Q}_p/\mathbb{Z}_p \overset{p}{\to} \mathbb{Q}_p/\mathbb{Z}_p \to 0$). If K contains a primitive p-th root of unity, this is only possible, if K has only one prime $\mathfrak{p}_0$ above p, see [4] 3.3. In particular, G is equal to the decomposition group for $\mathfrak{p}_0$, and we may use the same arguments as in the local case (considering again $G_K^{ab} \oplus \tilde{Y}_{S'_\infty}$).

<u>3.5. COROLLARY.</u> *If K/k is a p-extension (and Leopoldt's conjecture is true for K and p), let $d = \dim H^1(G_k)$ and $r = \dim H^2(G_k)$ be the numbers of generators and relations of the pro-p-group G_k, respectively. Then $r = d-r_2-1 = d_G(\mathrm{Tor}(G_K^{ab})^*)$ = p-rank of $\mathrm{Tor}(G_k^{ab})$, and there is an exact sequence*

(29) $$0 \to \mathbb{Z}_p[G]^r \to R_{d,p}^{ab} \to G_K^{ab} \to 0 .$$

<u>Proof</u>. The equality $1-d+r = \chi(G_k) = -r_2$ was shown by Tate [24], $H^2(G_k,\mathbb{Q}_p/\mathbb{Z}_p) = 0$ implies $H^2(G_k)^* \cong \{x \in G_k^{ab} | px = 0\}$, see [5] 5.6., and $H^2(G_K,\mathbb{Q}_p/\mathbb{Z}_p) = 0$ implies $(G_K^{ab})^G \cong G_k^{ab}$, see [9] 2.3. Finally, for M a finite $\mathbb{Z}_p[G]$-module and G a p-group, one has $d_G(M^*)$ = p-rank of $M^*/I_p(G)M^* \cong (M^G)^*$.

<u>3.6. Examples and remarks</u>. a) The numbers d and r in 3.5. have been studied extensively by Koch in [17]. If s, resp. s', denotes the cardinality of S, resp. the subset $S' = \{\mathfrak{p} \in S | \mu_p \subseteq k\}$, one has

(30) $$\begin{aligned} s' \le r \le s' + c_p + r_1 + r_2 - 1 &\quad \text{for } \mu_p \not\subseteq k, \\ r = s + c_p - 1 &\quad \text{for } \mu_p \subseteq k, \end{aligned}$$

where c_p is the p-rank of the S-class group of k (quotient of Cl_k by the classes of the primes in S), and in both cases $r = s'-1$ for large S.

b) If K is a p-extension of $k = \mathbb{Q}$ ($p \ne 2$) and Leopoldt's conjecture is true for K and p (e.g., K abelian), there is an exact sequence

$$(31) \qquad 0 \to \mathbb{Z}_p[G]^{s'} \to R_{s'+1,p} \to G_K^{ab} \to 0,$$

with s' as in a) (use (30)).

c) If $k = \mathbb{Q}(\sqrt{-D})$ is imaginary quadratic and $p \geq 5$ does not divide the class number of k, then for $S = S_p$ the group $G_k = G_{k,S_p}$ is free on two generators by (30). So for any p-extension K of k which is unramified outside p the Leopoldt conjecture is true for K and p (by (20)) and there is an isomorphism

$$G_K^{ab} \cong R_{2,p}^{ab} .$$

The same is true for p = 3 if the localizations above $\mathbb{Q}_3$ do not contain μ_3.

4. The special case of $\mathbb{Z}_p$-extensions.

Let K/k, $G = \mathrm{Gal}(K/k)$, $S, \bar{K}, G_L = \mathrm{Gal}(\bar{K}/L)$ and the other notations be as in the beginning of §3. If k is a global field, assume that Leopoldt's conjecture is true for K and p, and that k is totally imaginary for p = 2.

4.1. THEOREM. *Let* G_p *be a p-Sylow group of* G *and* K_p *be the fixed field of* G_p. *If* G_p *is cyclic, the following assertions are equivalent:*

i) $G^{ab} \cong M' \oplus R$ *with* M' *cohomologically trivial and* R *torsion free.*

ii) The extension K/K_p *is embeddable in a* $\mathbb{Z}_p$*-extension.*

Proof. In a decomposition i), the $\mathbb{Z}_p[G]$-module R has the property (*), so as G_p-module $R \cong \mathbb{Z}_p \oplus P$ with P projective, as follows from 1.2.b) and (3). The projection $G_K^{ab} \twoheadrightarrow \mathbb{Z}_p$ induces an isomorphism in the cohomology under G_p, so there is a commutative diagram

$$(32) \qquad \begin{array}{ccccccccc} 1 & \to & G_K^{ab} & \to & G_{K_p}/[G_K,G_K] & \to & G_p & \to & 1 \\ & & \downarrow & & \downarrow & & \| & & \\ 1 & \to & \mathbb{Z}_p & \to & \mathbb{Z}_p & \to & G_p & \to & 1 \end{array} ,$$

which shows ii).

On the other hand, if there is a diagram (32), we can solve the embedding problem

(33)
$$\begin{array}{ccccccccc} & & & & & & G_k & & \\ & & & & & \swarrow & \downarrow & & \\ 1 & \to & R^{ab}_{m,p} & \to & E & \to & G & \to & 1 \end{array}$$

(i.e., the dotted arrow making the diagram commutative exists), where E corresponds to an element of $H^2(G,R^{ab}_{m,p})$ which under the restriction map goes to that element of $H^2(G_p,R^{ab}_{m,p})$, which corresponds to the lower sequence in (32) via some G_p-isomorphism $R^{ab}_{m,p} \cong \mathbb{Z}_p \oplus \mathbb{Z}_p[G_p]^{m-1}$ (G generated by m elements). Indeed, the solvability may be checked on G_p by a theorem of Hoechsmann, and there it is solvable by assumption. (In fact one has to look at the induced problems with kernel $R^{ab}_{m,p}/p^r R^{ab}_{m,p}$ for all r to have finite modules and then use the fact that G_k is finitely generated).

We get a map $G^{ab}_K \to R^{ab}_{m,p}$, which induces an isomorphism in cohomology (because it does in dimensions i = 1,2,3). Adding a suitable map $\mathbb{Z}_p[G]^r \to R^{ab}_{m,p}$, we get a surjective map

$$G^{ab}_K \oplus \mathbb{Z}_p[G]^r \twoheadrightarrow R^{ab}_{m,p} \quad ,$$

whose kernel Q must be cohomologically trivial. Therefore the corresponding exact sequence splits, as $R^{ab}_{m,p}$ is torsion free, so

$$R^{ab}_{m,p} \oplus Q \cong G^{ab}_K \oplus \mathbb{Z}_p[G]^r \quad ,$$

which shows i) by the Krull-Schmidt theorem.

<u>4.2. Remark</u>. If G_p has d generators, d > 1, consider the statements

ii)' K/K_p can be embedded in a $\hat{F}_d$-extension, $\hat{F}_d$ the free pro-p-group on d generators.

iii) The embedding problem

(34)
$$\begin{array}{ccccccccc} & & & & & & G_{K_p} & & \\ & & & & & & \downarrow & & \\ 1 & \to & \hat{R}^{ab}_d & \to & \hat{F}_d/[\hat{R}_d,\hat{R}_d] & \to & G_p & \to & 1 \end{array}$$

is solvable.

Then i) $\Longleftrightarrow$ iii) $\Longleftarrow$ ii)', and iii) $\Longrightarrow$ ii)' for local fields by a result of Lur'e [18], compare [14] for the case of p-groups.

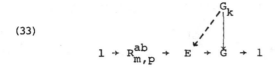

By 1.2.b) and 2.1. the modules M' and R in i) are determined by $Tor(M') = Tor(G^{ab}_K)$, $\mathbb{Q}_p \otimes M'$ and $\mathbb{Q}_p \otimes R$. But $\mathbb{Q}_p \otimes G^{ab}_K$ is known, and M' and R are uniquely defined up to projectives, so for (p-Sylow groups

embeddable in) $\mathbb{Z}_p$-extensions the $\mathbb{Z}_p[G]$-structure of G_K^{ab} is completely determined by $\mathrm{Tor}(G_K^{ab})$. We illustrate this first by completely determining the structure in the local case.

For this we also allow K/k to be infinite, in which case G_k^{ab} is a module over the completed group ring $\mathbb{Z}_p[[G]] = \varprojlim \mathbb{Z}_p[G/U]$, where U runs over all open normal subgroups of G. The relation module for G may then be described by $R_{m,p}^{ab} = R_{m,p}^{ab}(G) = \varprojlim R_{m,p}^{ab}(G/U)$, starting from a homomorphism $F_m \to G$ with dense image, which induces exact sequences $1 \to R_m(U) \to F_m \to G/U \to 1$ for all U. Another description is $R_{m,p}^{ab}(G) = \hat{R}_m^{ab} \otimes_{\hat{\mathbb{Z}}} \mathbb{Z}_p$, where $1 \to \hat{R}_m \to \hat{F}_m \to G \to 1$ is a presentation by a free profinite group $\hat{F}_m$ on m generators.

4.3. THEOREM. *Let* k *be of degree* n *over* $\mathbb{Q}_p$ *and* K/k *be a Galois extension such that* K/K_p *is a* $\mathbb{Z}_p$*-extension or embeddable in a* $\mathbb{Z}_p$*-extension, where* K_p *is the fixed field of a* p*-Sylow group of* G.

a) G *has two generators.*

b) If K/K_p *is cyclotomic and of finite degree,*

$$G_K^{ab} \cong \mu_K(p) \oplus \mathbb{Z}_p[\![G]\!]^n \oplus \mathbb{Z}_p. \tag{35}$$

c) If K/K_p *is cyclotomic and of infinite degree,*

$$G_K^{ab} \cong \mathbb{Z}_p(1)^{\delta} \oplus \mathbb{Z}_p[\![G]\!]^n , \tag{36}$$

where $\mathbb{Z}_p(1)^{\delta}$ *is the Tate module of* $\mu_K(p)$ ($\mathbb{Z}_p(1)^{\delta} = \varprojlim \mu_{p^r}$ *for* $\mu_p \subseteq K$, $= 0$ *for* $\mu_p \not\subseteq K$).

d) If K/K_p *is not cyclotomic,*

$$G_K^{ab} \oplus \mathbb{Z}_p[\![G]\!] \cong M' \oplus R_{2,p}^{ab} \oplus \mathbb{Z}_p[\![G]\!]^{n-1} , \tag{37}$$

with M' *given by the exact sequence*

$$0 \to \mathbb{Z}_p[\![G]\!] \to \mathbb{Z}_p[\![G]\!]^2 \to M' \to 0 \tag{38}$$

$$1 \mapsto (x-g,\ 1+(q-1)\lambda),$$

where: x *generates the* p*-Sylow group,* q *is the order of* $\mu_K(p)$, $g \in \mathbb{Z}_p$ *with* $\zeta^x = \zeta^g$ *for all* $\zeta \in \mu_K(p)$, *and* λ *is the idempotent of* $\mathbb{Z}_p[\![G_0]\!]$ *which belongs to the action on* $\mu_K(p)$; *here* G_0 *is a maximal* p'*-subgroup (i.e., with order prime to* p).

__Proof__. Let L_0 (resp. L_1) be the fixed field of the inertia (resp. ramification) group and α: $\mathrm{Gal}(L_1/k) \to (\mathbb{Z}_p/p^s\mathbb{Z}_p)^{\times}$, $0 \le s \le \infty$, be the character of the operation on $Gal(K/L_1)$. Then $Gal(L_1/k)$ has two generators σ, τ, where τ generates $Gal(L_1/L_0)$ and σ can be chosen such that $\alpha(\sigma)$ generates the image of α. If τ^r generates $Ker\ \alpha \cap \langle\tau\rangle$, G is generated by $x\tau^r$ and σ (where σ and τ are suitable liftings in G), because the order of τ is prime to p and $x\tau^r = \tau^r x$.

If L_0^p is the maximal p-extension of k in L_0, the order of $Gal(L_1/L_0^p)$ is prime to p, and G_0 can be chosen as the image of a section of $Gal(K/L_0^p) \twoheadrightarrow Gal(L_1/L_0^p)$.

By taking limits, c) follows from b), and in b) we may assume G to be finite (by a compactness argument we may take compatible isomorphisms (35), for which the transition maps on $\mathbb{Z}_p$ are just multiplication with the group index). Now $\mu_K(p)$ is cohomologically trivial for cyclotomic K/K_p, and $\mathbb{Z}_p$ is a module with the property (*) for G, because the p-Sylow group maps isomorphically onto the maximal p-quotient. Therefore $G_K^{ab} \cong \mu_K(p) \oplus \mathbb{Z}_p \oplus P$ with projective P, which must be free by Swan's theorem.

For d) we may again restrict to finite groups and then only have to check that M' is the cohomologically trivial module with $Tor(M') = \mu_K(p) = Tor\ G_K^{ab}$ and $\mathbb{Q}_p \otimes M' = \mathbb{Q}_p[G]$. By 2.2. we only need to show that

$$\mathbb{Z}_p[G]^2 \to \mathbb{Z}_p[G] \to \mu_K(p)^* \to 0$$

$$\begin{matrix} (1,0) \mapsto x^{-1}-g, \\ (0,1) \mapsto 1+(q-1)\lambda^+ \end{matrix} \quad , \ 1 \mapsto \text{ generating element,}$$

is exact. This is easy, using the fact that x and G_0 generate G.

<u>4.4. Remarks</u>. a) If $g_0 = (G_0:1)$ is finite and $\beta: G_0 \to (\mathbb{Z}/p\mathbb{Z})^\times \subseteq \mathbb{Z}_p^\times$ is the character describing the operation on $\mu_K(p)$, one has $\lambda = n_0^{-1} \Sigma \beta(\rho)^{-1}\rho$, where the sum runs over all $\rho \in G_0$. For infinite G_0 one takes the limit of these elements for finite quotients.

b) The case $G_0 = 1$ has been studied by Iwasawa in [11] and the split case (i.e., G is the product of $\mathbb{Z}_p$ and G_0) by Dummit in [7]. They also get b) and c) but instead of d) an exact sequence $0 \to G_K^{ab} \to \mathbb{Z}_p[[G]]^n \to \mu_K(p) \to$ which cannot exist in the non-split case, because then $\mathbb{Q}_p \otimes G_K^{ab}$ is not free.

c) For $n \geq 2$ one may cancel one $\mathbb{Z}_p[[G]]$ in (37) and so get an explicit formula for G_K^{ab}. If the group G is given, it is easy to determine $R_{2,p}^{ab}$ and a free summand of $M' \oplus R_{2,p}^{ab}$ for $n = 1$. For example, in the split case $R_{2,p}^{ab} \cong \mathbb{Z}_p[[G]] \oplus \mathbb{Z}_p$ for $[K:K_p] < \infty$ and $R_{2,p}^{ab} \cong \mathbb{Z}_p[[G]]$ for $[K:K_p] = \infty$.

For global fields 4.1. immediately implies

4.5. PROPOSITION. *If the p-Sylow subextension of* K/k *is embeddable in a* $\mathbb{Z}_p$*-extension and* K *is a totally real number field,*

$$G_K^{ab} \cong \mathrm{Tor}(G_K^{ab}) \oplus \mathbb{Z}_p, \tag{39}$$

and $\mathrm{Tor}(G_K^{ab})$ *is cohomologically trivial.*

Now let k be an arbitrary finite extension of $\mathbb{Q}$ and $K = \bigcup_n K_n$ be the cyclotomic Γ-extension, $K_n = k(\mu_{p^{n+1}})$ and $\Gamma = \mathrm{Gal}(K/k)$. Let $\Gamma_n = \mathrm{Gal}(K/K_n)$ and assume that Leopoldt's conjecture with respect to p is true for all K_n (e.g. k abelian). We want to relate the $\mathbb{Z}_p[[\Gamma]]$-module $X_1 = G_K^{ab}$ (usually considered for $S = S_p$, i.e., $X_1 = \mathrm{Gal}(M/k)$, where M is the maximal abelian p-extension of K unramified outside p) and $X_3 = \mathrm{Gal}(L'/K)$, where L' is the maximal abelian p-extension of K, which is unramified and in which every prime splits completely.

By Tate's duality theorem we get an exact sequence

$$0 \to \mu_{K_n}(p) \to \prod_{\mathfrak{p}\in S(K_n)} \mu_{K_{n,\mathfrak{p}}}(p) \xrightarrow{\psi} \mathrm{Tor}(G_{K_n}^{ab}) \to R_1(K_n) \to 0 \tag{40}$$

where $R_1(K_n)$ is the kernel of the map

$$H^1(G_{K_n}, \mu_K(p)) \to \prod_{\mathfrak{p}\in S(K_n)} H^1(G_{K_{n,\mathfrak{p}}}, \mu_{K_{n,\mathfrak{p}}}(p)) \tag{41}$$

induced by the restriction maps (compare [21]2.5.ii), $H^2(G_K,\mathbb{Q}_p/\mathbb{Z}_p) = 0$ implies $H^2(G_K,\mathbb{Z}_p) \cong \mathrm{Tor}(G_K^{ab})^*$, ψ is then given by the reciprocity map). By taking limits we get an exact sequence

$$0 \to \mu_K(p) \to \bigoplus_{\mathfrak{p}\in S(K)} \mu_{K_\mathfrak{p}}(p) \to \varinjlim_n \mathrm{Tor}(G_{K_n}^{ab}) \to \mathrm{Hom}(X_3,\mu_K(p)) \to 0, \tag{42}$$

and, by dualizing and setting $X_4 = (\varinjlim_n \mathrm{Tor}(G_{K_n}^{ab}))^*$ (the limit being taken via the transfer maps), the exact sequence

$$0 \to X_3(-1) \to X_4 \to \prod_{\mathfrak{p}\in S(K)} \mathbb{Z}_p(1) \to \mathbb{Z}_p(1) \to 0, \tag{43}$$

where $M(n)$ denotes the n-th Tate twist of a $\mathbb{Z}_p[[\Gamma]]$-module M (as in [6]). Let $\Delta = \mathrm{Gal}(K_0/k)$, $d = (\Delta:1)$, and e_i be the idempotent in $\mathbb{Z}_p[\Delta]$ belonging to the i-th power of the cyclotomic character, $0 \leq i \leq d-1$.

We then may split $X_1 (X_3,\dots)$ into the direct sum of the $e_i X_1$ $(e_i X_3,\dots)$ and consider these as modules under $\Lambda = \mathbb{Z}_p[\![\Gamma_o]\!]$.

Suppose now that $e_{1-i}X_3$ is known (and so also $(e_{1-i}X_3)(-1) = e_{-i}(X_3(-1)))$ and suppose further that we can calculate $e_{-i}X_4$ from (43) (e.g., if S(K) contains just one prime). Then we can get $e_i X_1$ as follows: Choose a minimal presentation

$$(44) \qquad \Lambda^{\ell_i} \xrightarrow{(\alpha_{rs})} \Lambda^{m_i} \to e_{-i}X_4 \to 0,$$

and take the transpose as in 2.2. to get an exact sequence

$$(45) \qquad 0 \to \Lambda^{m_i} \xrightarrow{(\alpha_{sr}^+)} \Lambda^{\ell_i} \to M_i \to 0,$$

(M_i defined by exactness). Then there is an isomorphism

$$(46) \qquad e_i X_1 \cong M_i \oplus \Lambda^{d_i},$$

where

$$(47) \qquad d_i = m_i - \ell_i + \begin{cases} r_1 + r_2 & \text{for d even and i odd,} \\ r_2 & \text{else.} \end{cases}$$

Indeed, we have $(e_{-i}X_4)_{\Gamma_m} = (e_i \varinjlim_n \mathrm{Tor}(G_{K_n}^{ab})^{\Gamma_m})^* = (e_i \mathrm{Tor}(G_{K_m}^{ab}))^*$ for the module of coinvariants under Γ_m, using the fact that the transfer induces an isomorphism $\mathrm{Tor}(G_{K_n}^{ab}) \xrightarrow{\sim} \mathrm{Tor}(G_{K_{n+1}}^{ab})^{\Gamma_n}$ if $H^2(G_{K_{n+1}}, \mathbb{Q}_p/\mathbb{Z}_p) = 0$. So by 2.2. $(M_i)_{\Gamma_m}$ is cohomologically trivial with torsion module isomorphic to $e_i \mathrm{Tor}(G_{K_m}^{ab})$. The same is true for $(e_i X_1)_{\Gamma_m}$, as follows from the spectral sequence

$$(48) \qquad H^i(\Gamma_m, H^j(G_K, \mathbb{Q}_p/\mathbb{Z}_p)) \Longrightarrow H^{i+j}(G_{K_m}, \mathbb{Q}_p/\mathbb{Z}_p).$$

Therefore by 2.1. these modules only differ by projective $\mathbb{Z}_p[\Gamma_o/\Gamma_n]$-modules, whose structure is easily calculated knowing the structure of $e_i \mathbb{Q}_p \otimes G_{K_m}^{ab}$. Passing to the limit we obtain (46).

Bibliography.

1. Artin, E. and Tate, J., Class field theory, Harvard 1961.

2. Borevič, Z.I. On the group of principal units of a normal p-extension of a regular local field, Proc. Math. Inst. Steklov 80 (1965), 31-47.

3. Borevič, Z.I. and El Musa, A.J., Completion of the multiplicative group of p-extensions of an irregular local field, J. Soviet Math. 6, 3 (1976), 6-23.

4. Brumer, A., Galois groups of extensions of number fields with given ramification, Michigan Math. J. 13 (1966), 33-40.

5. Brumer, A., Pseudocompact algebras, profinite groups and class formations, J. Algebra 4 (1966), 442-470.

6. Coates, J., p-adic L-functions and Iwasawa's theory, in Algebraic Number Fields (Durham Symp. 1975, ed. A. Fröhlich), 269-353. Academic Press, London 1977.

7. Dummit, D., An extension of Iwasawa's Theorem on Finitely Generated Modules over Power Series Rings, Manuscripta Math. 43(1983), 229-259.

8. Gruenberg, K.W. Relation modules of finite groups, conf. board of math. sciences 25, AMS, Providence 1976.

9. Haberland, K., Galois Cohomology of Algebraic Number Fields, VEB Deutscher Verlag der Wissenschaften, Berlin 1978.

10. Iwasawa, K., On Galois groups of local fields, Trans. Amer. Math. Soc. 80 (1955), 448-469.

11. Iwasawa, K., On $\mathbb{Z}_\ell$-extensions of algebraic number fields, Ann. of Math. (2) 98(1973), 246-326.

12. Jannsen, U., Über Galoisgruppen lokaler Körper, Invent. Math. 70 (1982), 53-69.

13. Jannsen, U. and Wingberg, K., Die p-Vervollständigung der multiplikativen Gruppe einer p-Erweiterung eines irregulären $\mathfrak{p}$-adischen Zahlkörpers, J. reine angew. Math. 307/308 (1979), 399-410.

14. Jannsen, U. and Wingberg, K., Einbettungsprobleme und Galoisstruktur lokaler Körper, J. reine angew. Math. 319 (1980), 196-212.

15. Jannsen, U. and Wingberg, K., Die Struktur der absoluten Galoisgruppe $\mathfrak{p}$-adischer Zahlkörper, Invent. Math. 70 (1982), 71-98.

16. Kawada, Y., Class formations, Proc. Symp. Pure Math. 20 (1971), 96-114.

17. Koch, H., Galoissche Theorie der p-Erweiterungen, VEB Deutscher Verlag der Wissenschaften / Springer Berlin-Heidelberg-New York 1970.

18. Lur'e, B.B., Problem of immersion of local fields with a non abelian kernel, J.Soviet Math. 6, no. 3 (1976), 298-306.

19. Lyndon, R.C., Cohomology theory of groups with a single defining relation, Ann. of Math. (2) 53 (1950), 650-665.

20. Nguyen-Quang-Do, T., Sur la structure galoisienne des corps locaux et la théorie d'Iwasawa II, J. reine angew. Math. 333 (1982), 133-143.

21. Schneider, P., Über gewisse Galoiscohomologiegruppen, Math. Z. 168 (1979), 181-205.

22. Serre, J-P., Cohomologie galoisienne, Lecture Notes in Math. 5, Springer Verlag, Berlin-Heidelberg-New York 1964.

23. Swan, R., Induced representations and projective modules, Ann. of Math. (2) 71(1960), 522-578.

24. Tate, J., Duality theorems in Galois cohomology over number fields Proc. Intern. Congress Math. 1962, Stockholm 1963, p. 288-295.

25. Wingberg, K. Die Einseinheitengruppe von p-Erweiterungen regulärer p-adischer Zahlkörper als Galoismodul, J. reine angew. Math. 305 (1979), 206-214.

VALUES OF ZETA-FUNCTIONS AT NON-NEGATIVE INTEGERS

S. Lichtenbaum*
Department of Mathematics
Cornell University
Ithaca, N.Y. 14853

The general problem that we want to consider is the computation of values of zeta-functions $\zeta(X,s)$ of schemes X of finite type over Spec $\mathbb{Z}$ at non-negative integral values of s in terms of cohomological or K-theoretical invariants of X. The behavior of $\zeta(X,s)$ near $s = 0$ is closely related to the étale cohomology of the constant sheaf $\mathbb{Z}$, and the behavior of $\zeta(X,s)$ near $s = 1$ is closely related to the étale cohomology of the sheaf G_m. However, for positive integral values of $s \geq 2$, there are no appropriate sheaves known, and it is likely that none exist. Instead, it is necessary to consider complexes of sheaves.

At present, we cannot define these complexes, but it is possible to predict a great many properties which they should have, and the number of relationships which should exist with results already known or conjectured is impressive, and leads to a fascinating picture.

In this paper, we will mainly be interested in smooth projective varieties over a finite field, in order to avoid the complicating factors caused by the presence of infinite primes. In the first two sections, we discuss the cases when $s = 0$ and $s = 1$. Then we go on in §3 to describe the "axioms" that the aforementioned complexes should satisfy (for an arbitrary scheme). In §4, we show that the axiom we call "Hilbert's Theorem 90" implies both the classical generalization of Hilbert's Theorem 90 and the Mercuriev-Suslin "Hilbert Theorem 90 for K_2". In §5, we discuss the relationship to similar complexes of Zariski sheaves whose existence has been conjectured by Beilinson. In §6, we explain how these complexes should give rise to very general duality theorems, and in §7 we come back to the case of varieties over finite fields and their zeta-functions. We conclude in §8 with some miscellaneous remarks.

---

*This research was partially supported by NSF grants. The author would also like to thank the I.H.E.S. and the University of Paris (Orsay) for their hospitality during the academic year 1982-83, when much of this work was done.

It should be emphasized here that we and Beilinson had independently conjectured the existence of complexes of sheaves satisfying various "axioms". Since Beilinson was working in the Zariski topology and we were working in the étale topology, these "axioms" were not identical; still there were many close similarities. The present paper was written after the author learned of Beilinson's work, and has been influenced in many respects by Beilinson's ideas. We would also like to thank L. Breen, B. Mazur, and C. Soulé for many helpful conversations.

1. Values of zeta-functions at $s = 0$.

In this and the next section, let k be a finite field with $q = p$ elements, and let X be a smooth, projective, geometrically connected scheme over k of dimension d. Let $Z(X,t)$ be the rational function such that $\zeta(X,s) = Z(X,q^{-s})$.

When $s = 0$, so $t = 1$, $\zeta(X,s)$ has a simple pole. Let $H^i(X,F)$ denote the étale cohomology groups of X with values in the sheaf F. Let F now be the constant sheaf $\mathbb{Z}$. Then the $H^i(X,\mathbb{Z})$ are zero for i large, and finite for $i \neq 0, 2$. $H^0(X,\mathbb{Z}) = \mathbb{Z}$ and $H^2(X,\mathbb{Z})$ is the dual of the finitely-generated rank one abelian group $CH_0(X)$ consisting of zero-cycles on X modulo rational equivalence.

Then the following formula holds:

$$\lim_{t\to 1}(1-t)Z(X,t) = \pm\chi(X,\mathbb{Z}),$$

where

$$\chi(X,\mathbb{Z}) = \# \frac{\#H^0(X,\mathbb{Z})_{tor}}{\#H^1(X,\mathbb{Z})} \cdot \frac{\#H^2(X,\mathbb{Z})_{cotor}\,\#H^4(X,\mathbb{Z})\ldots}{\#H^3(X,\mathbb{Z})\,\#H^5(X,\mathbb{Z})\ldots} .$$

(See Milne [M2] for a proof of essentially this result.) $H^0(X,\mathbb{Z})_{tor}$ and $H^1(X,\mathbb{Z})$ are zero, but the formula is more appealing with them than without. As will become apparent from later generalizations, there should also be a regulator term $R_0(X)$ equal to the 1×1 determinant of the intersection pairing on zero-cycles x d-cycles (modulo numerical equivalence) on X. However, since there is always a zero-cycle of degree 1, this regulator term is always 1.

2. Values of zeta-functions at $s = 1$.

In order to deduce a formula analogous to the one in §1, it is now necessary to assume that $H^2(X,G_m)$ is finite. (It is certainly plausible that this is always true.)

Under this assumption, the $H^i(X,G_m)$ are zero for i large, and finite for $i \neq 1, 3$. $H^1(X,G_m) = \mathrm{Pic}(X)$ is a finitely-generated abelian group and $H^3(X,G_m)$ is the $\mathbb{Q}/\mathbb{Z}$-dual of a finitely-generated abelian group C isogenous to one-cycles on X modulo numerical equivalence. Define a regulator term $R_1(X)$ as follows: Let $\alpha_1 \dots \alpha_r$ be a basis of $H^1(X,G_m)$ modulo torsion. Let $\beta_1 \dots \beta_r$ be a basis of C modulo torsion. Intersection induces a pairing $\langle\ ,\ \rangle$ from $H^1(X,G_m) \times C \to \mathbb{Q}$, and we define $R_1(X)$ to be $\det\langle \alpha_i, \beta_j \rangle$.

Then r is equal to the order $a_1(X)$ of the pole of $Z(X,t)$ at $t = q^{-1}$ and the following formula holds:

$$\lim_{t \to q^{-1}} (1-qt)^{a_1(X)} Z(X,t) = \pm \frac{\chi(X,G_a)}{\chi(X,G_m)} .$$

Here $\chi(X,G_a) = q^{\chi(X,O_X)}$ and

$$\chi(X,G_m) = \frac{\#H^0(X,G_m)\ \#H^2(X,G_m) \dots\ \ R_1(X)}{\#H^1(X,G_m)_{tor}\ \#H^3(X,G_m)_{cotor}\ \#H^5(X,G_m) \dots} .$$

The proof of this formula is given up to p-torsion in [L3] and again completed by Milne in [M2].

3. The complex $\Gamma(r)$.

If we wish to go beyond $s = 0$ and $s = 1$ to the general case where s may be any non-negative integer, it appears that it is not possible to have sheaves which play an analogous role to that of $\mathbb{Z}$ or G_m. Instead, we are forced to consider complexes of sheaves in the étale topology. The very existence of these complexes is at the moment hypothetical, but the hypothetical properties of these hypothetical complexes present a fascinating picture, well worth investigating.

Since all of the properties we would like our complex to have are well-defined in the derived category, we place ourselves in that context.

Let X be a scheme, and let $D(X)$ be the derived category of the category of all complexes of étale sheaves of abelian groups on X. We will regard sheaves as complexes which are zero outside of degree zero. We conjecture that for each non-negative integer r there exist a complex $\Gamma(r)$ in $D(X)$ with the following properties:

(0) $\Gamma(0) = \mathbb{Z}$. $\Gamma(1) = G_m[-1]$.

(1) For $r \geq 1$, $\Gamma(r)$ is acyclic outside of $[1,r]$.

(2) Let α_* be the functor which assigns to every étale sheaf on X the associated Zariski sheaf. Then the Zariski sheaf $R^{q+1}\alpha_*\Gamma(q) = 0$. (We will refer to this as "Hilbert's Theorem 90" for reasons to be explained shortly.)

(3) Let n be a positive integer prime to all residue field characteristics of X. Then there exists a triangle in $D(X)$ of the form

$$\begin{array}{ccc} & \mathbb{Z}/n\mathbb{Z}(r) & \\ \swarrow & & \nwarrow \\ \Gamma(r) & \xrightarrow{n} & \Gamma(r) \end{array},$$

where $\mathbb{Z}/n\mathbb{Z}(r)$ denotes the r-fold Tate twist of $\mathbb{Z}/n\mathbb{Z}$, so that $\mathbb{Z}/n\mathbb{Z}(1) = \mu_n$, $\mathbb{Z}/n\mathbb{Z}(2) = \mu_n \otimes \mu_n$, etc. This triangle of course gives rise to a long exact sequence of cohomology:

$$\ldots \to H^i(X,\Gamma(r)) \xrightarrow{n} H^i(X,\Gamma(r)) \to H^i(X,\mathbb{Z}/n\mathbb{Z}(r)) \to H^{i+1}(X,\Gamma(r)) \to \ldots$$

(4) There are product mappings $\Gamma(r) \overset{L}{\otimes} \Gamma(s) \to \Gamma(r+s)$, which induce maps on cohomology:

$$H^i(X,\Gamma(r)) \otimes H^j(X,\Gamma(s)) \to H^{i+j}(X,\Gamma(r+s)).$$

(5) The cohomology sheaves $\mathcal{H}^i(X,\Gamma(r))$ are isomorphic to the étale sheaves $Gr^r_\gamma K^{et}_{2r-i}(X)$ up to torsion involving primes $\leq (r-1)$. [Here Gr_γ is the gradation corresponding to Soulé's γ-filtration on higher K-theory. See [S1]. This isomorphism should come from an Atiyah-Hirzebruch spectral sequence which degenerates up to torsion involving primes $\leq (r-1)$.]

(6) If F is a field, $H^r(F,\Gamma(r))$ is canonically isomorphic to the Milnor K-groups $K^M_r(F)$ defined in [Mi].

We note here that these "axioms" are quite strong; for example, they imply the Mercuriev-Suslin theorem [MS] that $K_2(F)/nK_2(F) \simeq H^2(F,\mu_n \otimes \mu_n)$ for n prime to the characteristic of F. Axiom (3) yields the exactness of

$$H^2(F,\Gamma(2)) \xrightarrow{n} H^2(F,\Gamma(2)) \to H^2(F,\mu_n \otimes \mu_n) \to H^3(F,\Gamma(2)).$$

Axiom (6) identifies $H^2(F,\Gamma(2))$ with $K_2^M(F)$ which is the same as $K_2(F)$, and Axiom (2) in this case says that $H^3(F,\Gamma(2)) = 0$.

4. "Hilbert's Theorem 90".

If we specialize Axiom 2 to the case where F is a field and $r = 1$, it says that $H^2(F,\Gamma(1)) = 0$. This is $H^1(F,G_m)$, and its vanishing is exactly Emma Noether's generalization of Hilbert's Theorem 90. In this section, we wish to show that Axiom 2 is also a generalization of the "Hilbert Theorem 90 for K_2" [MS, 14.1] of Mercuriev and Suslin, in the presence of Axioms 1 and 6.

Let L be a finite Galois extension of F with group G. We start with the usual Hochschild-Serre spectral sequence:

$$H^p(G,H^q(L,\Gamma(2))) \Rightarrow H^{p+q}(F,\Gamma(2)).$$

We first observe that Axiom 1 implies $H^q(L,\Gamma(2)) = 0$ for $q < 1$. Looking at the exact sequence of terms of low degree, identifying $H^2(F,\Gamma(2))$ and $H^2(L,\Gamma(2))$ with $K_2(F)$ and $K_2(L)$ respectively, and using that $H^3(F,\Gamma(2)) = H^3(L,\Gamma(2)) = 0$, we obtain the exact sequences

(1) $0 \to H^1(G,H^1(L,\Gamma(2))) \to K_2(F) \to K_2(L)^G \to H^2(G,H^1(L,\Gamma(2))) \to 0$

(2) $0 \to H^1(G,K_2(L)) \to H^3(G,H^1(L,\Gamma(2)))$.

If G is cyclic, we may identify $H^1(G,K_2(L))$ with $H^{-1}(G,K_2(L))$ and $H^3(G,H^1(L,\Gamma(2)))$ with $H^1(G,H^1(L,\Gamma(2)))$ and conclude that the induced map α

$$H^{-1}(G,K_2(L)) \stackrel{\alpha}{\to} H^1(G,H^1(L,\Gamma(2)))$$

is injective.

But this map is also described by the diagram

$$\begin{array}{ccccc}
 & & & & \mathrm{Ker}\,N \twoheadrightarrow H^{-1}(G,K_2(L)) \\
 & & & & \downarrow \\
 & & K_2(L) & \stackrel{\sim}{\to} & K_2(L) \\
 & & \downarrow N^* & & \downarrow N \\
0 \to H^1(G,H^1(L,\Gamma(2))) & \to & K_2(F) & \to & K_2(L)^G
\end{array}$$

More precisely, let x be in $H^{-1}(G,K_2(L))$, and let y in $K_2(L)$ be such that $Ny = 0$ and y represents x. Then $\alpha(x)$ is given by N^*y, which evidently is in $H^1(G,H^1(L,\Gamma(2)))$. But now if $N^*y = 0$, the injectivity of α implies that $x = 0$, i.e. that y lies in the image of $(\sigma - 1)$, σ a generator of G. This is exactly the Mercuriev-Suslin "Theorem 90 for K_2".

5. Beilinson's Complex.

A.A. Beilinson has independently [Be] conjectured the existence of a complex (which we will denote by $\Gamma_B(r)$) in the derived category of <u>Zariski</u> sheaves on X satisfying axioms similar to those we hope hold true for $\Gamma(r)$. More precisely, he would like:

$(0)^*$ $\Gamma_B(0) = \mathbb{Z}$. $\Gamma_B(1) = \mathbb{G}_m[-1]$.

$(1)^*$ For $r \geq 1$, $\Gamma_B(r)$ is acyclic outside of $[1,r]$.

$(3)^*$ $\Gamma_B(r) \overset{L}{\otimes} \mathbb{Z}/n = \tau_{\leq r} R\alpha_* \mathbb{Z}/n\mathbb{Z}(r)$ if n is invertible on X.

$(5)^*$ $Gr_\gamma^r(K_j X) \cong H^{2r-j}(X, \Gamma_B(r))$ up to torsion and probably up to "small factorials".

$(6)^*$ For X smooth, $H^r(X, \Gamma_B(r))$ is $\underline{\underline{K}}_r^M(X)$.

The connection between Beilinson's complexes and ours should be given by $\Gamma_B(r) = \tau_{\leq r} R\alpha_* \Gamma(r)$. (Recall that if $\dot{A} = (A_n, d_n\colon A_n \to A_{n+1})$ is a complex, $\tau_{\leq n}\dot{A}$ is the complex B_n defined by $B_m = A_m$ for $m < n$, $B_n = \ker d_n$, $B_m = 0$ for $m > n$, with the obvious differentials. It is immediate that $\tau_{\leq n}$ induces a map on the derived category. Warning: This notation is that used by Deligne in [D], but disagrees with that used by Hartshorne in [H]. (Hartshorne uses $\sigma_{\leq n}$ instead of $\tau_{\leq n}$, and assigns a different meaning to $\tau_{\leq n}$.)

Note that, if we define $\Gamma_B(r)$ to be $\tau_{\leq r} R\alpha_* \Gamma(r)$, then $(0) \Rightarrow (0)^*$ since $\alpha_* \mathbb{Z} = \mathbb{Z}$ and $\alpha_* \mathbb{G}_m = \mathbb{G}_m$. Also, $(1) \Rightarrow \Gamma_B(r)$ is acyclic in degrees < 1, and the truncation operator assures acylicity in degrees $> r$ so we have $(1)^*$.

$(3)^*$ is equivalent to asserting that

$$\begin{array}{ccc} & \tau_{\leq r} R\alpha_* \mathbb{Z}/n\mathbb{Z}(r) & \\ \swarrow & & \nwarrow \\ \Gamma(r) & \xrightarrow{n} & \Gamma(r) \end{array}$$

is a triangle in the derived category, and this follows immediately from (2), (3), and the following lemma, whose proof is straightforward.

LEMMA. If

$$\begin{array}{ccc} & Z & \\ \swarrow & & \nwarrow \\ X & \rightarrow & Y \end{array}$$

is a triangle and $H^{r+1}(X) = 0$, then

$$\begin{array}{ccc} & \tau_{\leq r}Z & \\ \swarrow & & \nwarrow \\ \tau_{\leq r}X & \rightarrow & \tau_{\leq r}Y \end{array}$$

is a triangle.

$(5)^*$ would follow from a strengthening of (5) and $(6)^*$ for fields follows from (6). $(6)^*$ in general would follow from a strengthened (6), namely

$$R^r\alpha_*\Gamma(r) = \underline{\underline{K}}^M_r(X).$$

Beilinson also says that the $\Gamma_B(r)$ should satisfy "something like Gillet's axioms" (see [G], §1), but we leave this question for a later time.

6. Duality.

The complexes $\Gamma(r)$ should give rise to a vast duality theorem, generalizing and clarifying simultaneously many seemingly unrelated results. We treat here only the case of varieties over a field, although this should be but a small part of the picture.

Let k be a complete local field of dimension n. Let V be a proper, smooth, and geometrically connected variety of dimension d over k. Then $H^{2d+n+2}(V,\Gamma(d+n))$ should be canonically isomorphic to $\mathbb{Q}/\mathbb{Z}$ and the cohomology pairings $H^i(V,\Gamma(r)) \times H^j(V,\Gamma(s)) \rightarrow H^{2d+n+2}(V,\Gamma(d+n))$ which exist when $i+j = 2d+n+2$, $r+s = d+n$, $i,j,r,s \geq 0$, should induce dualities in some appropriate sense of the term.

We make this somewhat vague expression precise in the case when $n = 0$, i.e., varieties over finite fields. Then $H^i(V,\Gamma(r))$ should be finite except when $i = 2r$ or $2r+2$. Letting $j = 2d+2-i$ and $s = d-r$, we should have:

a) The pairings are perfect dualities of finite abelian groups if $i \neq 2r$ or $2r+2$.

b) $H^{2r}(V,\Gamma(r))$ should be finitely-generated and the natural map $H^{2d+2-2r}(V,\Gamma(d-r))$ to $\mathrm{Hom}(H^{2r}(V,\Gamma(r)),\mathbb{Q}/\mathbb{Z})$ induced by the pairing should be an isomorphism.

The general duality theorem is known in the following special case

1) If $n = 0$, $d = 0$, it's trivial.
2) If $n = 0$, $d = 1$, it is "unramified global class field theory" for curves over finite fields.
3) If $n = 0$, $d = 2$, $r = s = 1$, it is the statement that $H^i(V,G_m)$ is dual to $H^{4-i}(V,G_m)$ for surfaces over finite fields. For this is to be true, we must assume that $H^2(V,G_m)$ is finite (the Tate conjecture for divisors on V) and in this case it is essentially proven by Artin and Tate in [T2] "up to p" and by Milne in [M1] for the p-part.
4) If $n = 0$, d arbitrary, $r = 0$, $s = d$, $i = 2$ and $j = 2d$, this duality should give the "unramified class field theory" of Kato and Saito. For this to happen, it must be true that $H^{2d}(V,\Gamma(d))$ is isomorphic to zero-cycles on V modulo rational-equivalence.
5) If $n = 1$, $d = 0$, we obtain local class-field theory.
6) If $n = 1$, $d = 1$, $r = 1$, $s = 1$, we obtain the duality theorem proved by the author in [L1], ($H^i(V,G_m)$ is dual to $H^{3-i}(V,G_m)$ for V a curve over a p-adic field).
7) If $d = 0$, n is arbitrary, $r = 0$, $s = n$, $i = 2$, $j = n$, we obtain Kato's generalization [K] of local class-field theory using Milnor K-theory.

7. Values of zeta-functions and $\Gamma(r)$.

We are now in a position to predict the correct generalization of the results in paragraphs 1 and 2. So again, let k be a finite field with $q = p^f$ elements, let X be projective, smooth, and geometrically connected over k, and let d be the dimension of X. Then we conjecture that the following should hold:

1) The hypercohomology groups $H^i(X,\Gamma(r))$ are zero for large i.
2) $H^{2r}(X,\Gamma(r))$ is a finitely-generated abelian group.
3) $H^i(X,\Gamma(r))$ is finite for $i \neq 2r$, $2r+2$.
4) $H^{2d+2}(X,\Gamma(d))$ is canonically isomorphic to $\mathbb{Q}/\mathbb{Z}$.
5) The pairing described in Axiom 4:

$$H^i(X,\Gamma(r)) \times H^{2d+2-i}(X,\Gamma(d-r)) \to H^{2d+2}(X,\Gamma(d)) \simeq \mathbb{Q}/\mathbb{Z}$$

for $0 \leq r \leq d$, $0 \leq i \leq 2d+2$ is a "duality pairing" in the sense that

If $i \neq 2r$, it induces an isomorphism of $H^i(X,\Gamma(r))$ with the $\mathbb{Q}/\mathbb{Z}$-dual of $H^{2d+2-i}(X,\Gamma(d-r))$. We note that if $i = 2r$, then $H^{2d-2r+2}(X,\Gamma(d-r))$ is identified with the $\mathbb{Q}/\mathbb{Z}$-dual of the finitely-generated abelian group $H^{2r}(X,\Gamma(r))$.

It follows from 2) and 5) (and the well-known fact that $H^2(X,\mathbb{Z})$ (which is the dual of $\pi_1(X)_{ab}$) is isogenous to $\mathbb{Q}/\mathbb{Z}$) that $H^{2d}(X,\Gamma(d))$ is a finitely-generated abelian group of rank one, and so has a "degree map" to $\mathbb{Z}$. (By the results of Kato and Saito, $H^{2d}(X,\Gamma(d))$ must in fact be the group of zero-cycles on X modulo rational equivalence.)

6) The groups $H^{2r}(X,\Gamma(r))$ and $H^{2d-2r}(X,\Gamma(d-r))$ have the same rank $m(r)$.

Let $\alpha_1 \dots \alpha_{m(r)}$ be a basis (modulo torsion) for the finitely-generated group $H^{2r}(X,\Gamma(r))$ and $\beta_1 \dots \beta_{m(r)}$ be a basis (modulo torsion) for the finitely-generated group $H^{2d-2r}(X,\Gamma(d-r))$. Then we may define the regulator $R_r(X)$ to be $\det\langle\alpha_i,\beta_j\rangle$ where $\langle\ ,\ \rangle$ is the composite of the natural pairing into $H^{2d}(X,\Gamma(d))$ with the "degree map".

7) $m(r)$ = the order $a(r)$ of the pole of $Z(X,t)$ at $t = q^{-r}$.

8) $\lim\limits_{t\to q^{-r}} (1-q^r t)^{a(r)} Z(X,t) = \pm q^{\chi(X,O_X,r)} \chi(X,\Gamma(r))$, where $\chi(X,\Gamma(r))$ and $\chi(X,O_X,r)$ are defined as follows:

$$\chi(X,\Gamma(r)) =$$

$$\frac{\#H^0(X,\Gamma(r))\dots\#H^{2r}(X,\Gamma(r))_{tor}\#H^{2r+2}(X,\Gamma(r+2))_{cotor}\#H^{2r+4}(X,\Gamma(r))\dots}{\#H^1(X,\Gamma(r))\quad \#H^3(X,\Gamma(r))\dots \qquad R_r(X)}$$

and

$$\chi(X,O_X,r) = \sum_{\substack{0\le i\le r\\ 0\le j\le d}} (-1)^{i+j}(r-i)^{hij},\quad hij = \dim H^j(X,\Omega^i).$$

(The definition of $\chi(X,O_X,r)$ is due to Milne [M2].)

Observe that the mysterious group "C" in the formula for $\lim\limits_{t\to q^{-1}} (1-qt)^{a_1(X)} Z(X,t)$ now assumes its rightful identity as $H^{2d-2}(X,\Gamma(2d-2))$.

These conjectures about varieties over finite fields are, of course, related to each other and to the "axioms". In fact, if we neglect p-torsion, the finiteness conjectures 2) and 3) together with the "axioms" should imply 1), 4), 5), 6), 7), and 8) by the methods of [L2]. In order to include the p-torsion as well, using Milne's method in [M2], we would need a "p-axiom" similar to the "prime-to-p" Axiom 3

Note also that even for $r = 1$, we do not know that $H^3(X,\Gamma(1))$ is finite. The special case of conjecture 3) of this section, which asserts the finiteness of $H^{2r+1}(X,\Gamma(r))$, generalizes this, and so should be of a higher order of difficulty to prove than the other "axioms" and conjectures.

8. The case of number fields.

Even though we have restricted our attention up until now to the case of zeta-functions of varieties defined over finite fields, the complexes $\Gamma(r)$ certainly ought to have significant connections with zeta-functions in the number field case as well. We cite two examples:

In the case when $X = \operatorname{Spec} \mathcal{O}_F$, $\mathcal{O}_F$ being the ring of integers in a number field F, the author had previously conjectured [L2] a formula relating $\zeta(X,-n)$ to $K_{2n}(X)$ and $K_{2n+1}(X)$, for $n \geq 1$. It now seems likely that the correct groups are $H^2(X,\Gamma(n+1))$ and $H^1(X,\Gamma(n+1))$, respectively. Of course, these groups are closely connected with the corresponding K-groups and may even be equal to them up to 2-torsion. But it is interesting to observe that $H^1(\operatorname{Spec} \mathbb{Z}, \Gamma(2))$ is presumably $Gr_\gamma^2 K_3(\mathbb{Z})$. $K_3(\mathbb{Z})$, which has order 48, ([LS]), is made up of two pieces $Gr_\gamma^2 K_3(\mathbb{Z})$ and $Gr_\gamma^3 K_3(\mathbb{Z})$. Now $Gr_\gamma^3 K_3(\mathbb{Z}) = Gr_\gamma^3 K_3(\mathbb{Q}) = K_3^M(\mathbb{Q})$, which has order 2 ([BT]), so $Gr_\gamma^2 K_3(\mathbb{Z})$ has order 24. Since $H^2(\operatorname{Spec} \mathbb{Z}, \Gamma(2))$ is presumably $K_2(\mathbb{Z})$, which has order 2, we obtain a formula for $\zeta(-1) = -\frac{1}{12}$, namely $\pm \#H^2(\operatorname{Spec} \mathbb{Z}, \Gamma(2))/\#H^1(\operatorname{Spec} \mathbb{Z}, \Gamma(2))$, which seems likely to be correct, even when 2-torsion is taken into account.

In the case when X is regular, connected, of dimension d, and of finite type over $\operatorname{Spec} \mathbb{Z}$, it seems likely that the order of the pole of $\zeta(X,s)$ at $s = d-n$, $n > 0$ is given by $\sum_{i=1}^{2n} (-1)^i$ rank $H^i(X,\Gamma(n))$. When $n = 1$, this becomes rank $H^1(X,\mathcal{O}_X^*)$ - rank $H^0(X,\mathcal{O}_X^*)$, and this formula was conjectured by Tate in [T1]. If X is a projective variety over a finite field, this agrees with the conjectures in §7, which assert that $H^i(X,\Gamma(n))$ is finite for $i < 2n$, and that the order of the pole is given by the rank of $H^{2n}(X,\Gamma(n))$. If $X = \operatorname{Spec}$ then presumably we have, up to torsion, that

$$H^1(X,\Gamma(n)) = Gr^n_\gamma K_{2n-1}(\mathcal{O}_F) = K_{2n-1}(\mathcal{O}_F)$$

and $H^i(X,\Gamma(n))$ is finite for $i \neq 1$. The formula then predicts a zero of $\zeta(X,s)$ at $s = 1-n$ of order equal to the rank of $K_{2n-1}(\mathcal{O}_F)$. This is a well-known result of Borel [B]. Our formula should also be equivalent to the conjecture made by Soulé in his talk [S2] at the Warsaw International Congress, which asserts that the order of the pole of $\zeta(X,s)$ at $s = d-n$ is given by $\sum_{i=1}^{2n} (-1)^i \text{ rank } Gr^n_\gamma K_{2n-i}(X)$.

References

[Be] A. A. Beilinson, Letter to C. Soulé, November 1, 1982.

[Bo] A. Borel, Stable real cohomology of arithmetic groups, Ann. Sci. ENS, 7 (1974) 235-272.

[BT] H. Bass and J. Tate, The Milnor ring of a global field; in "Algebraic K-Theory II", Springer Lecture Notes in Math. 342, 1973.

[D] P. Deligne, Théorie de Hodge, II, Publ. Math. I.H.E.S., no. 40 (1974) 5-57.

[G] H. Gillet, Riemann-Roch theorems for higher K-theory, Adv. in Math. 40 (1981) 203-289.

[H] R. Hartshorne, Residues and Duality, Springer Lecture Notes in Math. 20, 1966.

[K] K. Kato, A generalization of local class field theory by using K groups I and II. Journal of the Fac. of Sci. Univ. Tokyo 26 (1979) 303-376 and 27 (1980) 603-683.

[L1] S. Lichtenbaum, Duality theorems for curves over p-adic fields, Inv. Math. (7) 1969, 120-136.

[L2] S. Lichtenbaum, Values of zeta-functions, étale cohomology, and algebraic K-theory, in Algebraic K-Theory II, Springer Lecture Notes in Math. 342, 1973.

[L3] S. Lichtenbaum, Zeta-functions of varieties over finite fields at s = 1, in Arithmetic and Geometry: Papers Dedicated to I.R. Shafarevich on the Occasion of His Sixtieth Birthday, Volume I, Birkhauser, 1983.

[LS] R. Lee and R. H. Szczarba, On $K_3(\mathbb{Z})$, Ann. of Math.

[M1] J. Milne, On a conjecture of Artin and Tate, Ann. of Math. 102 (1975) 517-533.

[M2] J. Milne, Values of zeta-functions of varieties over finite fields, preprint, 1983.

[Mi] J. Milnor, Algebraic K-theory and quadratic forms, Inv. Math. 9 (1970) 318-344.

[MS] A.S. Mercuriev and A.A. Suslin, K-cohomology of Severi-Brauer varieties and norm residue homomorphism, Izv. Akad. Nauk. SSSR Ser. Mat. (46) 1982, 1011-1046.

[S1] C. Soulé, Opérations en K-théorie algebrique, preprint, 1983.

[S2] C. Soulé, K-théorie et zéros aux points entiers de fonctions zéta Proceedings of the ICM, Warsaw, 1983.

[T1] J. Tate, On a conjecture of Birch and Swinnerton-Dyer and a geometric analogue, Séminaire Bourbaki no 306, 1965-66, W.A. Benjamin, Inc. (1966).

[T2] J. Tate, Algebraic cycles and poles of zeta-functions, in <u>Arithmetical Algebraic Geometry</u>, Harper and Row, New York, 1965.

:UCLIDEAN RINGS OF INTEGERS OF FOURTH DEGREE FIELDS

'. J. van der Linden
lathematisch Instituut
Iniversiteit van Amsterdam
loetersstraat 15
018 WB Amsterdam

NTRODUCTION

Let K be a totally complex quartic field of discriminant Δ . In 951 Cassels [4] proved an upper bound on Δ in the case that the ring f integers $\mathcal{O}$ of K is Euclidean (for the norm). In computing this pper bound Cassels made an error: In Lemma 16(iii) we should read k^2 nstead of k. We may improve the bound somewhat by replacing twice the onstant $4/\pi^2$ by $2/\pi\sqrt{3}$ in the same lemma 16. If we redo the computaion after these changes we get an upper bound equal to 230202117. This pper bound is too large to determine all Euclidean rings of this type.

In this paper we will improve upon Cassels' results, in the case hat K has a real quadratic subfield. This will lead to a complete etermination of all cyclic quartic totally complex fields, for which the ing of integers is Euclidean, cf. section 3. These are the two fields ith conductors 5 and 13. Apparently it was not yet known that the field f conductor 13 has a Euclidean ring of integers.

Cassels' upper bound also applies to rings with a Euclidean ideal lass. For the definition of Euclidean ideal class we refer to [8]. In ection 1 we give a definition for our case. Our improvements of Cassels' ounds do not apply to all rings with a Euclidean ideal class, but both or rings of integers of cyclic fields and for Euclidean rings they do. e will not give our proofs in full detail, for this we refer to [9]

We fix some notation. Throughout this paper we deal with a quartic otally complex field K of discriminant Δ with a real quadratic subield K_0 of discriminant Δ_0 . The relative discriminant of K/K_0 will e denoted by $\mathcal{D}$. The rings of integers of K and K_0 will be denoted y $\mathcal{O}$ and $\mathcal{O}_0$, respectively. The functions $N: K \to \mathbb{Q}$ and $N_0: K_0 \to \mathbb{Q}$ re the usual norm functions. The ideal norm functions are also denoted y N and N_0 respectively. We have $N(\alpha) = N(\alpha\mathcal{O})$ for $\alpha \in K^*$ and $N_0(\beta)| = N_0(\beta\mathcal{O}_0)$ for $\beta \in K_0^*$. We denote the class group and the class umber of a number field L by Cl(L) and h(L) respectively. For a field xtension $k \subset K$ the map $\iota: Cl(k) \to Cl(K)$ is given by $\iota[a] = [a\mathcal{O}]$.

§1 EUCLIDEAN IDEAL CLASSES

The concept of Euclidean ideal class was introduced by Lenstra [8] Here we give the definition for our particular case. We recall that $\mathcal{O}$ is a *Euclidean ring* if and only if the following property holds:

(1) For all $\alpha \in K$ there exists $\beta \in \alpha + \mathcal{O}$, such that $N(\beta) < 1$.

This suggests the following generalization. Let $\mathfrak{a}$ be a (fractional) $\mathcal{O}$ ideal. We call $\mathfrak{a}$ a *Euclidean ideal* if the following property holds:

(2) For all $\alpha \in K$ there exists $\beta \in \alpha + \mathfrak{a}$, such that $N(\beta) < N\mathfrak{a}$.

Clearly property (2) only depends on the ideal class of $\mathfrak{a}$. The ideal class of a Euclidean ideal will be called a *Euclidean ideal class*. Noti that $\mathcal{O}$ is a Euclidean ring if and only if the principal ideal class $[\mathcal{O}]$ is Euclidean.

If $\mathcal{O}$ has a Euclidean ideal class there are several restrictions o the class group of K :

<u>(3) PROPOSITION</u> *Let* $[\mathfrak{a}]$ *be a Euclidean ideal class of* $\mathcal{O}$. *Then*
(i) $[\mathfrak{a}]$ *is the only Euclidean ideal class of* $\mathcal{O}$*;*
(ii) $[\mathfrak{a}]$ *contains all integral* $\mathcal{O}$*-ideals of minimal norm* >1*;*
(iii) $Cl(K)$ *is cyclic with generator* $[\mathfrak{a}]$*;*
(iv) $h(K) \leq 6$.
<u>Proof</u> See [8], (1.5), (1.6) and (2.7). □

If K/k is a Galois extension for some subfield k, e.g. $k = K_0$, we ca say more:

<u>(4) LEMMA</u> *Suppose that* K/k *is a Galois extension with group* G. *If has a Euclidean ideal class then* G *acts trivially on* $Cl(K)$ *and* Index $[Cl(K):\iota Cl(k)] \mid n$, *where* n *is equal to the degree* $[K:k]$.
<u>Proof</u> The action of G maps Euclidean ideals upon Euclidean ideals. Hence by (3)(i) and (iii) G acts trivially on $Cl(K)$. If $[\mathfrak{a}]$ is the Euclidean ideal class we have

$$[\mathfrak{a}]^n = \prod_{\sigma\in G}[\sigma\mathfrak{a}] = [\prod_{\sigma\in G}\sigma\mathfrak{a}] \in \iota Cl(k).$$

Because $[\mathfrak{a}]$ generates $Cl(K)$ we get Index $[Cl(K):\iota Cl(k)] \mid n$. □

(5) PROPOSITION *Suppose that* $\mathcal{O}$ *has a Euclidean ideal class. Then*

(i) Index $[Cl(K):\iota Cl(K_0)] \mid 2$;

(ii) $h(K) \mid 4$ *if* $K/\mathbb{Q}$ *is a Galois extension;*

(iii) $h(K) \mid 2$ *if* $Gal(K/\mathbb{Q}) \simeq V_4$.

Proof Parts (i) and (ii) are derived from (4) with $k = K_0$ and $k = \mathbb{Q}$ respectively. For (iii) suppose in contrast that $h(K)=4$. Using Galois cohomology and the analytic class number formula, and taking (3), (4) and (5)(i) into account, we find that there remain only 4 possibilities for K. It can easily be shown that the rings of integers of these 4 fields do not have a Euclidean ideal class. For more details cf. [9] ch. 10. □

2 AN UPPER BOUND ON THE DISCRIMINANT

In this section we consider the space $U = \mathbb{C} \times \mathbb{C}$. The two subspaces $U_r = \mathbb{R} \times \mathbb{R}$ and $U_i = i\mathbb{R} \times i\mathbb{R}$ are orthogonal in U. We denote by $\pi_r: U \to U_r$ and $\pi_i: U \to U_i$ the orthogonal projections. Let Γ be a lattice in U, U_r or U_i. The *determinant* $\nu(\Gamma)$ is defined as the usual Haar measure of a fundamental domain of Γ.

The two complex embeddings of K give rise to an embedding $K \to U$ with dense image. From now on we identify K with its image. The subfield K_0 is a dense subset of U_r. Each $\mathcal{O}$-ideal is a lattice in U and each $\mathcal{O}_0$-ideal is a lattice in U_r, cf. [2] Kap. II §3 Satz 1.

By continuity we may extend the norm function N to U and the norm function N_0 to U_r. They are given by

$$N(x,y) = |x|^2|y|^2 \quad , \text{ for } x,y \in \mathbb{C}; \tag{6}$$
$$N_0(x,y) = |x|\,|y| \quad , \text{ for } x,y \in \mathbb{R}.$$

Analogously we define a norm function N_0 on U_i by

$$N_0(ix,iy) = |x|\,|y| \quad , \text{ for } x,y \in \mathbb{R}. \tag{7}$$

Let a be an $\mathcal{O}$-ideal. We denote $a_r = a \cap U_r$ and $a_i = a \cap U_i$. It can be shown that a_r and $\pi_r a$ are $\mathcal{O}_0$-ideals and hence they are lattices in U_r. Also a_i and $\pi_i a$ are lattices in U_i. The determinants of these lattices satisfy the following relations:

$$\nu(a) = \nu(a_r)\nu(\pi_i a) = \nu(\pi_r a)\nu(a_i) \ . \tag{8}$$

The determinant of an ideal can be computed as follows:

(9) $\nu(a) = \frac{1}{4} Na \sqrt{\Delta}$;
$\nu(b) = N_0 b \sqrt{\Delta_0}$.

Here a is an O-ideal and b is an O_0-ideal.

For each $t \in \mathbb{R}_{>0}$ we define a subset V_t of U by

(10) $V_t = \{x \in U: N(x) < t\}$.

We will use the sets V_t to study whether an ideal is Euclidean.

(11) PROPOSITION *Let a be an O-ideal. Define* $t(a) = \inf\{t \in \mathbb{R}_{>0} : U = a + V_t\}$.

(a) If $t(a) < Na$ then a is Euclidean.

(b) If $t(a) > Na$ then a is not *Euclidean.*

Proof For (a), if $t(a) < Na$ then clearly (2) holds. For (b) we use a generalization of a theorem of Barnes and Swinnerton-Dyer [1] thm. M. se also [8] (2.3). □

Notice that when it happens that $t(a) = Na$ we cannot use (11) to determine whether a is Euclidean.

Let a be an O-ideal. We construct $x_r \in U_r$ and $x_i \in U_i$ such tha $N_0(x_r - \alpha)$ and $N_0(x_i - \beta)$ are large for all $\alpha \in \pi_r a$ and all $\beta \in \pi_i$ Let $x \in U$ be such that $\pi_r(x) = x_r$ and $\pi_i(x) = x_i$. We will derive that $x \notin a + V_t$ for some large t depending on $\nu(\pi_r a)$ and $\nu(\pi_i a)$. If it happens that $t > Na$ we derive from (11)(b) that a is not Euclidean. For this we need the following lemma.

(12) LEMMA *Let $x_1, x_2, y_1, y_2, a, b \in \mathbb{R}_{>0}$ be such that $x_1 x_2 \geq a$ and $y_1 y_2 \geq b$. Then* $(x_1 + y_1)(x_2 + y_2) \geq (\sqrt{a} + \sqrt{b})^2$.

Proof Because $(\sqrt{x_1 y_2} - \sqrt{y_1 x_2})^2 \geq 0$ we have
$x_1 y_2 + y_1 x_2 \geq 2\sqrt{x_1 y_2 y_1 x_2} \geq 2\sqrt{ab}$, hence
$(x_1 + y_1)(x_2 + y_2) = x_1 x_2 + y_1 y_2 + x_1 y_2 + x_2 y_1 \geq (\sqrt{a} + \sqrt{b})^2$. □

(13) PROPOSITION *Let a be an O-ideal. There exists $x \in U$, such that $x \notin a + V_t$ for $t = (16 + 6\sqrt{6})^{-2}(\nu(\pi_r a) + \nu(\pi_i a))^2$.*

Proof Ennola [5] has shown that there exist $x_r \in U_r$ and $x_i \in U_i$ su that for all $\beta \in \pi_r a$ and all $\gamma \in \pi_i a$ we have

$$N_0(x_r - \beta) > (16 + 6\sqrt{6})^{-1} \nu(\pi_r a),$$
$$N_0(x_i - \gamma) > (16 + 6\sqrt{6})^{-1} \nu(\pi_i a).$$

Take $x \in U$ such that $\pi_r(x) = x_r$ and $\pi_i(x) = x_i$. For $\alpha \in a$ we have by (12):

$$N(x - \alpha) \geq (N_0(x_r - \pi_r(\alpha)) + N_0(x_i - \pi_i(\alpha)))^2$$
$$\geq (16 + 6\sqrt{6})^{-2}(\nu(\pi_r a) + \nu(\pi_i a))^2. \square$$

(14) PROPOSITION *Let* a *be an* $\mathcal{O}$*-ideal that is invariant under* $\mathrm{Gal}(K/K_0)$, *i.e.* $\sigma a = a$ *for all* $\sigma \in \mathrm{Gal}(K/K_0)$. *Denote*

$$q(a) = (Na)^{-\frac{1}{2}} N_0(a_r) \quad \text{and}$$
$$\kappa = 16(16 + 6\sqrt{6})^2.$$

If a *is Euclidean then*

$$\Delta_0 (q(a) + q(a)^{-1}\sqrt{N_0\mathcal{D}})^2 \leq \kappa .$$

Proof Because a is invariant under $\mathrm{Gal}(K/K_0)$ we have for all $\alpha \in a$

$$\pi_r(\alpha) = \frac{1}{2}\,\mathrm{Tr}(\alpha) \in \frac{1}{2}\, a_r,$$

where $\mathrm{Tr}: K \to K_0$ is the trace map. This shows that $\nu(\pi_r(a)) \geq \frac{1}{4}\nu(a_r)$. From (11), (13), (8) and (9) we derive that a is Euclidean only if

$$Na \geq (16 + 6\sqrt{6})^{-2}(\nu(\pi_r a) + \nu(\pi_i a))^2$$
$$\geq (16 + 6\sqrt{6})^{-2}\left(\frac{1}{4}\nu(a_r) + \frac{\nu(a)}{\nu(a_r)}\right)^2$$
$$= \kappa^{-1}\left(N_0 a_r \sqrt{\Delta_0} + \frac{Na}{N_0 a_r}\sqrt{\Delta_0 N_0 \mathcal{D}}\right)^2, \text{ i.e.}$$

$$\Delta_0 (q(a) + q(a)^{-1}\sqrt{N_0\mathcal{D}})^2 \leq \kappa . \square$$

(15) COROLLARY *If* $\mathcal{O}$ *has a Euclidean ideal that is invariant under* $\mathrm{Gal}(K/K_0)$, *then* $\Delta \leq 14206929$.

Proof The inequality of the means gives:

$$\Delta = \Delta_0 (N_0\mathcal{D})^2 \leq \left(\frac{\kappa}{16}\right)^2 = 14206929.9 \quad . \square$$

(16) Remark The result of (13) remains true if we replace a by the larger lattice $\pi_r a \times \pi_i a$. The quotient of the determinants of these lattices can be arbitrarily large. However, if a is invariant under $\mathrm{Gal}(K/K_0)$ this quotient is at most 4. This is what we used in (14). It appears that we have not obtained the best result. If we can get a result like (13), where we only need that U_r and U_i are 'almost' orthogonal in

some sense, we may prove a theorem like (14) for ideals in arbitrary complex quartic fields. We have not yet succeeded in such an approach.

To apply (14) we have to know whether there is a Euclidean ideal th is invariant under $\mathrm{Gal}(K/K_0)$. The following proposition lists several cases.

(17) PROPOSITION *Suppose that* $\mathcal{O}$ *has a Euclidean ideal class. This idea class contains an ideal invariant under* $\mathrm{Gal}(K/K_0)$ *if we are in one of the following cases:*

(a) $h(K)$ *is odd;*

(b) The integral $\mathcal{O}$*-ideal of minimal norm lies over an inert or ramifying prime in* K/K_0 *;*

(c) $K/\mathbb{Q}$ *is cyclic.*

Proof Let σ be the generator of $\mathrm{Gal}(K/K_0)$. In case (a) let a be a Euclidean ideal, then $(a\sigma a)^{\frac{1}{2}(h(K)+1)} \in [a]$ is invariant under $\mathrm{Gal}(K/K_0)$ In case (b) let a be the prime ideal of minimal norm. It is invariant because it lies over an inert or ramifying prime. It is Euclidean by (3)(ii).

In case (c) let τ be a generator of $\mathrm{Gal}(K/\mathbb{Q})$ and let a be a Euclidean ideal. Because $[\tau a] = [a]$ we have $\tau a = \alpha a$ for some $\alpha \in K$ with $N\alpha = 1$. Using Hilbert 90 we find $\beta \in K$ with $\alpha = \beta\tau\beta^{-1}$. Then $c = \beta a$ is invariant under $\mathrm{Gal}(K/K_0)$. □

Notice that when $\mathcal{O}$ is a Euclidean ring the bounds of (14) and (15 apply by (17)(a). In this case the bound of (14) becomes $\Delta_0(1 + \sqrt{N_0\mathcal{D}})^2 \leq \kappa$.

§3 CYCLIC EXTENSIONS OF $\mathbb{Q}$

In this section we consider the case that $K/\mathbb{Q}$ is cyclic. We denot by f the conductor of K and by f_0 the conductor of K_0. We have $\Delta_0 = f_0$ and $\Delta = f_0 f^2$. Let χ be a quartic character that generates th character group of K.

Suppose that $\mathcal{O}$ has a Euclidean ideal class. By (17)(c) we know th the bound of (14) applies to K.

(18) PROPOSITION *Suppose that* $K/\mathbb{Q}$ *is cyclic. If* $\mathcal{O}$ *has a Euclidean ideal class then* K *is one of the fields listed in table 1.*

Proof Because $h(K)\,|\,4$ we know by genus theory that at most 2 finite primes ramify in $K/\mathbb{Q}$. In fact if $h(K) = 1$ only one prime ramifies in $K/\mathbb{Q}$ and if $h(K)=2$ exactly one prime ramifies in $K_0/\mathbb{Q}$ and two in K/K_0.

:able 1 *Cyclic quartic totally complex fields for which the ring of in-* :*egers may have a Euclidean ideal class. When* $h(K) = 1$ *or* 2 *the field* *s completely determined by the conductors* f *and* f_0 *, when* $h(K) = 4$ *e also need the value of* $\chi(3)$ *to determine* K.

0	f	h(K)	f_0	f	h(K)	
5	5	1	5	85	2	
8	16	1	8	48	2	
3	13	1	8	80	2	
9	29	1	13	65	2	
7	37	1	13	104	2	
3	53	1	17	119	2	
1	61	1	40	80	4	$\chi(3) = -1$
5	40	2	85	85	4	$\chi(3) = -1$
5	65	2				

inally if $h(K) = 4$ exactly two primes ramify in $K/\mathbb{Q}$, both totally.

First suppose that $h(K) = 1$. If f_0 is even we have $f_0 = 8$ and $= 16$, so we may suppose that f_0 is odd. In this case we have $f =$ $_0 = p$, where p is the prime that ramifies in $K/\mathbb{Q}$. Because K is tot- lly complex we have $p \equiv 5 \bmod 8$. Then the norm of the relative discri- inant $N_0\mathcal{D}$ is equal to p. In (14) we may take $a = 0$, with $q(0) = 1$. ence we have $p(1 + \sqrt{p})^2 \leq \kappa$, i.e. $p \leq 109$. By using the analytic class umber formula or existing tables of class numbers, e.g. [6], we find hat $h(K) = 1$ only if $p \in \{5, 13, 29, 37, 53, 61\}$.

In the cases that $h(K) = 2,4$ the proof goes along similar lines. ince $q(a)$ is equal to the square root of a product of norms of differ- nt primes ramifying in K/K_0 only a few cases have to be considered, .g. if 2 is totally ramified in $K/\mathbb{Q}$ we may take $q(a) = \sqrt{2}$ because prime of norm 2 is in the Euclidean ideal class. For more details we efer to [9], ch. 10. □

A straightforward computation shows that for all fields in table 1, xcept for those with conductors 5 and 13, there is no Euclidean ideal lass. This can be done by showing that for a given integral ideal a in he class that contains all integral ideals of minimal norm there exists $\in 0$ such that (2) is not satisfied. In all cases we only need an ideal ith either $Na \leq 20$ or $a | \Delta$.

It was already known to Kummer [7] that the ring of integers of the ield with conductor 5 is Euclidean. Now we show that the ring of integers of the field with conductor 13 is Euclidean as well.

Let K be the quartic field of conductor 13. Let σ be a generato of $\mathrm{Gal}(K/\mathbb{Q})$. The ring of integers of K is equal to $\bigoplus_{i=0}^{3} \mathbb{Z}\sigma^i\beta$, wher $\beta = \frac{1}{4}(-1 + \sqrt{13} + \sqrt{-26 + 6\sqrt{13}}) = \zeta + \zeta^3 + \zeta^9$ for some primitive 13-th roo of unity ζ. The unit $\eta = \frac{1}{2}(3 + \sqrt{13}) = -\beta - \sigma\beta - \sigma^2\beta - 2\sigma^3\beta$ is a fundamental unit of $\mathcal{O}$.

The proof that $\mathcal{O}$ is Euclidean is analogous to that of Ojala in [10] for the ring $\mathbb{Z}[\zeta_{16}]$: we divide the fundamental domain $\{x = \Sigma_{i=0}^{3} a_i\sigma^i\beta: a_i \in \mathbb{R}, |a_i| \leq \frac{1}{2}\}$ of $\mathcal{O}$ in $\mathbb{C} \times \mathbb{C}$ into parallelepipeds of the form $\{x = \Sigma_{i=0}^{3} a_i\sigma^i\beta : a_i \in \mathbb{R}, r_i/10 \leq a_i \leq (r_i+1)/10\}$, for all quadruples (r_0, r_1, r_2, r_3) with $r_i \in \mathbb{Z}$ and $-5 \leq r_i \leq 4$. For each of these parallelepipeds P we processed the following algorithm with help of a computer.

Step 1 Let $\mu = \Sigma_{i=0}^{3} m_i\sigma^i\beta$ be the center of P and let T be the set of all $\alpha \in \mathcal{O}$ of the form $\alpha = \Sigma_{i=0}^{3} a_i\sigma^i\beta$ with $|a_i - m_i| \leq 1$ for $0 \leq i \leq$ For each $\alpha \in T$ we check whether $P \subset \alpha + V_1$, cf. (10). If indeed $P \subset \alpha + V_1$ we stop, otherwise we go to step 2.

Step 2 Suppose that $\eta\mu = \Sigma_{i=0}^{3} e_i\sigma^i\beta$. Let T_1 be the set of all $\alpha \in$ of the form $\alpha = \Sigma_{i=0}^{3} a_i\sigma^i\beta$ with $|a_i - e_i| \leq 1$. For each $\alpha \in T_1$ we check whether $\eta P \subset \alpha + V_1$. If this is the case we stop, otherwise we perform a similar step with η replaced by η^{-1}. If we also do not find that $\eta^{-1}P \subset \alpha + V_1$ for some α we go to step 3.

Step 3 Divide P in 16 parallelepipeds of half size by cutting each edge in two halves. For each of the new parallelepipeds we return to step 1.

Actually we performed each step for all P before going to the ne: step. Because the process terminates for all P we know that $\mathcal{O}$ is Euclidean by (11). In fact after performing step 1 for the third time tl process stopped. The checks in steps 1 and 2 whether some parallelepiped P is in $\alpha + V_1$ for some $\alpha \in \mathcal{O}$ can be done by checking whether all vertices of P are in some convex set contained in $\alpha + V_1$. For instai we used sets of the form $C_1 \times C_2$ in $\mathbb{C} \times \mathbb{C}$, where C_i is a disc in (

Combining the results above we get the following theorem

(19) THEOREM *Suppose that $K/\mathbb{Q}$ is cyclic. Then $\mathcal{O}$ has a Euclidean id class if and only if the conductor of K is equal to 5 or 13. In both cases the ring $\mathcal{O}$ itself is Euclidean.* □

§4 V_4-EXTENSIONS OF $\mathbb{Q}$

If $K/\mathbb{Q}$ is a V_4-extension for which $\mathcal{O}$ has a Euclidean ideal class then $h(K) \leq 2$ by (5)(iii). If $h(K) = 2$ it may occur that the Euclidean ideal class, i.e. the non-principal class, does not contain an ideal that is invariant under $Gal(K/K_0)$. However, the fields with $h(K) = 2$ and $\Delta \leq 232020117$ are all listed in [3]. For these fields we checked whether the non-principal ideal class contains an ideal invariant under $Gal(K/K_0)$ and whether (17)(b) holds. The fields for which there is a Euclidean ideal class that does not contain an ideal invariant under $Gal(K/K_0)$ are among the remaining fields. All these remaining fields have $\Delta \leq 7958041$. Because this value is less than the bound of (15) and the discriminant of a V_4-extension of $\mathbb{Q}$ is a square we obtain the following result.

<u>(20) THEOREM</u> *Let* $K/\mathbb{Q}$ *be a* V_4*-extension. If* $\mathcal{O}$ *has a Euclidean ideal class, then* $\Delta \leq 14205361$. □

Notice that the notion 'ring with Euclidean ideal class' is not equal to 'Euclidean ring' in the case of V_4-extensions. Indeed the ring of integers of $\mathbb{Q}(\sqrt{-3}, \sqrt{13})$ has a non-principal Euclidean ideal class, cf. [8] (2.7).

ACKNOWLEDGEMENT

The author was supported by the Netherlands Organization for the Advancement of Pure Research (Z.W.O.).

REFERENCES

[1] BARNES, E.S. - H.P.F. SWINNERTON-DYER, *The inhomogeneous minima of binary quadratic forms (II)*, Acta Math. <u>88</u> 279-316 (1952).

[2] BOREWICZ, S. - I.R. ŠAFAREVIČ, *Zahlentheorie*, German translation of Russian original by H. Koch, Basel: Birkhäuser Verlag (1966).

[3] BUELL, D.A. - H.C. WILLIAMS - K.S. WILLIAMS, *On the imaginary bicyclic biquadratic fields with class number 2*, Math. Comput. <u>31</u> 1034-1042 (1977).

[4] CASSELS, J.W.S., *The inhomogeneous minimum of binary quadratic, ternary cubic and quaternary quartic forms*, Proc. Camb. Philos. soc. <u>48</u> 72-86/519-520 (1952).

[5] ENNOLA, V., *On the first inhomogeneous minimum of indefinite binary quadratic forms and Euclid's algorithm in real quadratic fields*, Ann. Univ. Turku, Ser. AI Tom. 28 (1958).

[6] HASSE, H., *Über die Klassenzahl abelscher Zahlkörper*, Berlin: Akademie Verlag (1952).

[7] Letter from KUMMER to KRONECKER, October 2 1844, KUMMER Collected papers I, Berlin-Heidelberg-New York: Springer Verlag 87-91 (1975)

[8] LENSTRA jr., H.W., *Euclidean ideal classes*, Astérisque 61 121-131 (1979).

[9] LINDEN, F.J. van der,*Euclidean rings with two infinite primes*, Ph. D.-thesis, to appear in spring 1984.

[10] OJALA, T., *Euclid's algorithm in the cyclotomic field* $\mathbb{Q}(\zeta_{16})$, Math. Comput. 31 268-273 (1977).

EQUATIONS OVER FUNCTION FIELDS

R.C. Mason,
Gonville & Caius College,
Cambridge,
U.K.

The last four years have seen considerable progress made towards the effective resolution of many problems associated with equations over function fields. The equations concerned are the direct analogues of certain families of Diophantine equations over number fields, and in order to make a comparison between the cases of function fields and number fields, let us start with a review, necessarily brief, of the relevant researches in the latter, before proceeding to a discussion of the problems associated with the former and of the results prior to the new approach. This new approach is based absolutely on an original fundamental inequality, and the main purport of this note is to announce this inequality and to introduce some of the consequences which flow therefrom.

In 1909 the first major contribution towards a systematic study of Diophantine equations was made by Thue [9]. He proved that whenever f is a binary form, irreducible over $\mathbb{Q}$ and of degree at least 3, the equation $f(x,y) = 1$ has only finitely many integer solutions. Thue's result was a direct consequence of his studies on the approximation of algebraic numbers by rationals. This approach was developed by Siegel, who in 1925 proved that the hyperelliptic equation $y^2 = g(x)$, where g denotes a square-free polynomial with integer coefficients and of degree at least 3, has only finitely many solutions in integers. Siegel's researches in this area culminated in his illustrious paper of 1929, in which he proved that any polynomial equation in two variables with integer coefficients has only finitely many integer solutions, provided that the equation represents a curve of positive genus. A further analysis, of equations of genus zero in his paper, thus provided a complete characterisation of those equations in two variables which possess only finitely many integer solutions.

Unfortunately there is one serious reservation to be entertained towards each of these theorems, namely that they are ineffective in failing to provide bounds on the solutions. Although they do establish that each of the various equations has only finitely many solutions in integers, they furnish no means of actually determining the solutions. The problem of providing an effective resolution of general families of Diophantine equations remained outstanding until 1968, when Baker [1], having previously established an inequality to linear forms in the logarithms of algebraic numbers, applied his work to give a new proof of Thue's theorem of 1909 which was effective. Baker's result thereby reduced to a finite amount of computation the problem of determining all the solutions in integers of the Thue equation $f(x,y) = 1$. Following this, Baker established explicit bounds on integer solutions of the hyperelliptic equation. In 1970 he discovered with Coates [2] a simpler and effective proof of the general theorem of Siegel which applies to any equation of genus 1, and a similar agreement obtains when the equation represents a curve of genus 0 with at least 3 infinite valuations. However, Baker's proof does not appear to extend easily to equations of any higher genus, and an effective resolution of the important problem of determining the integer points on curves of genus 2 or more remains unsolved. In fact for such equations a more powerful result than Siegel's has just been discovered: Faltings has reported at this conference on his proof of the celebrated Mordell conjective, that any curve of genus 2 or more possess only finitely many rational points. Unfortunately Faltings' theorem is believed to be ineffective.

We shall be concerned with the analogy over function fields of these results over number fields. Let us denote by k an algebraically closed field of characteristic zero presented explicitly (see [3]) and by $k(z)$ the rational function field over k. Let us consider as an illustration the Thue equation $F(X,Y) = 1$, where F is a separable binary form with coefficients in the rational function field, and of degree at least 3. Corresponding to the integers there is the polynomial ring $k[z]$, so the natural object of concern is the set of solutions X,Y in $k[z]$ of $F(X,Y) = 1$. In contrast to the classical Thue equation discussed above, there are four questions which may be posed concerning this equation and its polynomial solutions. Do the solutions have bounded degrees? If so, can such bounds be determined effectively? Are there but finitely many solutions? Can all the solutions be determined effectively? The pluricity of problems here arises from the fact that since the ground field k is infinite, bounds on the degrees of the solutions do not of themselves enable the solutions to be determined, nor do they imply that only finitely many solutions exist. The first question was resolved in the affirmative by Gill in 1930, when he activated an analogue of Thue's method relating to the approximation of algebraic functions by rational function Sadly this was ineffective, and in 1973 Osgood [6] established an effec tive version of Gill's theorem by a different method, relating to earli researches of Kolchin on algebraic differential equations. Osgood's technique was considerably developed by Schmidt, who calculated explicit bounds on the degrees. Actually Schmidt's extension displayed much greater power than this, for he succeeded in establishing explicit bounds on the degrees of all the solutions in $k(z)$, not just in $k[z]$, when F has degree 5 or more. This work was later enlarged to include certain classes of hyperelliptic and superelliptic equations, of which more below. Concerning the third of the four questions, a criterion for the existence of infinitely many solutions may be deduced from the Manin-Grauert theorem when F has degree at least 4. This theorem had established for function fields the analogue of Faltings' result for number fields, so that any equation of genus 2 or more has only finitely many solutions in $k(z)$, provided that it is not birationally equivalent to an equation with coefficients in the ground field k.

In a recent paper [3] the final, central and subsuming question four is resolved by establishing a simple algorithm for the determination of all polynomial solutions of the Thue equation $F(X,Y) = 1$. The criterion for the existence of infinitely many solutions and an improved bound for their degrees are but immediate consequences of the analysis. Furthermore, the algorithm is exceedingly efficient, and permits many equations to be solved directly, even without resorting to machine computation. The attack also contrasts with earlier approaches in relying not on studies of differential equations; instead the crux is an inequality whose motivation lies in the study of linear forms in the logarithms of algebraic functions: thus the style may be regarded as analogous to Baker's approach on Diophantine equations over number fields.

The result in fact refers to a more general situation than that already described: it is concerned with the Thue equation over an arbitrary finite extension K of the rational function field $k(z)$. Let $\mu,\alpha_1,\ldots,\alpha_n$ be elements of $\mathcal{O}$, the ring of elements integral over $k[z]$, with μ non-zero and $\alpha_1,\alpha_2,\alpha_3$ distinct. The main theorem established in this context is as follows.

THEOREM. All the solutions X,Y in $\mathcal{O}$ of the Thue equation

$$(X-\alpha_1Y)\ldots(X-\alpha_nY) = \mu \qquad (1)$$

may be determined effectively. In general there will be only finitely many solutions. The only instance to the contrary occurs when (1) can be transformed by a substitution

$$X = \alpha x + \beta y, \quad Y = \gamma x + \delta y \qquad (\alpha,\beta,\gamma,\delta \in \mathcal{O},\ \alpha\delta \neq \beta\gamma)$$

into an equation $f(x,y) = 1$, where f has coefficients in the ground field k: in this case there will be the infinite family derived from *the solutions x,y in k of $f(x,y) = 1$, but otherwise only finitely many solutions X,Y in 0 of (1).*

We shall now examine the fundamental inequality on which the proof of this theorem is based. Let us begin by recalling some preliminaries concerning valuations on function fields (see [4]). Associated with the extension K/k is a non-singular projective curve C defined over k, whose points may be identified with non-archimedean additive valuations on K. Each such valuation v also has associated with it a local parameter z_v: each element of f possesses a Laurent expansion in powers of z_v, and $v(f)$ is just the order of vanishing of this series. The usual sum formula

$$\sum_v v(f) = 0 \qquad (f \neq 0)$$

holds. The height of an element is defined by

$$H(f) = -\sum_v \min(0,v(f)),$$

that is, the number of poles of f, counted according to multiplicity; the height of a finite set is defined similarly, and the height of a polynomial is just the height of its set of coefficients. Each Laurent expansion may be differentiated with respect to its local parameter z_v: the result is denoted by $\frac{df}{dv}$. In view of the existence of a global derivation on K, the sum $\sum_v v(\frac{df}{dv})$ is independent of the choice of f in $K \setminus k$: accordingly we define

$$2g - 2 = \sum_v v(\frac{df}{dv}) \qquad (f \in K \setminus k); \qquad (2)$$

g is termed the genus of K/k. We also record the following elementary observations concerning the derivative of a Laurent series:

$$v(\frac{df}{dv}) = v(f) - 1 \text{ if } v(f) \neq 0, \; v(\frac{df}{dv}) \geq 0 \text{ if } v(f) \geq 0. \qquad (3)$$

We may now state and prove the fundamental inequality.

LEMMA. Let S denote a finite set of valuations on K, and suppose that U and V are non-zero elements of K all of whose poles and zeros lie within S, and with $U + V = 1$. Then either U and V are both elements of k, and so $H(U) = 0$, or

$$H(U) \leq \#S + 2g - 2,$$

where $\#S$ denotes the number of elements of S.

Proof. We may assume that U and V are not elements of k. Let S_1,S_2,S_3 denote the sets of valuations v at which $v(U)<0$, $v(U)>0$, $v(V)>0$ respectively. Thus S_1,S_2 and S_3 are disjoint subsets of S, and if v lies in some S_i we denote by m_v the positive integer

$-v(U), v(U), v(V)$ according as $i = 1,2,3$. In view of (3), we obtain $v(\frac{dU}{dv}) = -m_v - 1$ $(v \in S_1)$, $v(\frac{dU}{dv}) = m_v - 1$ $(v \in S_2 \cup S_3)$, $v(\frac{dU}{dv}) \geq 0$ $(v \notin S_1 \cup S_2 \cup S_3)$. The genus formula (2) now yields.

$$2g - 2 \geq -\sum_{S_1} m_v + \sum_{S_2 \cup S_3} m_v - \#S,$$

from which the required result follows, since by the sum formula $H(U) = \sum_{S_i} m_v$ for $i = 1,2,3$.

This result may be interpreted as a lower bound for linear forms with integer coefficients in the logarithms of units in algebraic function fields. This was indeed the original motivation for the approach, but, after distillation and crystallisation of argument it proved possible to avoid completely both reference to the logarithms of functions and to the consequent studies of the theory of units in function fields. As a corollary to the lemma, it follows that all the possibilities for U may be determined effectively: apart from those in k they are finite in number. The proof of this corollary is achieved by a reduction of the problem to that of solving systems of linear equations over k.

In order to exploit the lemma in the solution of the Thue equation (1), the following identity is considered:

$$(X-\alpha_1 Y)(\alpha_2-\alpha_3) + (X-\alpha_2 Y)(\alpha_3-\alpha_1) + (X-\alpha_3 Y)(\alpha_1-\alpha_2) = 0,$$

that is, we choose $U = (X-\alpha_1 Y)(\alpha_2-\alpha_3)/(X-\alpha_3 Y)(\alpha_2-\alpha_1)$. If S denotes the set of valuations v at which $v(z) < 0$, $v(\mu) > 0$, or $v(\alpha_i-\alpha_j) > 0$ for some $1 \leq i < j \leq 3$, then both U and V have all their zeros and poles in S. The lemma and its corollary thus lead to the algorithm for the complete determination of the solutions. In addition, we obtain that

$$H(X,Y) \leq 8H + 2g + r - 1,$$

where H denotes the height of (1) and r denotes the number of distinct poles of Z. It is to be remarked that the dependence on H provides a stark contrast with the analogous case of number fields, when the bounds of Baker [1] are of exponential complexity, thereby rendering actual computations laborious. The bound here reflects that the algorithm enables many equations to be solved explicitly even without machine computation. Finally, the result represents an improvement on a previous bound obtained by Schmidt in this context, $89H+212g+r-1$, which he obtained using analyses of differential equations. It is also of interest to note that Győry has recently employed these results, together with those of Baker, to prove that if $F(x,y) = 1$ is a Thue equation with coefficients in some finitely generated, possibly transcendental, extension of $\mathbb{Q}$, then the equation has only finitely many solutions in any subring finitely generated over $\mathbb{Z}$.

The fundamental inequality has also served prominently in the resolution [4] of the hyperelliptic equation

$$Y^2 = (X-\beta_1)\dots(X-\beta_m) \tag{4}$$

over function fields, where $\beta_1,\dots,\beta_m$ are $m(\geq 3)$ distinct elements of $\mathcal{O}$. The first stage in the solution is the proof that, as X,Y run over the solutions in $\mathcal{O}$ of (4), the field L generated over K by the square roots ξ_1,ξ_2,ξ_3 of $X-\beta_1$, $X-\beta_2$, $X-\beta_3$ respectively, has only a finite number of possibilites. This follows easily from the well known theorem that, if J denotes the Jacobian variety associated with K, then the 2-torsion subgroup of J is finite, of order 4^g. The next stage is the effective determination of the range of possibilities for such a field: the technique here involves the construction of 2-division points on J, and this has been shown to lead to the solution of a system of polynomial equations over k in at most $2g+1$ unknowns; the systems arising are constructed so that they have only finitely many solutions, and thus all their solutions may be computed. The determination of the fields having been achieved, the fundamental inequality is applied with $U = (\xi_2-\xi_3)/(\xi_2-\xi_1)$. Since $(\xi_i-\xi_j)(\xi_i+\xi_j) = \beta_j-\beta_i$ for each i,j it follows that U and V have their poles and zeros within the set S consisting of valuations on L at which $v(z) <$ or $v(\beta_i-\beta_j) > 0$ for some $1 \leq i < j \leq 3$. The lemma and its corollary now lead to an algorithm for the determination of all the solutions of (4) in $\mathcal{O}$. In general there will be only finitely many solutions, the only instance to the contrary occurring when (4) may be transformed by a substitution

$$X = \alpha x + \beta, \quad Y = \gamma y \qquad (\alpha,\beta,\gamma\in\mathcal{O},\ \alpha\gamma\neq 0)$$

into an equation $y^2 = f(x)$, where f has coefficients in k; in this case there will be the infinite family derived from the solutions in k of $y^2 = f(x)$, but otherwise only finitely many solutions in $\mathcal{O}$. The analysis also yields

$$H(X) \leq 26H + 8g + 4(r-1),$$

where H denotes the height of (4); this may be compared with Schmidt's bound $10^6(H+g+r)$.

In a forthcoming book [5, Chapters IV,V], the complete resolutions of equations of genera 0 and 1 are attained, provided that as usual in the former case the associated curve possesses at least 3 infinite valuations. The bounds calculated on the heights of the solutions are, as for the Thue and hyperelliptic equations above, linear functions of the height of the equation concerned. For example, if $F(X,Y)$ denotes an absolutely irreducible polynomial over K of genus one, height H and total degree d, then the solutions in $\mathcal{O}$ of $F(X,Y) = 0$ satisfy

$$H(X,Y) \le 1500(d+1)^{d^2+11}H + 8d(2g+r-1)$$

The fundamental inequality applies for fields of positive characteristic p, provided that U is not a p-th power in K. This apparently innocuous exceptional case actually gives rise to a much greater variety in the sets of solutions to equations. For example, if $X = x$, $Y = y$ is any solution in $\mathcal{O}$ of the Thue equation $XY(X+Y) = 1$, then so is $X = x^{p^i}$, $Y = y^{p^i}$ for any positive integer i: the set of solutions may therefore possess such a family of unbounded height. It turns out that [5, Chapter VI], for the general Thue and hyperelliptic equation, the set of solutions in $\mathcal{O}$ may be divided into a finite number of such families, possibly together with those corresponding to the solutions in k of an equation defined over k, as in the case of characteristic zero above, and at most finitely many other exceptional solutions. Furthermore, the various families and the exceptional set may all be determined effectively.

Another application of the fundamental inequality [5, Chapter VII] has been the establishment of bounds on the heights of all the solutions of Thue and superelliptic equations, not just the integral solutions. Let us suppose that F is a polynomial in X of height H and degree n which splits separably in K. Then all the solutions in K of the superelliptic equation $Y^m = F(X)$ satisfy

$$H(X) \le 35H + 8g$$

provided $n = 2, m \ge 5$ or $n \ge 3, m \ge 4$. If $n \ge 3$, then the fundamental inequality is applied with $U=(X-\alpha_1)(\alpha_2-\alpha_3)(\alpha_2-\alpha_1)$, where $F(X) = \alpha_0 \prod_{i=1}^{n} (X-\alpha_i)$. Success is obtained since at each pole or zero of U or V, with a fixed number of exceptions, the corresponding order of vanishing is a multiple of m: thus $\#S \le 3H(U)/m + N$, where N is the number of exceptional valuations. Since $H(U) \le \#S + 2g - 2$ by the lemma, a bound follows provided that $m \ge 4$. If $n = 2$ we choose $U = (X-\alpha_1)/(\alpha_2-\alpha_1)$. Schmidt's method of differential equations is at present slightly more powerful in this direction, since he has shown [8] that bounds may be determined in principle for the solutions in $k(z)$ also when $m = 3, n \ge 5$ and $m = 2, n \ge 17$. For the Thue equation, if $F(X,Y)$ is a separable binary form over K with height H, then any solution in K of $F(X,Y) = 1$ satisfies

$$H(X,Y) \le 59H + 14g$$

provided F has degree at least 4. This latter condition is essential, since otherwise the equation would represent a curve of genus 0 or 1, which may possess points in K of arbitrarily large height.

In conclusion we shall discuss two further aspects of the fundamental inequality, one retrospective and the other prospective. If we consider the special case when K is the rational function field k(z), then we obtain the following remarkable and beautiful original result.

COROLLARY. Let A, B and C be coprime polynomials in $k[z]$ *with* $A + B + C = 0$*, and such that their product ABC has N distinct zero in k. Then either A, B and C are all constants, or*

$$\max(\deg A, \deg B, \deg C) < N.$$

Here k is any algebraically closed field of characteristic zero, although the assumption of algebraic closure may be removed, provided that N is then replaced by the largest degree of a separable factor in k[z] of ABC, and the assumption of characteristic zero may also be omitted, provided that A, B and C are not all perfect characteristic powers. This corollary to the fundamental inequality may be applied to obtain some familiar results in the subject as immediate sub corollaries: k is any field of characteristic zero.

(i) The Fermat equation $X^n + Y^n = Z^n$ has no non-trivial solution in rational functions if $n \geq 3$. For we may assume that X,Y and Z are coprime polynomials, with the maximum of their degrees M say. Then by the corollary either $XYZ = 0$, or $M = 0$, or $nM < 3M$, as required.

(ii) Using the same method, we conclude that any polynomial solution X,Y of the Mordell equation $Y^2 = X^3 - A$ satisfies $\deg X = 0$ or $\deg X \leq 3 \deg A - 2$. This is an old result of Davenport.

(iii) If A, B, C are fixed polynomials, then all the solutions in coprime polynomials of $AX^a + BY^b + CZ^c = 0$ have bounded degrees provided $1/a + 1/b + 1/c < 1$. This is a recent result of Dwork.

Finally, in a very recent development the fundamental inequality has been generalised to the following result.

THEOREM. Suppose that S is a finite set of valuations on K, and $U_1, \ldots, U_n$ *are non-zero elements all of whose poles and zeros lie in S and with* $U_1 + \ldots + U_n = 1$*. Suppose further that there is no proper subset I of* $\{1, \ldots, n\}$ *with* $\sum_{i \in I} U_i = 1$*. Then*

$$H(U_1, \ldots, U_n) \leq 4^{n-1}(\# S + 2g).$$

By a remarkable coincidence, Evertse has also reported at this conference that he has recently established the ineffective analogue of this result for number fields. The theorem has profound consequences for norm form equations. For example, if L denotes a finite extension of K, and M denotes a free non-degenerate $\mathcal{O}$-module in L, then it follows that all the solutions of

$$\mathrm{Norm}_{L/K}(x) = c$$

may be determined effectively for each c in K.

I would like to express my grateful thanks to St. John's College, Cambridge, of which I was latterly a Fellow, for their generous financial assistance in enabling me to attend this conference.

REFERENCES

[1] Baker, A. Contributions to the theory of Diophantine equations: I On the representation of integers by binary forms, Philos. Trans. Roy. Soc. London Ser.A263(1968), 173-191.

[2] Baker, A. and Coates, J. Integer points on curves of genus 1, Proc. Camb. Philos. Soc. 67(1970), 595-602.

[3] Mason, R.C. On Thue's equation over function fields, J. London Math. Soc. Ser. 2 24 (1981), 414-426.

[4] Mason, R.C. The hyperelliptic equation over function fields, Proc. Camb. Philos. Soc. 93(1983), 219-230.

[5] Mason, R.C. Diophantine equations over function fields, LMS Lecture Notes, Cambridge University Press, to appear.

[6] Osgood, C.F. An effective lower bound on the "Diophantine approximation" of algebraic functions by rational functions, Mathematika 20(1973), 4-15.

[7] Schmidt, W.M. Thue's equation over function fields, J. Austral. Math. Soc. Ser.A 25(1978), 385-422.

[8] Schmidt, W.M. Polynomial solutions of $F(x,y) = z^n$. Queen's Papers in Pure Appl. Math., 54(1980), 33-65

[9] Thue, A. Über Annäherungswerte algebraischer Zahlen, J. Reine Angew. Math. 135(1909), 284-305.

On Thue's principle and its applications

J. Mueller*
Department of Mathematics
Fordham University
Bronx, N.Y., 10458/U.S.A.

I. Introduction

Let α be a real algebraic number and let k be a real number field with $\alpha \notin k$. The effective measure of irrationality of α over k, $\mu_{eff}(\alpha,k)$ is defined to be the infimum of all μ's for which

$$|\alpha-\beta| > C(\alpha)\ H_k(\beta)^{-\mu}$$

holds for every $\beta \in k$ and some effectively computable $C(\alpha) > 0$. Here $H_k(\beta)$ is the field height of β i.e. the largest coefficient of the primitive integral polynomial with roots $\sigma(\beta)$, for all distinct embeddings $\sigma: k \rightarrow C$. We say $\mu_{eff}(\alpha,k)$ is non-trivial if $\mu_{eff}(\alpha,k) < \deg \alpha$.

The only non-trivial effective result valid for every algebraic number α and every number field k is given by Baker and Feldman [1], [7]. Unfortunately, their improvement is rather small. However, bette effective results are known for special numbers. For example, the celebrated cubic irrational $\sqrt[3]{2}$ is shown by G.V. Chudnovsky [6] to have measure of irrationality $\mu_{eff}(\sqrt[3]{2},Q) \leqq 2.4297...$ This number was first studied by Baker [2] in 1964, using the Padé approximation method.

This paper will use the Thue principle as in the recent papers [3], [4]. The basic strategy of this method consists in a precise quantitative statement which attaches an effective measure of irrationality $\mu_{eff}(\alpha,\beta)$ to any pair (α,β) in which β is an approximation in k to α. If the approximation is sufficiently good, the corresponding measure of irrationality is non-trivial. In favorable circumstance it can approach Roth's exponent 2.

In this talk we present some applications of Thue's principle including a new result in the following theorem

<u>THEOREM</u> Let α be a real algebraic number of degree $r \geq 3$. Let $\delta > 0$. For any $2 \leqq s < r$, there exists infinitely many real number fields k of degree s such that

(a) if $s = r-1$, then

(1) $$\mu_{eff}(\alpha,k) \leqq 2+\delta.$$

* Research partially supported by NSF grant MCS 82-05253

(b) if $2 \leqq s < r-1$, then

(2) $$\mu_{\text{eff}}(\alpha,k) \leqq \frac{4r}{s+3} + \delta .$$

Besides our formulation of Thue's principle, a theorem of Wirsing on approximation of real numbers by algebraic numbers of bounded degree plays an essential part in the proof.

I would like to thank Professor Enrico Bombieri for his helpful suggestions and advice. I would also like to thank the University of Geneva for its hospitality while this manuscript was prepared.

II. The Thue Principle

We shall first review two earlier versions of Thue's principle. Thue's original formulation [11] deals with rational approximations to an algebraic number, whereas Gelfond's version [8], Theorem 1, p. 22 deals with approximations of an algebraic number by algebraic numbers. It should be noted that Gelfond's construction is different from that of Thue and therefore his conclusion and Thue's are somewhat different In our notation, we can formulate Thue's and Gelfond's results as follows:

(A) Thue's version. Let α be an algebraic number of degree $r \geqq 3$. Then there is an effectively computable constant $q_0(\alpha,\delta,\varepsilon)$ such that if

(3) $$|\alpha - \frac{p}{q}| < q^{\frac{-r}{2}-1-\delta} \qquad q > q_0(\alpha,\delta,\varepsilon)$$

then

$$\mu_{\text{eff}}(\alpha,Q) < \frac{r}{2(1+\delta)} + 1 + \varepsilon .$$

A more precise formulation of Thue's result, which allows the explicit calculation of the constant q_0 can be found in Mahler's paper [9], Hilfsatz 3, p. 709. The value of q_0 thus obtained turned out to be very large and so far no examples of inequality (3) with such values of q_0 have been found. It should be noted that according to Roth's theorem, if q_0 tends to ∞ then the exponent in (3) should tend to 2. This indicates that (3) is an exceedingly strong hypothesis.

(B) Gelfond's version. Let α be an algebraic number of degree $r \geqq 3$. Let k be a number field, $\alpha \notin k$. Then there is an effectively computable constant $C(\alpha,k,\varepsilon)$ such that if there is a $\beta \in k$ with

(4) $|\alpha-\beta| < H_k(\beta)^{-\rho}$

and

$$H_k(\beta) > C(\alpha,k,\varepsilon)$$

then

$$\mu_{eff}(\alpha,k) \leq \frac{2r}{\rho} + \varepsilon.$$

We note that Gelfond's original result [8], Theorem 1, p. 22, had the additional restriction $\rho > \sqrt{2r}$. This restriction was subsequently removed by Hyyrö, [13].

The value of the constant $C(\alpha,k,\varepsilon)$ was calculated by Hyyrö [13] for th case in which k is the rational field $\mathbb{Q}$. We remark here that because the constant $C(\alpha,k,\varepsilon)$ depends on k and because in our theorem we take $k = \mathbb{Q}(\beta)$, where β satisfies an inequality such as (4), this version of Thue's principle cannot be used to prove our theorem.

Before stating our formulation of Thue's principle, we mention brie ly about Thue's method. It is known that this method is based on the construction of an auxiliary polynomial $P \in k[x_1,x_2]$ with $\deg_{x_i} P \leq d_i$ and having the following properties: (i) P vanishes at the point (α,α) to high order; (ii) P vanishes at the point (β,β') only to low order, where β,β' are approximations to α; and (iii) the height $H(P)$ of P is not too large, where $H(P)$ is the maximum of the heights of coefficients of P. It is in the verification of (ii) that Bombieri ingeniously applied Dyson's lemma and as a result he removed the restriction on the height of β which was present in earlier formulations and in that way cleared a major obstacle to finding explicit applications.

Let α be algebraic of degree r and let σ be $0 < \sigma < 1$. We define $A_1 = A_1(\alpha,\sigma;k)$ as the lowest upper bound of the constants A with the following property:

For every Θ, $\sqrt{\frac{2}{r+\sigma^2}} < \Theta < \sqrt{\frac{r+\sigma^2}{2}}$ and for every d_1,d_2

we can find $P \in k[x_1,x_2]$, $P \neq 0$, with integer coefficients in k, of degree $\deg_{x_i} P \leq d_i$, such that

$$\frac{\partial^{i_1+i_2}}{\partial x_1^{i_1}\,\partial x_2^{i_2}} P(\alpha,\alpha) = 0$$

for all i_1, i_2 with

$$\Theta^{-1}\frac{i_1}{d_1} + \Theta\frac{i_2}{d_2} < \sqrt{\frac{2}{r+\sigma^2}}$$

and such that

$$\frac{1}{[k:Q]} \log H(P) \leq \frac{A_1}{\deg\alpha} \log H(\alpha) d_1 + O(d_2),$$

uniformly for $d_1, d_2 \to \infty$. We can now state

<u>Thue's Principle</u> Let k be a real number field with $[k:Q]=s$ and let α be a real algebraic number of degree r, $r \geq 3$, $\alpha \notin k$. If $k = Q(\beta)$ and if $|\alpha-\beta| < 1$, then for each σ, $0<\sigma<1$, we have

$$(5) \qquad \mu_{eff}(\alpha,k) \leq \frac{2(r+\sigma^2)}{(1-\sigma)^2} \quad \frac{\log H(\beta) + A_1 \frac{s}{r} \cdot \log H(\alpha) + 2s}{\log \frac{1}{|\alpha-\beta|}}$$

Moreover, the right hand side of (5) is a bound for $\mu_{eff}(\alpha',k)$ for every element α' of $k(\alpha)$.

<u>Remarks</u> 1. What we state here as Thue's principle is a corollary of the so called Thue-Siegel principle in [4]. A derivation of this can be found in [5].

2. The assumption $k = Q(\beta)$ can be replaced by $\beta \in k$ if we also replace $\log H(\beta)$ by $\log H_k(\beta)$ in (5).

3. We call (α,β) an anchor pair if the right-hand side of (5) is strictly less than r.

In general, we have

$$A_1 \leq \frac{r}{\sigma^2} \left(1 + \frac{r}{\log H(\alpha)}\right) .$$

(This inequality is easily deduced from the bound in Lemma 6 and the remark on p. 284 of [3]).

If we substitute this bound, a simple estimation with $\sigma = \frac{1}{2}$ will give the following result which even if it is not as strong as the Thue principle, provides a more transparent upper bound for $\mu_{eff}(\alpha,k)$.

<u>COROLLARY</u> Let α be a real algebraic number of degree r, $r \geq 3$. If there exists a β in Q such that

(i) $H(\beta) \geq H(\alpha)^{1/b}$, and

(ii) $|\alpha-\beta| < H(\beta)^{-\rho}$,

then

$$\mu_{eff}(\alpha,Q) \leq \frac{(8r+2)(4sb+1)}{\rho} + \frac{sb(8r+4)^2}{\rho \log H(\alpha)} .$$

III. Applications

The following simple lemma is very useful in what follows.

LEMMA Let $p(x)$ be a polynomial with coefficients in Q and of degree r. Suppose $p(\beta) \neq 0$ with $\beta \in Q$, then $p(x)$ has a root α with

$$|\alpha-\beta| \leq r \left|\frac{p(\beta)}{p'(\beta)}\right|$$

To see this we write $p(x) = d \prod_{i=1}^{r} (x-\alpha_i)$ where the product is over all the roots α_i of $p(x)$. Our lemma follows easily by taking the logarithm derivative of p.

We now illustrate Thue's principle with some examples. Our first example is example 3 of [3]. That is, α is the positive root of the irreducible polynomial $p(x) = x^r - mx + 1$.
The irreducibility follows from the fact that for $|m| \geq 3$, α^{-1} is a so called P.V. number. Note that α^{-1} is a root of the polynomial $x^r - mx^{r-1} + 1$. By lemma, we have

$$|\alpha - m^{-1}| \leq rm^{-r-1}.$$

In the notation of Corollary, we have $s = 1$, $b = 1$, $\rho = r+1 - \frac{\log r}{\log m}$, then

$$\mu_{eff}(\alpha, Q) \leq \frac{40r+10}{\rho} + \frac{(8r+4)^2}{(\rho)\log m}$$

which implies $\mu_{eff}(\alpha,Q) \leq 40$ if $m > e^{3r^2}$.

Our second example is dealt with in detail in [4]. It is shown in [4], lemma 2 that if

$$\alpha = (1-\frac{h}{a})^{1/r}, \quad 1 \leq h < a.$$

then

$$A_1 \leq 1 + \frac{r}{\sigma^2 \log a}$$

Using this bound we obtain the following:

Let

$$p(x) = ax^r - (a-h);$$

by lemma 2, we have

$$|\alpha - 1| \leq \frac{h}{a},$$

and, choosing σ in an optimal way, we deduce from (5)

$$\mu_{eff}(\alpha,Q) \leq \frac{2 \log a}{\log a - \log h} + O\left(\frac{1}{(\log a)^{1/3}}\right)$$

We refer to [4] for further details.

Our final example is the theorem mentioned in the introduction. Recall the following result of Wirsing [12].

Wirsing's Theorem. Let α be real and not algebraic of degree $\leq s$. Then there exist infinitely many algebraic numbers β of degree $\leq s$ such that

$$(6) \qquad |\alpha-\beta| \ll H(\beta)^{-\frac{s+3}{2}} .$$

Moreover, if $s = r-1$ and if α is real algebraic of degree r, then there exist infinitely many β such that

$$|\alpha-\beta| \ll H(\beta)^{-r}.$$

The implicit constants depend on α and s and can be effectively determined.

It should be noted that Wirsing's result is more precise than our statement, but this simplified version suffices for our purpose.

To prove our theorem, we apply Thue's principle to the pair (α,β). We consider α as being fixed and β with very large height. Let us consider the case in which $s = r-1$. Then by choosing $\sigma = (\log H(\beta))^{-1/3}$ we have

$$\frac{r+\sigma^2}{(1-\sigma)^2} = r + O((\log H(\beta))^{-1/3}),$$

$$\log \frac{1}{|\alpha \ \beta|} = r \log H(\beta) + O(1) \quad ,$$

and since $s = r-1$

$$A_1 \cdot \frac{s}{r} \log H(\alpha) + 2s = O\left((\log H(\beta))^{2/3}\right),$$

where the constants involved in the $O(\dots.)$ symbol depend only on $r, H(\alpha)$. Now the Thue principle yields

$$\mu_{eff}(\alpha,k) \leq 2 + O\left(\frac{1}{(\log H(\beta))^{1/3}}\right)$$

with $k=Q(\beta)$, which is a more precise form of (1).

Finally we remark that the same procedure together with (6) will give a proof of (2). This completes the proof of our theorem.

IV. Concluding Remarks

As we have shown in previous sections, Thue's principle can be applied to obtain non-trivial measures of irrationality, either for each real algebraic number over certain number fields, or for some speci classes of numbers constructed from a given number field. We should mention that our first two examples are a special case of a more general class of numbers. To construct such numbers, we first define the so called Thue numbers.

DEFINITION Let η be an algebraic integer and $\eta_j (j=2,\ldots,r)$ the conjugates of η. We call η a Thue Number if $|\eta| > 1$ and $|\eta_j| < 1$ for each j, $2 \leq j \leq r$.

Let $F_1(x)$ and $F_2(x)$ be polynomials with rational integer coefficients and of degree s_1 and s_2. Let η be an algebraic number such that η^{-1} is a Thue number in k and let $G(x)$ be the defining polynomial of η. For $r > \max(s_1, s_2)$, we define

$$M(x) = x^{r-s_1}F_1(x)+G(x)F_2(x)$$

and

$$N(x) = x^{r-s_2}F_1(x) + \xi F_2(x),$$

where $\xi = 1+\eta$.

Our first and second examples are special cases of the above polynomials. Suppose that α is a real root of $M(x)=0$ or $N(x)=0$ with $F_1(x)$ and $F_2(x)$ restricted in certain ways, then we can again show, using the same method, that $\mu_{eff}(\alpha,k)$ is non-trivial. Details will be carried out in a separate paper.

V. Appendix

In the following proposition we will show that Wirsing's theorem (6) can be strengthened under the assumption that α is algebraic.

PROPOSITION Let α be a real algebraic number of degree $r \geq 3$ and le $\varepsilon > 0$ be given. Let $2 \leq s < r$. Then there exist infinitely many algebrai numbers β of degree s such that

$$|\alpha-\beta| < H(\beta)^{-s-1+\varepsilon}$$

Proof: Define $w_s^*(\alpha)$ to be the greatest lower bound for w's for which

$$|\alpha-\beta| < H(\beta)^{-w-1}$$

holds for infinitely many real algebraic numbers β of degree $\leq s$. Define $w_s(\alpha)$ to be the greatest lower bound of w's for which

$$|P(\alpha)| < H(P)^{-w}$$

holds for infinitely many polynomials $P(x)$ with rational integer coefficients and of degree $\leq s$. Wirsing [12] has shown that

$$w_s^*(\alpha) \geq \frac{w_s(\alpha)}{w_s(\alpha)-s+1}$$

for every real number α which is not algebraic of degree $\leq s$.

On the other hand, if α is algebraic of degree $r > s$, Schmidt's celebrated subspace theorem [10] implies $w_s(\alpha) = s$. By Wirsing's inequality, $w_s^*(\alpha) \geq s$ and the Proposition is proved.

We conclude with the remark that the most interesting case $s=r-1$ can be dealt with in an elementary way, with an explicit construction of numbers β such that

$$|\alpha-\beta| << H(\beta)^{-r}$$

For a detailed treatment of this remark, we refer the readers to [5].

References

1. A. Baker, The Theory of Linear Forms in Logarithms. In "Transcendence Theory: Advances and Applications", A. Baker & D.W. Masser ed., New York-London 1977.

2. A. Baker, Rational approximations to $\sqrt[3]{2}$ and other algebraic numbers Quart. J. Math. Oxford 15(1964), 375-383.

3. E. Bombieri, On the Thue-Siegel-Dyson Theorem, Acta Mathematica 1 (1982), 255-296.

4. ____________ and J. Mueller, On effective measures of irrationality for $\sqrt[n]{\frac{a}{b}}$ and related numbers, J. reine angew. Math. 342 (1983

5. ____________ and J. Mueller, Remarks on the approximation to an algebraic number by algebraic numbers, to appear.

6. G.V. Chudnovsky, On the method of Thue-Siegel, Annals of Math. 117 (1983), 325-382.

7. N.I. Feldman, An effective refinement of the exponent in Liouville's theorem (Russian), Izv. Akad. Nauk 35 (1971), 973-990 Also Math USSR Izv. 5 (1971), 985-1002.

8. A.O. Gelfond, Transcendental and algebraic numbers, English translation by L.F. Boron, New York 1960.

9. K. Mahler, Zur Approximation algebraischer Zahlen. I. (Über den größten Primteiler binärer Formen.), Math. Annalen 107, 1933, 691-730.

10. W.M. Schmidt, Diophantine Approximation, Springer Lecture Notes in Mathematics, 785.

11. A. Thue, Über Annäherungswerte algebraischer Zahlen, J. reine angew. Math. 135(1909), 284-305.

12. E. Wirsing, Approximation mit algebraischen Zahlen beschränkten Grades, J. reine angew. Math. 206 (1961), 66-77.

13. S., Hyyrö, Über rationale Näherungswerte algebraischer Zahlen, Ann. Acad. Sci. Fenn. Ser. A I. Math. 376 (1965), 1-15.

FORMATIONS DE CLASSES ET MODULES D'IWASAWA

T. Nguyen-Quang-Do

U.E.R. de Mathématiques.
Université Paris VII

Il est bien connu (voir par exemple [11], [8]) que les méthodes de cohomologie galoisienne peuvent s'appliquer avec fruit à la théorie wasawa. Le but de cet exposé et d'en faire un usage intensif (di- sion cohomologique, formations de classes, dimension projective...) r obtenir, à propos des modules d'Iwasawa standard, un certain ore de résultats qui ne semblent pas accessibles par la théorie clas- ue. Voici quelques exemples de ces résultats :

Soit p un nombre premier. Soit k une extension finie de $\mathbb{Q}$ (resp. $\mathbb{Q}_p$). On désigne par K_∞ la composée de toutes les $\mathbb{Z}_p$-extensions de k, par Γ le groupe de Galois de K_∞/k. On considère le module d'Iwasawa ndard X_{K_∞} (pour une définition précise, voir §O) sur l'algèbre de upe complète $\Lambda = \mathbb{Z}_p[[\Gamma]]$. Alors :

(i) Le Λ-module X_{K_∞} est un sous-module d'un Λ-module de dimension projective inférieure ou égale à 1.

(ii) Il résulte de (i) qu'en particulier, X_{K_∞} n'a pas de sous-module pseudo-nul non nul.

(iii) Soit $r_2(k)$ (resp. n_k) le nombre de places complexes de k (resp. le degré de k sur $\mathbb{Q}_p$). Le Λ-rang de X_{K_∞} est égal à $r_2(k)$ (resp. n_k)

Dans les deux cas, local et global, ce rang s'interprète comme t égal à une caractéristique d'Euler-Poincaré.

(iv) Dans certains cas favorables, la série caractéristique de X_{K_∞} t être calculée par des méthodes empruntées à la topologie algé- que (calcul différentiel de Fox).

Tous les résultats précédents restent valables en prenant pour K_∞ une $\mathbb{Z}_p$-extension multiple de k qui vérifie une certaine forme faible de la conjecture de Leopoldt.

O - DEFINITIONS ET NOTATIONS

Soient p un nombre premier et k une extension finie de $\mathbb{Q}$ (resp. de $\mathbb{Q}_p$). Si p=2 et k est un corps de nombres, on supposera en plus que k contient les racines quatrièmes de l'unité. Dans toute la suite, S désignera un ensemble fini de places de k contenant les places au-dessus de p et les places à l'infini. Pour toute extension K/k, on continuera à noter S le prolongement de S à K, si aucune confusion n'est possible L'extension K/k est appelée S-ramifiée si elle est non ramifiée en toute place en dehors de S. Soit Ω la pro-p-extension S-ramifiée maximale (resp. la pro-p-extension maximale) de k. On désigne par G_k le groupe de Galois $Gal(\Omega/k)$ et par X_k l'abélianisé de G_k. Les propriétés suivantes sont bien connues (voir [5] et [14]) :

- la dimension cohomologique de G_k est inférieure ou égale à 2 : $cd\ G_k \leq 2$

- soient $d = \dim H^1(G_k,\mathbb{Z}/p\mathbb{Z})$ et $r = \dim H^2(G_k,\mathbb{Z}/p\mathbb{Z})$ respectivement les nombres minimaux de générateurs et de relations de G_k. La *caractéristique d'Euler-Poincaré* de G_k sera définie par : $\chi(G_k) = -1 + d - r$. Si $r_2(k)$ (resp. n_k) est le nombre de places complexe de k (resp. le degré de k sur $\mathbb{Q}_p$), on sait que $\chi(G_k) = r_2(k)$ (resp. n_k

Dans la suite, nous utiliserons l'homologie des groupes profinis, dont nous rappelons maintenant les bases (voir [2]).

Soit G un pro-p-groupe, et soit Λ l'algèbre de groupe complète $\mathbb{Z}_p[[G]]$. Dans la catégorie $\mathcal{C}$ des Λ-modules pseudo-compacts (en fait, dans la suite on ne considérera que des Λ-modules compacts, qui seront des groupes de Galois) on définit un "produit tensoriel complet" $T(A,B) = A \hat{\otimes} B$ qui possède les bonnes propriétés ([2], section 1). En particulier, on peut définir les foncteurs dérivés $Tor_n^{\Lambda}(A,B)$ de $T(A,B)$ en utilisant des résolutions projectives de A ou B. En abrégé, on notera : $Tor_n^G(A,B)$. Pour tout $A \in \mathcal{C}$, on définit l'homologie de A par $H_n(G,A) = Tor_n^G(A,\mathbb{Z}_p)$ ([2], 4.2) et l'on vérifie que $H_n(G,A) = \varprojlim H_n(G/U,A/I(U).A)$, où U parcourt les sous-groupes distingués ouverts de G et I(U) est l'idéal d'augmentation de $\mathbb{Z}_p[[U]]$. Cette définition est donc compatible avec celle de [14], problème 4, I-55. Il en résulte en particulier que pour tout Λ-module compact A, $H_n(G,A) = H^n(G,A^*)^*$, où (*) désigne le dual de Pontryagin.

1 - MODULE DES RELATIONS ET CONJECTURE DE LEOPOLDT

Nous commençons par quelques rappels de résultats purement algébriques. On considère une présentation libre de rang m de G, c'est-à-dire une suite exacte $1 \to R_m \to F_m \to G \to 1$, où F_m est le pro-p-groupe libre à m générateurs. Le $\mathbb{Z}_p[[G]]$-module $R_m^{ab}(G) = R_m/[R_m,R_m]$ (avec l'action naturelle de G) est appelé un *module de relations* de G. C'est un invariant de G à équivalence projective près, dont les propriétés sont bien connues (voir par exemple [11], chap. II). Pour la commodité du lecteur, nous rappelons quelques-unes de ces propriétés qui nous seront utiles dans la suite

PROPOSITION 1.1 :

On a une suite exacte de $\mathbb{Z}_p[[G]]$*-modules :*

$$0 \to R_m^{ab}(G) \to \mathbb{Z}_p[[G]]^m \to \mathbb{Z}_p[[G]] \to \mathbb{Z}_p \to 0$$

(c'est la "résolution de Lyndon").

Preuve :

Considérons la suite exacte d'augmentation $0 \to I(F_m) \to \mathbb{Z}_p[[F_m]] \to \mathbb{Z}_p \to 0$ où $I(F_m)$ est l'idéal d'augmentation. Comme le pro-p-groupe F_m est libre, on sait ([2], 5.1) que $I(F_m)$ est un $\mathbb{Z}_p[[F_m]]$-module libre de rang m. Le début de la suite exacte Tor sur l'algèbre $\mathbb{Z}_p[[R_m]]$ ([2], 5.2) s'écrit :

$$0 \to \mathrm{Tor}_1^{R_m}(\mathbb{Z}_p,\mathbb{Z}_p) \to I(F_m)/I(R_m).I(F_m) \to \mathbb{Z}_p[[F_m]]/I(R_m).\mathbb{Z}_p[[F_m]]$$

$$\to \mathbb{Z}_p \to 0, \text{ c'est-à-dire}$$

$$0 \to R_m^{ab}(G) \to \mathbb{Z}_p[[G]]^m \to \mathbb{Z}_p[[G]] \to \mathbb{Z}_p \to 0 \qquad \text{QED}$$

COROLLAIRE 1.2 :

Soient $n \geq m \geq \dim H^1(G,\mathbb{Z}/p\mathbb{Z})$. *Alors* $R_n^{ab}(G) \simeq R_m^{ab}(G) \times \mathbb{Z}_p[[G]]^{n-m}$

Preuve : si G est fini, on applique le lemme de Schanuel et le théorème de Krull-Schmidt à la suite exacte 1.1 ; si G est infini, on passe à la limite projective en utilisant un argument de compacité.

COROLLAIRE 1.3 :

Supposons G *fini. Alors* $\operatorname{rang}_{\mathbb{Z}_p} R_m^{ab}(G) = |G|(m-1)+1$
(c'est la "formule de Schreier")

Revenons maintenant à l'arithmétique pour montrer une suite exacte qui sera fondamentale dans la suite

THEOREME 1.4 (voir [12] ou [11], chap. VI).

On reprend les définitions et notations du §0. Soit K/k *une extension galoisienne (finie ou non) contenue dans* Ω, *et soit* $G = \mathrm{Gal}(K/k)$. *On a une suite exacte de* $\mathbb{Z}_p[[G]]$*-modules :*

$$0 \to \Delta(K) \to \mathbb{Z}_p[[G]]^r \to R_d^{ab}(G) \to X_K \to 0, \text{ où } \Delta(K) = H_2(G_K,\mathbb{Z}_p)$$

(cette suite exacte a été étudiée indépendamment par Uwe Jannsen - voir son exposé à ces "Journées Arithmétiques").

Preuve : Considérons une présentation libre minimale de $G_k : 1 \to W \to F_d \to G_k \to 1$ (rappelons que $d = \dim H^1(G_k,\mathbb{Z}/p\mathbb{Z})$). Soit R l'image réciproque de G dans F_d (voir figure)

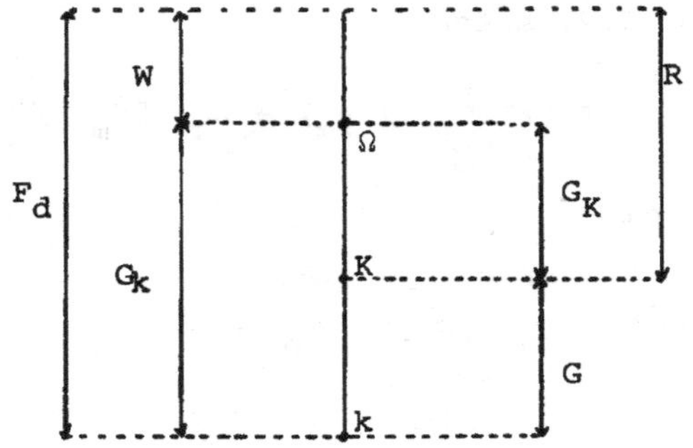

La suite exacte d'inflation-restriction associée à l'extension de groupes $1 \to W \to R \to G_K = \mathrm{Gal}(\Omega/K) \to 1$ s'écrit, puisque le pro-p-groupe R est libre :

$$0 \to H^1(G_K,\mathbb{Q}_p/\mathbb{Z}_p) \to H^1(R,\mathbb{Q}_p/\mathbb{Z}_p) \to H^0(G_K,H^1(W,\mathbb{Q}_p/\mathbb{Z}_p)) \to H^2(G_K,\mathbb{Q}_p/\mathbb{Z}_p) \to 0,$$

soit, par dualité :

$$0 \to H_2(G_K,\mathbb{Z}_p) \to H_0(G_K,W^{ab}) \to R^{ab} \to X_K \to 0$$

Comme $\mathrm{cd}\, G_k \leq 2$, le $\mathbb{Z}_p[[G_k]]$-module W^{ab} est libre sur les images d'un système minimal de relations de W (i.e. un système minimal de F_d-générateurs de W) (voir [2], 5.3). Il en résulte que $W^{ab} \simeq \mathbb{Z}_p[[G_k]]^r$ et que $H_0(G_K,W^{ab}) \simeq \mathbb{Z}_p[[G]]^r$ QED

Le module $\Delta(K)$ est $\mathbb{Z}_p$-libre. Soit $\delta(K) = \mathrm{rang}_{\mathbb{Z}_p} \Delta(K)$. Dans le cas local, $\Delta(K) = 0$ car $\mathrm{scd}\, G_k = 2$. Dans le cas global, et si l'extension K/k est finie, $\delta(K)$ mesure le "défaut" de la conjecture de Leopoldt pour K. Plus précisément :

<u>COROLLAIRE 1.5</u> :

Supposons que l'extension K/k *est finie. Alors* $\mathrm{rang}_{\mathbb{Z}_p} X_K = 1+r_2(K)+\delta(K)$ (resp. $1+n_K$). *En particulier, dans le cas global,* K *vérifie la conjecture de Leopoldt (pour p) si et seulement si* $\Delta(K) = 0$

<u>Preuve</u> : En utilisant la suite exacte du théorème 1.4 et la formule de Schreier (corollaire 1.3), un simple calcul de $\mathbb{Z}_p$-rangs nous donne : $\mathrm{rang}_{\mathbb{Z}_p} X_K = 1 + \chi(G_K) + \delta(K)$. Dans le cas local, $\delta(K) = 0$ car $\mathrm{scd}\, G_K = 2$. Dans le cas global, on sait ([7], 2.3) que la conjecture de Leopoldt pour p et K équivaut à l'égalité $\mathrm{rang}_{\mathbb{Z}_p} X_K = 1+r_2(K)$ QED

<u>Remarque</u> : Dans [8], 4.3, Kuz'min montre directement l'isomorphisme de $\Delta(K)$ avec le noyau de l'homomorphisme naturel $U_K \otimes \mathbb{Z}_p \to \prod_{\mathfrak{p} \in S} K^*_{\mathfrak{p}} \otimes \mathbb{Z}_p$, où U_K est le groupe des unités de K.

Pour des raisons techniques, nous allons maintenant associer à toute extension K/k contenue dans Ω un module Y_K plus commode à manier que le module X_K.

DEFINITION 1.6 :

Comme précédemment, considérons une présentation libre minimale de G_k, i.e. une suite exacte $1 \to W \to F_d \to G_k \to 1$. Soit K/k une extension contenue dans Ω. Posons $Y_{K/k} = H_o(\Omega/K, I(G_k))$, où $I(G_k)$ est l'idéal d'augmentation de l'algèbre $\mathbb{Z}_p[[G_k]]$. En abrégé, nous écrirons Y_K au lieu de $Y_{K/k}$.

PROPOSITION 1.7 :

Soit K/k *une extension galoisienne contenue dans* Ω, *de groupe de Galois* G. *On a deux suites exactes de* $\mathbb{Z}_p[[G]]$*-modules :*

$0 \to X_K \to Y_K \to I(G) \to 0$ *(où* $I(G)$ *est l'idéal d'augmentation de* $\mathbb{Z}_p[[G]]$*) et* $0 \to \Delta(K) \to \mathbb{Z}_p[[G]]^r \to \mathbb{Z}_p[[G]]^d \to Y_K \to 0$

Preuve : Considérons la suite exacte d'augmentation :
$0 \to I(G_k) \to \mathbb{Z}_p[[G_k]] \to \mathbb{Z}_p \to 0$. Comme dans la démonstration de la résolution de Lyndon (proposition 1.1), en prenant l'homologie par rapport au groupe $G_K = Gal(\Omega/K)$, on obtient immédiatement la première suite exacte $0 \to X_K \to Y_K \to \mathbb{Z}_p[[G]] \to \mathbb{Z}_p \to 0$.

Pour montrer la seconde suite exacte, on considère la résolution de Lyndon $0 \to W^{ab} \simeq \mathbb{Z}_p[[G_k]]^r \to \mathbb{Z}_p[[G_k]]^d \to I(G_k) \to 0$. En prenant l'homologie par rapport à G_K, on obtient la suite exacte :
$0 \to H_1(\Omega/K, I(G_k)) \to \mathbb{Z}_p[[G]]^r \to \mathbb{Z}_p[[G]]^d \to Y_K \to 0$. Mais il est clair que $H_1(\Omega/K, I(G_k)) \simeq H_2(\Omega/K, \mathbb{Z}_p) = \Delta(K)$ QED

L'avantage du module Y_K est que, par définition même, si L est un sur-extension de K contenue dans Ω, on a $Y_K \simeq H_o(L/K, Y_L)$.

Pour terminer ce paragraphe, nous considérons quelques propriétés fonctorielles du module $\Delta(K)$. Nous nous placerons dans la situation suivante : L/k est une extension galoisienne finie contenue dans Ω, K/k une sous-extension galoisienne, $E = Gal(L/k)$, $G = Gal(K/k)$, $H = Gal(L/K)$

PROPOSITION 1.8 :

On a un isomorphisme $\Delta(K) \simeq \Delta(L)^H$
Preuve : On peut écrire la suite exacte du théorème 1.4 pour les ex-

tensions K et L respectivement, puis faire une chasse dans le diagramme. On peut aussi utiliser le module dualisant : soit D le module dualisant de G_K ([14], chap. I). Par définition, on a un isomorphisme fonctoriel $H^2(G_K, \mathbb{Q}_p/\mathbb{Z}_p)^* \simeq H^0(G_K, \mathrm{Hom}(\mathbb{Q}_p/\mathbb{Z}_p, D))$.

Si H est fini, G_L est un sous-groupe ouvert de G_K, donc admet aussi D comme module dualisant ([14], I, propos. 1.8). Il est alors immédiat que $\Delta(K) \simeq \Delta(L)^H$

QED

<u>COROLLAIRE 1.9</u> :

Supposons l'extension L/k *finie. Alors si* L *vérifie la conjecture de Leopoldt,* K *la vérifie également*

<u>Problème</u> : démontrer une réciproque du corollaire 1.9

<u>PROPOSITION 1.10</u> :

L'homomorphisme de transfert induit un homomorphisme $j = j_{L/K}$ *de* X_K *dans* X_L *tel qu'on ait la suite exacte :*

$$0 \to H^1(H, \Delta_L) \to X_K \xrightarrow{j} X_L^H \to H^2(H, \Delta_L) \to 0$$

<u>Preuve</u> : On a un triangle commutatif :

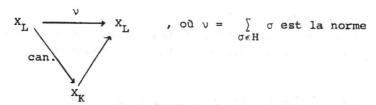

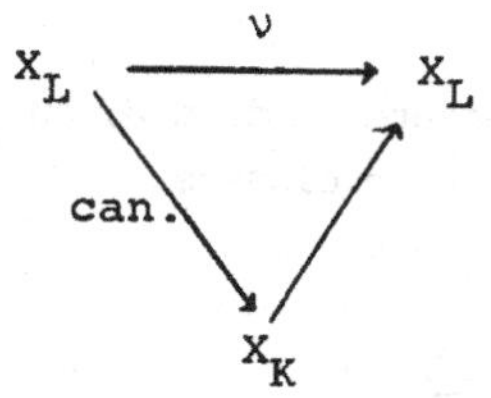

, où $\nu = \sum_{\sigma \in H} \sigma$ est la norme

Considérons le $\mathbb{Z}_p[H]$-module $Y_L = Y_{L/K}$ (cette fois-ci, le corps de base est K).

La norme induit la suite exacte bien connue :

$$0 \to \hat{H}^{-1}(H, Y_L) \to H_0(H, Y_L) \xrightarrow{\nu} H^0(H, Y_L) \to \hat{H}^0(H, Y_L) \to 0$$

D'après 1.7, $\hat{H}^i(H, Y_L) \simeq \hat{H}^{i+2}(H, \Delta_L)$ pour tout $i \in \mathbb{Z}$. Egalement d'après 1.7, on a une suite exacte $0 \to X_L \to Y_L \to I(H) \to 0$, d'où $X_L^H = Y_L^H$ (car $I(H)^H = 0$) et $X_K = Y_K$ (en faisant $H = (1)$). La proposition en découle

QED

On remarquera l'analogie du résultat 1.9 avec les propriétés des groupes de classes d'idéaux ([7], thm. 1.2). L'isomorphisme

Ker $j \simeq H^1(H,\Delta_L)$ a été également montré par G. Gras ([4], thm. II.2) par des méthodes non cohomologiques

COROLLAIRE 1.11 :

(i) Ker $j_{L/K}$ *est inclus dans le sous-module de torsion de* X_K

(ii) X_K *et* X_L^H *ont même* $\mathbb{Z}_p$-rang

Enfin, donnons une propriété technique qui nous servira dans la suite. Les extensions L/k et K/k sont galoisiennes, contenues dans Ω, mais ne sont plus supposées finies.

PROPOSITION 1.12 :

On a une suite exacte $H_o(H,\Delta_L) \to \Delta_K \to H_1(H,Y_L) \to 0$

Preuve : On écrit la suite exacte du théorème 1.4 pour K et L respectivement, et on fait une chasse dans le diagramme QED

COROLLAIRE 1.13 :

Supposons que $H = \mathrm{Gal}(L/K)$ *soit isomorphe à* $\mathbb{Z}_p$ *et que* $\Delta(L) = 0$. *Alors* $\Delta(K) \simeq X_L^H$.

Justification du titre :

Si $\Delta(K) = 0$ pour toute extension finie K/k contenue dans Ω, on voit facilement (1.4 et 1.10) que les X_K constituent une formation de classes.

2. AUTOUR DE LA CONJECTURE FAIBLE DE LEOPOLDT

Nous cherchons maintenant à étendre les résultats du corollaire 1.5 aux extensions infinies

DEFINITION 2.1 :

Soit K/k une extension *infinie* contenue dans Ω. On dira que K vérifie la *conjecture faible* de Leopoldt (pour p) si $\Delta(K) = 0$, i.e. $H^2(G_K,\mathbb{Q}_p/\mathbb{Z}_p) = 0$ (cette propriété est toujours vérifiée dans le cas local)

THEOREME 2.2 :

Soit k_∞^{cyc} *la* $\mathbb{Z}_p$*-extension cyclotomique de* k. *Alors toute extension* K_∞/k, *contenant* k_∞^{cyc} et *contenue dans* Ω, *vérifie la conjecture faible de Leopoldt*

<u>Preuve</u> : Pour $K_\infty = k_\infty^{cyc}$, c'est un théorème bien connu d'Iwasawa ([7], thm. 1.7 ; pour une démonstration cohomologique, voir [13], lemma 7). Pour montrer le cas général, il suffit visiblement de supposer que K_∞ est une extension finie de k_∞^{cyc}. D'après des propriétés standard des limites inductives, il existe alors une extension k_n/k dans la tour canonique de k_∞^{cyc}/k et une extension finie K/k_n telle que K_∞ est la $\mathbb{Z}_p$- extension cyclotomique de K. QED

<u>COROLLAIRE 2.3</u> : (voir aussi [8], thm. 4.3)

On a $\mathrm{scd}\ \mathrm{Gal}(\Omega/k_\infty^{cyc}) = 2$

<u>Preuve</u> : Comme $\mathrm{cd}\ G_k \leq 2$ et comme $\mathrm{Gal}(\Omega/k_\infty^{cyc})$ est un sous-groupe fermé de G_k, on a $\mathrm{cd}\ \mathrm{Gal}(\Omega/k_\infty^{cyc}) \leq 2$. Le résultat à prouver est alors équivalent au fait que $H^2(N, \mathbb{Q}_p/\mathbb{Z}_p) = 0$ pour tout sous-groupe ouvert N, ce qui a été démontré dans le théorème 2.2 QED

<u>DEFINITION 2.4</u> :

On dira qu'un pro-p-groupe G est *rangé* si l'algèbre $\Lambda = \mathbb{Z}_p[[G]]$ possède un corps des fractions $Q(\Lambda)$ (G est alors forcément infini ou trivial). Le Λ-rang d'un Λ-module X sera alors défini comme d'habitude par : $\mathrm{rang}_\Lambda X = \dim X \underset{\Lambda}{\otimes} Q(\Lambda)$

<u>Exemples 2.5</u> :

Les pro-p-groupes suivants sont rangés :

(i) G = (1)

(ii) $G = \mathbb{Z}_p^N$ (N copies de $\mathbb{Z}_p$). En fait, dans ce cas, $\Lambda = \mathbb{Z}_p[[G]]$ est un domaine factoriel

(iii) G est un pro-p-groupe qui est analytique, sans torsion. Alors possède un corps des fractions (non commutatif en général) (voir [6], 1.5).

<u>PROPOSITION 2.6</u> :

Soit K_∞/k *une extension galoisienne infinie contenue dans* Ω, *telle que* $G = \mathrm{Gal}(K_\infty/k)$ *soit un pro-p-groupe rangé. Posons* $\Lambda = \mathbb{Z}_p[[G]]$. *Alors* $\mathrm{rang}_\Lambda X_{K_\infty} = r_2(k) + \mathrm{rang}_\Lambda \Delta(K_\infty)$ (*resp.* n_k). *En particulier dans le cas global,* K_∞ *vérifie la conjecture faible de Leopoldt si et seulement si* $\mathrm{rang}_\Lambda X_{K_\infty} = r_2(k)$.

Preuve : D'après la suite exacte du théorème 1.4, on a : $\mathrm{rang}_\Lambda X_{K_\infty} = \mathrm{rang}_\Lambda R_d^{ab}(G) - r + \mathrm{rang}_\Lambda \Delta(K_\infty)$. D'après la résolution de Lyndon, $\mathrm{rang}_\Lambda R_d^{ab}(G) = d-1$, d'où $\mathrm{rang}_\Lambda X_{K_\infty} = \chi(G_k) + \mathrm{rang}_\Lambda \Delta(K_\infty)$. Comme $\Delta(K_\infty)$ est un sous-module de Λ^r, $\mathrm{rang}_\Lambda \Delta(K_\infty) = 0$ si et seulement si $\Delta(K_\infty) = 0$ QED

COROLLAIRE 2.7 :

Soit k *un corps de nombres. Soit* K_∞/k *une extension galoisienne contenue dans* Ω, *dont le groupe de Galois* G *est un pro-p-groupe rangé. Si* K_∞ *contient la* $\mathbb{Z}_p$*-extension cyclotomique* k_∞^{cyc} *de* k, *alors* $\mathrm{rang}_\Lambda X_{K_\infty} = r_2(k)$, *où* $\Lambda = \mathbb{Z}_p[[G]]$

Preuve : C'est immédiat, d'après 2.6 et 2.2 QED

Le corollaire précédent contient et unifie divers résultats de Greenberg ([3], proposition 2), Harris ([6], thm. 3.9), etc... On peut se demander s'il reste valable en remplaçant k_∞^{cyc} par n'importe quelle $\mathbb{Z}_p$-extension vérifiant la conjecture faible de Leopoldt. La réponse est affirmative. Plus précisément :

THEOREME 2.8 :

Soit K_∞/k *une extension galoisienne contenue dans* Ω, *et soit* L/k *une sous-extension galoisienne de* K_∞/k. *On suppose que les groupes de Galois* $\mathrm{Gal}(K_\infty/k)$ *et* $\mathrm{Gal}(L/k)$ *sont rangés. Si* $\Delta(L) = 0$, *alors* $\Delta(K_\infty) = 0$

Preuve : Faisons intervenir les modules Y_L et Y_{K_∞} (voir définition 1.6). Nous avons deux suites exactes : $0 \to \Delta(K_\infty) \to \Lambda^r \xrightarrow{\phi} \Lambda^d \to Y_{K_\infty} \to 0$ et $0 \to \Delta(L) \to \Lambda_1^r \xrightarrow{\phi_1} \Lambda_1^d \to Y_L \to 0$, où Λ et Λ_1 sont les algèbres complètes sur $\mathbb{Z}_p$ de $\mathrm{Gal}(K_\infty/k)$ et $\mathrm{Gal}(L/k)$ respectivement. Par définition, $Y_L = H_0(K_\infty/L, Y_{K_\infty})$, donc la matrice de ϕ_1 n'est autre que l'image de la matrice de ϕ par la projection naturelle $\Lambda \to \Lambda_1$. Par hypothèse, $\Delta(L) = 0$ i.e. ϕ_1 est injectif. Par suite, la matrice de ϕ_1 possède un sous-déterminant non nul d'ordre r (pour une théorie des déterminants dans un corps gauche, voir par exemple [1], chap. VI). Le sous-déterminant correspondant dans la matrice de ϕ est aussi non nul, i.e. ϕ est injectif et $\Delta(K_\infty) = 0$ QED

Comme conséquences immédiates, nous retrouvons des propriétés connues concernant les $\mathbb{Z}_p$- extensions d'un corps de nombres

COROLLAIRE 2.9 : (résulte aussi de 2.2)

Soit k *un corps de nombres. La composée de toutes les* $\mathbb{Z}_p$*-extensions de* k *vérifie la conjecture faible de Leopoldt.*

COROLLAIRE 2.10 :

Supposons que k *vérifie la conjecture de Leopoldt (ordinaire). Alors toute* $\mathbb{Z}_p$*-extension de* k *vérifie la conjecture faible de Leopoldt.*

Dans la même direction que ce corollaire, mais sans supposer que k vérifie la conjecture de Leopoldt, on a le résultat suivant (comparer à Greenberg, [3], fin du §3).

THEOREME 2.11 :

Soit k *un corps de nombres, et soit* $\mathcal{E}(k)$ *l'ensemble des* $\mathbb{Z}_p$*-extensions de* k*, considéré comme un espace projectif sur* $\mathbb{Z}_p$*. Alors l'ensemble des* $\mathbb{Z}_p$*-extensions de* k *qui vérifient la conjecture faible de Leopoldt contientle complémentaire d'une variété linéaire projective de* $\mathcal{E}(k)$*. En particulier, il est partout dense dans* $\mathcal{E}(k)$

Preuve : Rappelons d'abord la structure canonique de $\mathbb{Z}_p$- espace projectif de $\mathcal{E}(k)$. Soient K_∞ la composée de toutes les $\mathbb{Z}_p$ -extensions de k et $\Gamma_\rho = \mathrm{Gal}(K_\infty/k)$. On a $\rho \geq 1+r_2(k)$, avec égalité si et seulement si k vérifie la conjecture de Leopoldt. Le groupe Γ_ρ est le pro-p-groupe abélien libre à ρ générateurs. Toute $\mathbb{Z}_p$ -extension de k est déterminée par le noyau d'un épimorphisme $\alpha : \Gamma_\rho \to \Gamma_1$, où Γ_1 est le pro-p-groupe abélien libre à un générateur. Une fois qu'on a choisi des bases de Γ_ρ et Γ_1, un tel épimorphisme α est donné par un ρ-uple $a = (a_1,\dots,a_\rho) \in \mathbb{Z}_p^\rho$. En outre, α induit un homomorphisme d'anneaux $\Lambda_\rho = \mathbb{Z}_p[[\Gamma_\rho]] \to \Lambda_1 = \mathbb{Z}_p[[\Gamma_1]]$, qu'on désignera par Π_a. On a le lemme suivant de Monsky :

LEMME 2.12 : ([9], lemme 2.2)

Soit $f \in \Lambda_\rho$, $f \neq 0$*. Il existe des applications linéaires non nulles* $h_1,\dots, h_m : \mathbb{Z}_p^\rho \to \mathbb{Z}_p$ *telles que, si* $a \in \mathbb{Z}_p^\rho$ *et* $h_i(a) \neq 0$ *pour* $1 \leq i \leq m$*, alors* $\Pi_a(f) \neq 0$

La fin de la démonstration du théorème 2.11 est maintenant claire:

D'après le corollaire 2.9, on a une suite $0 \to \Lambda_\rho^r \overset{\phi}{\to} \Lambda_\rho^d \to Y_{K_\infty} \to 0$
Pour toute $\mathbb{Z}_p$- extension k_∞/k, de groupe de Galois Γ_1, on a une suite exacte $\Lambda_1^r \xrightarrow{\phi_1} \Lambda_1^d \to Y_{k_\infty} \to 0$, avec $Y_{k_\infty} = H_0(K_\infty/k_\infty, Y_{K_\infty})$. Donc $\phi_1 = \Pi_a(\phi)$, et il suffit d'appliquer le lemme à un mineur d'ordre r de dét ϕ.

Problème : démontrer une réciproque du théorème 2.8.

3. DIMENSION PROJECTIVE ET MODULESPSEUDO-NULS

Dans tout ce paragraphe, K_∞/k désignera une $\mathbb{Z}_p^N$-extension, i.e. une extension galoisienne telle que $\Gamma_N = \mathrm{Gal}(K_\infty/k)$ soit isomorphe à un produit de N copies de $\mathbb{Z}_p$. L'algèbre complète $\mathbb{Z}_p[[\Gamma_N]]$sera notée Γ_N. Si aucune confusion n'est possible, on écrira simplement Γ et Λ au lieu de Γ_N et Λ_N. Le résultat principal que nous voulons démontrer est le suivant :

THEOREME 3.1 :

Soit K_∞/k une $\mathbb{Z}_p^N$ -extension. Supposons que K_∞ vérifie la conjecture faible de Leopoldt. Alors X_{K_∞} n'a pas de sous-module pseudo-nul non nul

(Comparer à [3], proposition 5)

COROLLAIRE 3.2 :

X_{K_∞} n'a pas de sous-module pseudo-nul non nul dans chacun des cas suivants :
(i) k *vérifie la conjecture de Leopoldt (ordinaire)*

ou

(ii) K_∞ *est la composée de toutes les $\mathbb{Z}_p$-extensions de* k

Avant de procéder à la démonstration, rappelons d'abord quelques définitions et résultats concernant les dimensions homologiques sur Λ (voir [2], §3). Pour tout Λ-module compact A, la *dimension projective* (ou dimension homologique) de A, notée $\mathrm{hd}_\Lambda A$, est le plus petit entier n tel qu'on puisse trouver une résolution projective
$0 \to P_n \to P_{n-1} \to \ldots \to P_0 \to A \to 0$ (ou une résolution libre - c'est la même chose car Λ est local).

La *dimension globale* de Λ, notée gld Λ, est le sup des $hd_\Lambda A$. Comme Λ est isomorphe à l'anneau des séries formelles $\mathbb{Z}_p[[T_1,\dots,T_N]]$, on sait que gld $\Lambda = N+1$.

Les propriétés suivantes seront utilisées librement :

PROPOSITION 3.3 :

(i) *Soit* $0 \to B \to A \to C \to 0$ *une suite exacte de* Λ*-modules. Alors* $hd_\Lambda A \le \max(hd_\Lambda B, hd_\Lambda C)$, *avec égalité, sauf peut être quand* $hd_\Lambda C = 1 + hd_\Lambda B$.

En particulier, si $hd_\Lambda B < hd_\Lambda A$, *alors* $hd_\Lambda A = hd_\Lambda C$ *; si* $hd_\Lambda B > hd_\Lambda A$, *alors* $hd_\Lambda C = hd_\Lambda B + 1$

(ii) *Soit* $U \in \Lambda$, *n'ayant pas de diviseur de zéro dans un* Λ*-module* A. *Alors* $hd_\Lambda A = hd_{\Lambda/U} A/U.A$

(voir par exemple [15], chap. IV).

Preuve du théorème 3.1 :

Comme K_∞ vérifie la conjecture faible de Leopoldt, on a une suite exacte de Λ-modules $0 \to \Lambda^r \to \Lambda^d \to Y_{K_\infty}$ (voir déf. 1.6 et propos. 1.7), i.e. hd $Y_{K_\infty} \le 1$. De plus, la suite exacte $0 \to X_{K_\infty} \to Y_{K_\infty} \to I_\Lambda \to 0$ où I_Λ est l'idéal d'augmentation de Λ (propos. 1.7), montre que X_{K_∞} et Y_{K_∞} ont même Λ-torsion. Il suffit donc de montrer que Y_{K_∞} n'a pas de sous-module pseudo-nul non nul. C'est une propriété purement algébrique : *un* Λ*-module* A *de dimension projective inférieure ou égale à* 1 *n'a pas de sous-module pseudo-nul non nul.* La démonstration originelle de l'auteur utilisait des calculs fastidieux de déterminants. Mais, comme l'a fait remarquer J. Oesterlé, le résultat cherché découle simplement de ce que (voir [15], chap. IV) la hauteur des idéaux premiers associés à A est majorée par $hd_\Lambda A$ QED

Nous allons maintenant calculer la dimension projective de X_{K_∞} et de certains de ses sous-modules.

PROPOSITION 3.4 :

Soit K_∞/k *une* $\mathbb{Z}_p^N$*-extension vérifiant la conjecture faible de Leopoldt. Alors :*

(i) *si* $N \leq 2$, $hd_\Lambda X_{K_\infty} \leq 1$

(ii) *si* $N > 2$, $hd_\Lambda X_{K_\infty} = N-2$

<u>Preuve</u> : La suite exacte du théorème 1.4 : $O \to \Lambda^r \to R_d^{ab}(\Gamma) \to X_{K_\infty} \to O$ montre, d'après 3.3 (i), que $hd_\Lambda X_{K_\infty} = hd_\Lambda R_d^{ab}(\Gamma)$, sauf peut être si $R_d^{ab}(\Gamma)$ est libre, auquel cas $hd\, X_{K_\infty} \leq 1$. Comme $\Gamma \simeq \mathbb{Z}_p^N$, on a $cd\, \Gamma = N$, donc $hd_\Lambda \mathbb{Z}_p = N$ ([2], corol. 4.4). La résolution de Lyndon montre alors que $hd_\Lambda R_d^{ab}(\Gamma) = N-2$ si $N \geq 2$ et $hd_\Lambda R_d^{ab}(\Gamma) = O$ si $N = 1$ QED

<u>PROPOSITION 3.5</u> :

Soit K_∞/k *une* $\mathbb{Z}_p$ *ou* $\mathbb{Z}_p^2$*-extension, et soit* tX_{K_∞} *le sous-module de* Λ*-torsion de* X_{K_∞}. *Supposons que* k *vérifie la conjecture de Leopoldt. Alors* $hd\, tX_{K_\infty} \leq 1$.

<u>Preuve</u> : Comme k vérifie la conjecture de Leopoldt, toute $\mathbb{Z}_p$-extensio (simple ou multiple) de k vérifie la conjecture faible de Leopoldt (2.10). C'est vrai en particulier pour K_∞, donc $hd_\Lambda X_{K_\infty} \leq 1$

(i) Si $N = 1$, alors *tout sous-module de* X_{K_∞} *est de dimension projectiv inférieure ou égale à* 1 (cela résulte immédiatement de 3.3.i) et de $gld\, \Lambda = 2$). Donc $hd_\Lambda tX_{K_\infty} \leq 1$

(ii) Si $N = 2$, soit k_∞/k une $\mathbb{Z}_p$-extension contenue dans K_∞. Appelons un générateur de $Gal(K_\infty/k_\infty)$. Comme K_∞ et k_∞ vérifient la conjecture faible de Leopoldt, $\gamma-1$ n'a pas de diviseur de zéro dans tX_{K_∞} (prop. 1.12). Mais il est clair que $tX_{K_\infty}/_{(\gamma-1)} \cdot tX_{K_\infty}$ est un sous-module de X_{k_∞} donc est de dimension projective ≤ 1 d'après i). En appliquant 3.3 i), on trouve alors que $hd_\Lambda tX_{K_\infty} \leq 1$ QED

<u>Problème</u> : Calculer $hd\, tX_{K_\infty}$ pour $N > 2$

Pour finir, voici une généralisation d'un résultat de Kuz'min ([8],th 6.

<u>PROPOSITION 3.6</u> :

Soit K_∞/k *une* $\mathbb{Z}_p^N$*-extension vérifiant la conjecture faible de Leopoldt. Alors* $Ext_\Lambda^1(Y_{K_\infty}/\mathbb{Z}/p\mathbb{Z}) \simeq H^2(G_k,\mathbb{Z}/p\mathbb{Z})$. *Si* $N = 1$, *on peut rempla cer* Y_{K_∞} *par* X_{K_∞} *dans cet isomorphisme.*

<u>Preuve</u> : Appliquons à la suite exacte $O \to \Lambda^r \to \Lambda^d \to Y_{K_\infty} \to O$ le foncteu $Hom_\Lambda(.,\mathbb{Z}/p\mathbb{Z})$. Nous obtenons une suite exacte :

$0 \to Hom_\Lambda(Y_{K_\infty},\mathbb{Z}/p\mathbb{Z}) \to Hom_\Lambda(\Lambda^d,\mathbb{Z}/p\mathbb{Z}) \to Hom_\Lambda(\Lambda^r,\mathbb{Z}/p\mathbb{Z}) \to Ext^1_\Lambda(Y_{K_\infty},\mathbb{Z}/p\mathbb{Z}) \to 0$

d'où, par dualité, une suite exacte :

$0 \to Ext^1_\Lambda(Y_{K_\infty},\mathbb{Z}/p\mathbb{Z})^* \to (\mathbb{Z}/p\mathbb{Z})^r \to (\mathbb{Z}/p\mathbb{Z})^d \to \bar{Y}_k \to 0$, où $(\bar{.})$ signifie le quotient modulo p. Mais l'analogue mod p de 1-4 et 1-7 est valable (il suffit de remplacer $\mathbb{Z}_p$ par $\mathbb{Z}/p\mathbb{Z}$) :

LEMME 3.7 :

Pour toute extension galoisienne K/k contenue dans Ω, de groupe de Galois G, on a deux suites exactes de $\mathbb{Z}/p\mathbb{Z}[[G]]$ -modules :

$0 \to H^2(G_K,\mathbb{Z}/p\mathbb{Z})^* \to \mathbb{Z}/p\mathbb{Z}[[G]]^r \to \mathbb{Z}/p\mathbb{Z}[[G]]^d \to \bar{Y}_K \to 0$ *et*

$0 \to \bar{X}_K \to \bar{Y}_K \to I_p(G) \to 0$ *(où $I_p(G)$ est l'idéal d'augmentation de* $\mathbb{Z}/p\mathbb{Z}[[G]]$).

L'isomorphisme $Ext^1_\Lambda(Y_{K_\infty},\mathbb{Z}/p\mathbb{Z}) \simeq H^2(G_k,\mathbb{Z}/p\mathbb{Z})$ résulte alors du lemme de Schanuel

QED

COROLLAIRE 3.8 :

Le module Y_{K_∞} est Λ-libre si et seulement si le pro-p-groupe G_k est libre. Si tel est le cas, $X_{K_\infty} \simeq R_N^{ab}(\Gamma) \times \Lambda^{d-N}$

Preuve : Pour tout Λ-module A compact, A est Λ-libre si et seulement si $Ext^1_\Lambda(A,C) = 0$ pour tout Λ-module discret simple C ([2], 3-1). Mais comme Γ est un pro-p-groupe, le seul Λ-module discret simple est $\mathbb{Z}/p\mathbb{Z}$. Donc Y_{K_∞} est Λ-libre si et seulement si $H^2(G_k,\mathbb{Z}/p\mathbb{Z}) = 0$. Si tel est le cas, $X_{K_\infty} \simeq R_d^{ab}(\Gamma)$ par définition même, donc $X_{K_\infty} \simeq R_N^{ab}(\Gamma) \times \Lambda^{d-N}$ d'après le corollaire 1-2

QED

4. EXEMPLES DE CALCULS EXPLICITES

Les hypothèses et les notations sont les mêmes que dans le §3. Pour tout Λ-module A, on appellera *série caractéristique* de A, notée F(A), la série caractéristique (définie à une unité près) du sous-module de Λ-torsion de A. Notre but est de calculer $F(X_{K_\infty})$ dans un certain nombre de cas favorables, en utilisant le théorème 1.4 et des méthodes empruntées à la topologie algébrique. De fait, le formalisme employé est exactement le même que celui qui sert à calculer le "polynôme d'Alexander" d'un noeud. Des exemples de tels calculs se trouvent déjà dans [1b].

Rappel 4.1 :

Rappelons la base du calcul différentiel de Fox. Soit $x_1,\dots,x_d$ un système de générateurs libres de F_d et soit $Dx_1 = 1-x_1,\dots,Dx_d = 1-x_d$ une base de l'idéal d'augmentation $I(F_d)$ $(\simeq \mathbb{Z}_p[[F_d]]^d)$. L'application $x_i \to 1-x_i$ induit une "dérivation" $D : \mathbb{Z}_p[[F_d]] \to I(F_d)$ définie de la façon suivante : pour tout mot w de F_d, $D(w) = \sum_{i=1}^{d} \frac{\partial w}{\partial x_i} Dx_i$, où les "dérivées partielles" $\frac{\partial w}{\partial x_i}$ sont calculées à partir de la forme réduite de w. Plus précisément, si $w = y_1 \dots y_m$, avec $y_i = x_i^{\pm 1}$ pour $1 \le i \le m$, on a : $\frac{\partial w}{\partial x_i} = \sum_j y_1 \dots y_{j-1} \frac{\partial y_j}{\partial x_i}$, avec $\frac{\partial y_j}{\partial x_i} = 0$ si $i \neq j$, $\frac{\partial x_i}{\partial x_i} = 1$, $\frac{\partial x_i^{-1}}{\partial x_i} = -x_i^{-1}$. Pour des détails, voir [4b] (où les calculs sont faits pour des groupes discrets, mais l'adaptation aux groupes profinis est aisée).

Retournons maintenant à la preuve et aux notations de 1.4 et 1.7. Nous avons construit une suite exacte

$H_0(\Omega/K_\infty,W^{ab}) \simeq \Lambda^r \overset{\phi}{\to} H_0(\Omega/K_\infty,H_0(W,I(F_d)) \simeq \Lambda^d \to Y_{K_\infty} \to 0$, où ϕ s'obtient à partir de D par passage au quotient. Plus précisément, notons $\sigma_1,\dots,\sigma_d$ les images de $x_1,\dots,x_d$ dans $\Gamma = Gal(K_\infty/k)$, $D\sigma_1,\dots,D\sigma_d$ les images de $Dx_1,\dots,Dx_d$ dans Λ^d et $\frac{\partial w}{\partial \sigma_i}$ les images de $\frac{\partial w}{\partial x_i}$. Pour tout $w \in W$, nous ferons un abus de langage en désignant encore par w l'image de w dans $H_0(\Omega/K_\infty,W^{ab})$. Alors $\phi(w) = \sum_i \frac{\partial w}{\partial \sigma_i} D\sigma_i$. L'homomorphisme ϕ est ainsi donné par une "matrice jacobienne" $M_\phi = (\frac{\partial w_j}{\partial \sigma_i})$, $1 \le i \le d$, $1 \le j \le r$, les w_j étant un système de F_d-générateurs de w (les "relations" de G_k). La connaissance de M_ϕ permet :

- de décrire Y_{K_∞} par générateurs et relations
- de calculer $F(X_{K_\infty}) = F(Y_{K_\infty})$: en effet, si ρ est le Λ-rang de Y_{K_∞}, la théorie de Fitting nous dit que $F(Y_{K_\infty})$ est le p.g.c.d. des mineurs non nuls d'ordre $(d-\rho)$ extraits de la matrice M_ϕ.

Pour illustrer la méthode que nous venons d'exposer, nous calculerons M_ϕ dans la situation suivante : K_∞ est la composée de toutes les $\mathbb{Z}_p$-extensions de k, $\Gamma = \mathrm{Gal}(K_\infty/k)$, $\Lambda = \mathbb{Z}_p[[\Gamma]]$

4.2 - Cas local :

Supposons que k est une extension de degré n de $\mathbb{Q}_p$. Soit q l'ordre du groupe des racines p-primaires de l'unité contenues dans k

(i) Si q=1, on sait que le pro-p-groupe G_k est libre, donc Y_{K_∞} est Λ-libre (coroll. 3.7)

(ii) Supposons $q \neq 1,2$ (le cas q=2 nécessite un calcul particulier). Alors G_k est décrit par (n+2) générateurs x_i et une relation, le "mot de Demuškin" :

$$w = x_1^q[x_1,x_2][x_3,x_4]\ldots[x_{n+1},x_{n+2}]$$

En tenant compte que $\sigma_1 = 1$ (car Γ est abélien sans torsion), on obtient, en posant $\frac{n}{2} = m$:

$$\phi(w) = qD\sigma_1 + \sum_{i=1}^{m+1} (1-\sigma_{2i})D\sigma_{2i-1} + \sum_{i=2}^{m+1} (\sigma_{2i-1}-1)D\sigma_{2i}.$$

La matrice M_ϕ est une matrice à (n+2) lignes et une colonne

$$M_\phi = \begin{pmatrix} q+1-\sigma_2 \\ 0 \\ 1-\sigma_4 \\ \sigma_3-1 \\ \vdots \\ 1-\sigma_{n+2} \\ \sigma_{n+1}-1 \end{pmatrix}$$

(Remarquons que σ_2 correspond à la $\mathbb{Z}_p$-extension cyclotomique de k)

Il en résulte en particulier que $F(X_{K_\infty}) = 1$, autrement dit X_{K_∞} *n'a pas de Λ-torsion.*

4.3 - Exemple global :

Dans le cas où k est un corps de nombres, on ne connait pas en général les relations du pro-p-groupe G_k. Dans certains cas, cependant, où les relations globales "proviennent des relations locales" (pour de précisions, voir [8b], chap. XI), on peut expliciter ces relations. Le hypothèses sur k étant assez restrictives ([8b], Satz 11-16), nous nou contenterons d'examiner un exemple numérique.

Soit $k = \mathbb{Q}$, $p=3$, $S = \{\infty, \mathfrak{P}_1, \mathfrak{P}_2, q, \mathfrak{P}\}$, où $\mathfrak{P}_1$, $\mathfrak{P}_2$ sont les diviseurs de 3 dans k, q est inerte, $q \equiv 1 \pmod 3$, $q \not\equiv 1 \pmod 9$, et $\mathfrak{P}$ est un diviseur premier non principal de k (le nombre de classes de k est 3). Alors G_k est décrit par quatre générateurs s_q, t_q, $s_{\mathfrak{P}}$, $t_{\mathfrak{P}}$ et deux relations $w_1 = t_q^{q-1}[t_q^{-1}, s_q^{-1}]$, $w_2 = t_{\mathfrak{P}}^{N(\mathfrak{P})-1}[t_{\mathfrak{P}}^{-1}, s_{\mathfrak{P}}^{-1}]$ ([8b], p.125). En désignant par σ_1 et σ_2 les images de s_q et $s_{\mathfrak{P}}$ dans Γ (les images σ_3 et et σ_4 de t_q et $t_{\mathfrak{P}}$ sont triviales), on obtient :

$$M_\phi = \begin{pmatrix} 0 & 0 \\ 0 & 0 \\ (q-1)+(\sigma_1^{-1}-1) & 0 \\ 0 & (N(\mathfrak{P})-1)+(\sigma_2^{-1}-1) \end{pmatrix}$$

$$\text{Donc } F(X_{K_\infty}) = F(Y_{K_\infty}) = [(q-1)+(\sigma_1^{-1}-1)][(N(\mathfrak{P})-1)+(\sigma_2^{-1}-1)]$$

En particulier, *la Λ-torsion de X_{K_∞} n'est pas nulle et l'invariant* $\lambda(X_{K_\infty})$, défini par $\lambda(X_{K_\infty}) = \sum_P \mathrm{ord}_P \overline{F(X_{K_\infty})}$, la somme étant étendue à tous les idéaux premiers P de la forme $\overline{\tau-1}$ de $\overline{\Lambda} = \Lambda/p\Lambda$ (cet invariant est noté ℓ_o dans [9]), *n'est pas nul.*

Il serait intéressant de trouver un exemple où X_K vérifie les deux propriétés précédentes, mais où S contient seulement les places archimédiennes et les places divisant p.

BIBLIOGRAPHIE

[1] E. ARTIN : *Geometric Algebra,* Interscience (1957)

[1b] BOREVIC-EL MUSA : Completion of the multiplicative group of p-extensions of an irregular local field. J. Soviet Math., 6.3 (1976), 6-23.

[2] A. BRUMER : Pseudo-compact algebras, profinite groups and class-

formations. *J. of Algebra,* 4 (1966), 442-470

[3] R. GREENBERG : On the structure of certain Galois groups. *Inventions Math.*, 47 (1978), 85-99

[4] G. GRAS : Groupe de Galois de la p-extension abélienne p-ramifiée maximale d'un corps de nombres. *J. reine angew. Math,* 333 (1982), 86-132

[4b] R.H. FOX : Free differential calculs. I. Derivation in the free group ring. *Ann. Math.*, 57, 3 (1953), 547-560.

[5] M. HARRIS : p-adic representations arising from descent on abelian varieties. *Compositio Math.*, 39, 2 (1975), 177-245.

[6] K. HABERLAND : *Galois cohomology of algebraic number fields.* Deutscher Verlag der Wissenschafter, Berlin (1978)

[7] K. IWASAWA : On $\mathbb{Z}_{\ell}$-extensions of algebraic number fields. *Ann. Math.*, 98 (1973), 246-326

[8] L.V. KUZ'MIN : On the Tate module of an algebraic number field. *Math. USSR Izv.*, 6 (1972), 263-321

[8b] H. KOCH : *Galoissche Theorie der p-Erweiterungen.* Deutscher Verlag der Wissenschafter, Berlin (1970)

[9] P. MONSKY : Some invariants of $\mathbb{Z}_p^d$-extensions. *Math. Ann.*, 255 (1981), 229-233

[10] O. NEUMANN : On p-closed number fields and an analogue of Riemann's existence theorem, in *Algebraic Number Fields* (A. Fröhkich, ed.), Acad. Press, London (1977), 625-647

[11] T. NGUYEN-QUANG-DO : *Sur la structure galoisienne des corps locaux et la théorie d'Iwasawa.* Thèse d'Etat, Orsay (1982)

[12] T. NGUYEN-QUANG-DO : Description par générateurs et relations de certains modules d'Iwasawa. *Sém. Théorie des Nombres Delange-Pisot-Poitou* (Fév. 1982)

[13] P. SCHNEIDER : Über gewisse Galoiscohomologie gruppen. *Math. Zeitschrift,* 168 (1979), 161-205

[14] J.P. SERRE : *Cohomologie Galoisienne.* Lectures Notes in Math. n°5, Springer (1964)

[15] J.P. SERRE : *Algèbre locale-Multiplicités.* Lectures Notes in Math. n°11, Springer (1965)

ON THE REMAINDER TERM OF THE PRIME NUMBER FORMULA AND THE ZEROS OF RIEMANN'S ZETA-FUNCTION

J. Pintz
Mathematical Institute of the
Hungarian Academy of Sciences
Budapest, 1053 Reáltanoda u. 13-15.
HUNGARY

1. The aim of the present paper is to treat further developments compared with the author's earlier contribution [10] concerning irregularities in the distribution of primes. First we shall consider the oscillation of $|\Delta(x)|$ where we define (p always runs through the primes)

$$\Delta(x) = \sum_{n\le x} \Lambda(n)-x = \sum_{p^m\le x} \log p-x . \tag{1.1}$$

Supposing the existence of an arbitrary zeta-zero $\rho_0 = \beta_0 + i\gamma_0$, Littlewood (1937) raised the problem of explicit Ω-estimation of $\Delta(x)$ in terms of ρ_0. At that time the only existing relation, due to Phragmén, was *ineffective*, asserting for any $\varepsilon>0$

$$\Delta(x) = \Omega(x^{\beta_0-\varepsilon}) . \tag{1.2}$$

More generally one may ask for lower bounds for

$$S(x) = \max_{0\le u\le x} |\Delta(u)| \tag{1.3}$$

and

$$D(x) = \frac{1}{x}\int_0^x |\Delta(u)|\,du . \tag{1.4}$$

Littlewood's problem was solved in 1950 by Turȧn [15, part I] who showed

(1.5) $$S(x) \geq x^{\beta_o} \exp(-c_1 \frac{\log x}{\log_2 x} \log_3 x)$$

for $x > \max(c_2, e_2(|\rho_o|)$ where $\log_\nu x$ and $e_\nu(x)$ denote the ν-times iterated logarithmic function and exponential function, resp., and, as in the sequel, the c_ν denote explicitly calculable positive constants (eventually depending on some parameters indicated in brackets). The same inequality was proved by S.Knapowski [5] for $D(x)$ in place of $S(x)$.

The explicit prime number formula,

(1.6) $$\Delta(x) = -\sum_\rho \frac{x^\rho}{\rho} + O(\log x)$$

(where $\rho = \beta + i\gamma$ always runs through the non-trivial zeta-zeros) suggests, however, a larger oscillation for $\Delta(x)$ than that furnished by Turȧn's result (1.5). Using also Turȧn's power-sum theory it is possible to show (slightly improving the result of [10, part I]) the following.

THEOREM 1. *If* $\zeta(\rho_o) = 0$, $\varepsilon > 0$ *and*

(1.7) $$Y > \max(c_3, (\frac{|\rho_o|}{\varepsilon})^8, \exp(\frac{4}{\varepsilon^2 |\rho_o|^2}))$$

then there exists a value

(1.8) $$x \in [Y, Y^{6\log|\gamma_o|+60}]$$

such that

(1.9) $$|\Delta(x)| > (1-\varepsilon)\frac{x^{\beta_o}}{|\rho_o|} .$$

This yields a very weak (but non-trivial) lower bound for $S(x)$. However, using an entirely different method we can prove:

THEOREM 2. *If* ρ_o *is a zeta-zero with multiplicity* ν, $x > e^{\gamma_o^2/20}$, $A(x) = 10^{-4} x/\log x$, *then*

$$(1.10)\qquad D(x) > \frac{1}{x}\int_{A(x)}^{x} |\Delta(u)|du > \frac{c_4|\zeta^{(\nu)}(\rho_0)|}{(\nu-1)!|\rho_0|^3}\, x^{\beta_0} - c_5$$

and a fortiori we have the same inequality for $S(X)$ *and* $\max_{A(x)\le u\le x} |\Delta(u)|$, *too.*

Choosing $\rho_0 = 1/2+i\cdot 14.13\ldots$, the first zero of $\zeta(s)$ over the real axis, with some extra trouble we can show

COROLLARY 1. *For every* $x \geq 1$ *we have*

$$(1.11)\qquad D(x) \geq \frac{\sqrt{x}}{400}\ .$$

We remark that every improvement of (1.11) with a non-constant factor would already disprove the Riemann Hypothesis (RH) since Cramer [2] showed in 1922 that RH implies

$$(1.12)\qquad D(X) \leq \left(\frac{1}{x}\int_{0}^{x}\Delta^2(u)du\right)^{1/2} \leq c_6 x^{1/2} \qquad (x > c_7)$$

(and with some numerical computation one can choose even $c_6 = 1$). Thus we have

COROLLARY 2. *Assume* RH. *Then for* $x > c_7$,

$$(1.13)\qquad \frac{\sqrt{x}}{400} \leq D(x) \leq \sqrt{x}\ .$$

If RH is true then one can easily infer from (1.11) and (1.12) a good lower bound for $|\Delta(x)|$ for a positive proportion of all positive numbers.

COROLLARY 3. *Assume* RH *and let* $|A|$ *denote the measure of the set* A. *Then for* $x > c_7$

$$(1.14)\qquad |\{0 \leq u \leq x;\ |\Delta(u)| > \frac{\sqrt{x}}{800}\}| > \frac{x}{800^2}\ .$$

If RH does not hold but there is a zero $\rho_0 = \Theta+i\gamma_0$ where

$$(1.15)\qquad \Theta = \limsup_{\zeta(\rho)=0} (\mathrm{Re}\ \rho)$$

then we have a phenomenon similar to (1.14) in the stronger form. This is expressed by

COROLLARY 4. *Under the above conditions we have*

$$(1.16)\qquad c_8(\rho_o)x^{\beta_o} \le D(x) < S(x) \le c_9(\rho_o)x^{\beta_o} .$$

Finally we mention another result, seemingly weaker than Theorem 2, which, however, has important applications in the problems discussed in the following section.

THEOREM 3. *If* $\zeta(\rho_o) = 0$, $x > \max(c_{10}, \exp(\sqrt{|\rho_o|})$, $B(x) = x\exp(-40\log_2^2 x)$ *then*

$$(1.17)\qquad D(x) > \frac{1}{x}\int_{B(x)}^{x} |\Delta(u)|du > x^{\beta_o}\exp(-60\log_2^2 x) .$$

2. Our further investigations deal with the assertion of Riemann [11]

$$(2.1)\qquad \Delta_1(x) \overset{\text{def}}{=} \pi(x) - \text{li}\,x \overset{\text{def}}{=} \sum_{p\le x} 1 - \int_0^x \frac{dt}{\log t} < 0 \qquad (x > 2)$$

stated without proof in 1859. Although generally believed to be true for more than 50 years (and checked up to $x = 10^7$) this was disproved by Littlewood [9] in 1914: he showed that $\Delta_1(x)$ infinitely often changes sign. His theorem was completely *ineffective* and it took more than 40 years to give the explicit upper bound $e_4(7.705)$ for the first sign change of $\Delta_1(x)$ (Skewes [12]).

S.Knapowski was the first who succeeded in furnishing a lower estimate for the number $V_1(Y)$ of sign changes of $\Delta_1(x)$ in the interval $[2, Y]$. Applying Turán's one-sided power-sum method he proved in 1961-62 [6, 7]

$$(2.2)\qquad V_1(Y) > c_{11}\log_4 Y \quad \text{for} \quad Y > c_{12}$$

and the weaker *ineffective* inequality

$$(2.3)\qquad V_1(Y) > \log_2 Y \quad \text{for} \quad Y > Y_1$$

where the Y_i denote ineffective absolute constants. These results were improved in 1974-76 by Knapowski and Turán [8]. They showed by Turán's power-sum method that (2.2) and (2.3) remain true with the functions $c_{13}\log_3 Y$ and $c_{14}\log^{1/4} Y/\log_2^4 Y$. The author was able to replace the above functions by $c_{15}\log^{1/2} Y/\log_2 Y$ and $c_{16}\log Y/\log_2^3 Y$ [10, parts III - IV] using also Turán's method.

Making use of (1.17) (better to say, its analogoue where $\Delta(u)$ is replaced by $\Delta_1(u)$) we can now show the *ineffective*

THEOREM 4. $\Delta_1(x)$ *changes sign in the interval*

$$(2.4) \qquad [Y\exp(-500\log_2^3 Y), Y] \quad \text{if} \quad Y > Y_2 .$$

This implies trivially the *ineffective* lower bound $c_{17}\log Y/\log_2^3 Y$ for $V_1(Y)$. But this can also be shown *effectively*.

THEOREM 5. $V_1(Y) > \dfrac{\log Y}{500\log_2^3 Y}$ *for* $Y > c_{18}$.

We remark that Theorems 4 and 5, unlike all the earlier *effective* results of this kind, were proved independently from Turán's method. A suitable *effective* result of type (2.4) needs, however, Turán's method.

THEOREM 6. $\Delta_1(x)$ *changes sign in the interval*

$$(2.5) \qquad [Y^{c_{19}}, Y] \quad \text{if} \quad Y > c_{20} .$$

Finally we remark that in a recent work J.Kaczorowski [4] announced the ineffective inequality $V_1(Y) > c_{21}\log Y$ for $Y > Y_3$. We also remark that Riemann's assertion (2.1) had not only empirical background but was supported also by some theoretical arguments. The assertion

$$(2.6) \qquad \int_1^x \Delta_1(u)du < 0 \quad \text{for} \quad x > x_0$$

is e.g. equivalent with RH. But it is interesting to note that there is a relatively simple averaging procedure such that the statement "$\pi(x)$ - li x is negative on the average" is true without any conditions.

THEOREM 7. $\int_{\cdot}^{\infty} \Delta_1(x)\exp(-(\log^2 x)/y)dx \to -\infty$ as $y \to \infty$.

3. All results of Sections 1 and 2 are based on the investigation of the zeros of $\zeta(s)$ (although in the formulations of Theorems 1-3 only one zero and in Theorems 4-7 no zero appears). The aim of the present section is to examine the connection of the order of magnitude of $|\Delta(x)|$ with the distribution of zeros of $\zeta(s)$.

A general theorem of this type was obtained by Ingham [3, Theorem 22]:

Suppose $(s = \sigma + it)$

(3.1) $$\zeta(s) \neq 0 \quad \text{for} \quad \sigma > 1-\eta(t)$$

where $\eta(t) \in C^1[2,\infty)$, $\eta'(t) \leq 0$, $\lim\limits_{t\to\infty} \eta'(t) = 0$, $\eta(t) \gg \log^{-1} t$.

Let $0 < \varepsilon < 1$ be fixed and

(3.2) $$\omega_{(\eta)}(x) = \inf_{t\geq 1} (\eta(t)\log x + \log t) \quad .$$

Then

(3.3) $$\Delta(x) \ll x\exp(-\frac{1}{2}(1-\varepsilon)\,\omega_{(\eta)}(x)) \quad .$$

This implies e.g. that in case of

(3.4) $$\zeta(s) \neq 0 \quad \text{for} \quad \sigma > 1 - \frac{c_{22}}{\log^{\alpha} t}, \quad t > t_0$$

one has

(3.5) $$\Delta(x) \ll x\exp(-c_{23}(\alpha)\log^{1/(1+\alpha)} x) \quad .$$

Turán [15, part II] was the first to show that the inverse implication (3.5) ⇒ (3.4) is also true (with a $c'_{22} < c_{22}/40$, however). His result was later extended to more general domains by W.Staś [14]. The author [10, part II] succeeded in showing that the factor 1/2 can be deleted in (3.3) and that the slightly stronger assumption

(3.5) $$\Delta(x) \ll x\exp(-(1+\varepsilon)\,\omega_{(\eta)}(x))$$

already implies (3.1) (for $t > t_0$) if $\eta(t) = g(\log t)$, where

(3.7) $$g(u) \in C^1(1,\infty), \quad g'(u) \nearrow 0 \quad \text{as} \quad u \to \infty \quad .$$

The above results suggest that perhaps there is a real function $\omega(x)$ depending in a simple way on the distribution of zeta-zeros (without using a hypothetical zero-free region in the formulation of the results) which describes the largest possible order of magnitude of $|\Delta(x)|$. In the favourite case we may hope that such a function $\omega(x)$ determines the functions $S(x)$ or $D(x)$ (see (1.3) - (1.4)) with considerable accuracy. It turns out that this is really possible by choosing

$$(3.8) \qquad \omega(x) \overset{\text{def}}{=} \min_{\rho} ((1-\beta)\log x+\log|\gamma|) = \log\frac{x}{Z(x)}$$

where

$$(3.9) \qquad Z(x) = \max_{\rho} \frac{x^{\beta}}{|\gamma|}$$

is, up to an insignificant factor $|\rho|/|\gamma|$, the modulus of the larges error term in the explicit formula (1.6).

TEHOREM 8. *Using the notations* (1.3) - (1.4), (3.8) *we have*

$$(3.10) \qquad \log\frac{x}{S(x)} \sim \log\frac{x}{D(x)} \sim \omega(x) \quad as \quad x \to \infty .$$

Theorem 8 includes

THEOREM 9. $\Delta(x) \ll x\exp(-(1-\varepsilon)\,\omega(x))$

and

THEOREM 10. $S(x) > D(x) \gg x\exp(-(1+\varepsilon)\,\omega(x))$

Consequently $\Delta(x) = \Omega(x\exp(-(1+\varepsilon)\,\omega(x)))$.

Taking into account that in case of (3.1) we have trivially $\omega(x) \geq \omega_{(\eta)}(x)$ (cf. (3.2)), Theorem 9 implies

COROLLARY 5. *If* $\eta(t)$ *is an arbitrary real function and* (3.1) *is true then*

$$(3.11) \qquad \Delta(x) \ll x\exp(-(1-\varepsilon)\,\omega_{(\eta)}(x)) .$$

We remark that Theorem 8 also implies

COROLLARY 6. *Using the notation* (1.15) *we have*

$$(3.12) \qquad \log S(x) \sim \log D(x) \sim \log Z(x) \sim \Theta\log x .$$

(3.12) is a sharpening of the well-known relation

$$(3.13) \qquad \Theta = \inf\{\vartheta \,;\, \Delta(x) = O(x^{\vartheta})\} .$$

Although (3.12) is equivalent with Theorem 8 if $\Theta < 1$, it is much weaker in the sense that the crucial case in proving Theorem 8 is just

$\Theta = 1$. Corollary 6 already follows from (1.5) and from the corresponding result for $D(x)$. In case of $\Theta = 1$ Corollary 6 yields only $\log(x/S(x)) = o(\log x)$ whilst Theorem 8 gives an asymptotic relation for this quantity, the function $\omega(x)$. Theorem 8 can be considered also as a far-reaching extension of Wiener's result [16] *interpreted in this context* as

$$(3.14)\qquad S(x) = o(x) \Leftrightarrow \lim_{x\to\infty} \omega(x) = \infty$$

(although the main point there was the method used).

Furthermore we remark that the fact that $\omega(x)$ *itself* describes the asymptotic behaviour of *both* functions $S(x)$ and $D(x)$ implies that the maximal value $S(x)$ cannot be much larger than the mean value $D(x)$, a phenomenon not discovered before. Namely, Theorem 8 yields

THEOREM 11. $S(x) \ll D(x)(x/D(x))^{\varepsilon}$.

The interesting feature of the above result is that, unlike in Theorems 8-10, no zeta-zeros occur in the formulation (however, a direct proof seems to be hopeless).

4. In what follows we shall sketch the proof of Theorem 8. (The details of proof, as well as the proofs of the other theorems will appear in a series of papers entitled "Irregularities of prime distributions".) According to the remark following Corollary 6 we shall restrict ourselves to the case $\Theta = 1$.

Concerning the upper estimate of $\Delta(x)$ we obtain by Carlson's density theorem [1]

$$(4.1)\qquad N(1-\varepsilon, T) = \sum_{\substack{\beta\geq 1-\varepsilon\\ |\gamma|\leq T}} 1 \ll_{\varepsilon} T^{4\varepsilon}$$

and so we have for every natural number n

$$(4.2)\qquad \sum_{\substack{\beta>1-\varepsilon\\ e^n<|\gamma|\leq e^{n+1}}} \frac{x^{\beta}}{|\gamma|} \ll_{\varepsilon} e^{4n\varepsilon} \max_{e^n<|\gamma|\leq e^{n+1}} \frac{x}{e^{((1-\beta)\log x+\log|\gamma|)}}$$

$$\leq e^{-n\varepsilon} \max_{e^n<|\gamma|\leq e^{n+1}} \frac{x}{e^{((1-\beta)\log x+\log|\gamma|)(1-5\varepsilon)}} .$$

This implies by $\omega(x) = o(\log x)$ $(\Leftrightarrow \Theta = 1)$

(4.3) $$S(x) \ll_{\varepsilon} x^{1-\varepsilon/2} + \sum_{n=1}^{\infty} e^{-n\varepsilon} \frac{x}{e^{(1-5\varepsilon)\omega(x)}} \ll_{\varepsilon} \frac{x}{e^{(1-5\varepsilon)\omega(x)}} .$$

The lower estimate in Theorem 8 is the consequence of the following Lemma, which can be proved by Turán's power-sum method.

LEMMA. *Let* $0 < \varepsilon < c_{24}$, $\zeta(\beta_o + i\gamma_o) = \zeta(1-\delta_o+i\gamma_o) = 0$, $\gamma_o > 0$ $\delta_o < \varepsilon^{10}$. *Then for* $x > \gamma_o^{1/\varepsilon^{10}}$ *we have*

(4.4) $$\bar{D}(x) = \frac{1}{x} \int_{x^{1-\delta_o}\gamma_o^{-\varepsilon}}^{x} |\Delta(u)|du \geq \frac{1}{(x^{\delta_o}\gamma_o)^{\varepsilon}} \cdot \frac{x^{\beta_o}}{\gamma_o} .$$

It is really easy to check that if $\Theta = 1 \Leftrightarrow \omega(x) = o(\log x)$ and $\omega(x) = \delta_o \log x + \log \gamma_o$ then the conditions of our Lemma are satisfie and so we obtain

(4.5) $$D(x) \geq \frac{1}{(x^{\delta_o}\gamma_o)^{\varepsilon}} \cdot \frac{x}{x^{\delta_o}\gamma_o} = \frac{x}{e^{\omega(x)(1+\varepsilon)}} .$$

In order to sketch the proof of the Lemma we introduce the notations $(\varepsilon_1 = \varepsilon/24)$

(4.6) $$\varphi = \delta_o \log x + \log \gamma_o = \alpha \log x , \quad k = 5\varepsilon_1^2 \alpha\mu ,$$

where the real number μ will be chosen later so as to satisfy

(4.7) $$\mu \in [\log x - 6\varepsilon_1\varphi , \log x - 5\varepsilon_1\varphi] .$$

Using the well-known relations

(4.8) $$\int_1^{\infty} \Delta(u) \frac{d}{du}(u^{-s})du = \frac{\zeta'}{\zeta}(s) + \frac{s}{s-1} \overset{\text{def}}{=} H(s) \qquad (\sigma > 1)$$

(4.9) $$(2\pi i)^{-1} \int_{(2)} e^{As^2+Bs} ds = (4\pi A)^{-1/2} \exp(-B^2/4A)$$

we obtain our basic identity

$$(4.10)\qquad U(\mu) \overset{\text{def}}{=} (2\pi i)^{-1} \int_{(2)} H(s+\rho_o)\, e^{ks^2+\mu s}\, ds$$

$$= (4\pi k)^{-1/2} \int_1^\infty \frac{\Delta(u)}{u^{1+\rho_o}} \exp\left(-\frac{(\mu-\log u)^2}{4k}\right)\left(-\rho_o+\frac{\mu-\log u}{2k}\right) du \quad .$$

Since the weight function gives small weights if we are far from e^μ, it is relatively easy to show that the simple estimate $\Delta(u) \ll u$ implies

$$(4.11)\qquad U(\mu) = \int_{e^{\mu-5\varepsilon_1\varphi}}^{e^{\mu+5\varepsilon_1\varphi}} + O(e^{-\varphi/5}) \ .$$

Further we obtain from (4.7) and from (4.10)

$$(4.12)\qquad \left| \int_{e^{\mu-5\varepsilon_1\varphi}}^{e^{\mu+5\varepsilon_1\varphi}} \right| \le \frac{\gamma_o}{\varepsilon_1}\left(\frac{e^{11\varepsilon_1\varphi}}{x}\right)^{1+\beta_o} \int_{xe^{-11\varepsilon_1\varphi}}^{x} |\Delta(u)|\, du$$

$$\le \frac{\gamma_o}{x^{\beta_o}}\, \varepsilon_1^{-1} e^{22\varepsilon_1\varphi} \cdot \bar{D}(x) \ .$$

In order to obtain a lower estimation for $|U(\mu)|$ by a suitable choice of μ we shift the line of integration in (4.10) to $\sigma = -1-\beta_o$, thereby obtaining

$$(4.13)\qquad U(\mu) = \sum_\rho e^{k(\rho-\rho_o)^2+\mu(\rho-\rho_o)} + O(x^{-1}) \ .$$

Further we can trivially estimate the total contribution of all zeros which do not belong to the set

$$(4.14)\qquad M = \{\rho;\ |\gamma-\gamma_o| \le \varepsilon_1^{-1}\ ,\quad \beta \ge \beta_o-\alpha\} \ .$$

Thus we get (cf.(4.6))

$$(4.15) \qquad E(\mu) \overset{\text{def}}{=} \sum_{\rho\in M} e^{\{5\varepsilon_1^2\alpha(\rho-\rho_0)^2+\rho-\rho_0\}\mu} = U(\mu)+O(e^{-\varphi/2}) \quad .$$

Now $|E(\mu)|$ can be estimated from below (under a suitable choice of μ) by the continuous form of the second main theorem of Turán's power-sum theory [13], which asserts for arbitrary complex numbers $a_1, \dots a_n$ with $\mathrm{Re}\,a_1=0$ and for any real positive b, d the inequality

$$(4.16) \qquad \max_{b\le v\le b+d} \Big| \sum_{i=1}^{n} e^{a_i v} \Big| \ge \Big(\frac{d}{8e(b+d)}\Big)^n \quad .$$

(For this form see e.g. [10, part I], Theorem A of the Appendix). In any application of this theorem a crucial role is played by the value of n . In our case the estimate of Korobov-Vinogradov

$$(4.17) \qquad \zeta(1-h+it) \ll t^{c_{25}h^{4/3}} \log^{c_{26}} t \qquad (t > 2)$$

and Jensen's inequality leads to

$$(4.18) \qquad 1 \le n = |M| \le \frac{c_{27}}{\varepsilon_1\alpha}(\alpha^{4/3}\log\gamma_0 + \log_2\gamma_0) \quad .$$

From (4.18) we can infer $n \le c_{28}\varepsilon_1^{-1}\alpha^{1/3}\varphi$ and applying (4.16) we obtain a value μ satisfying (4.7) such that

$$(4.19) \qquad |E(\mu)| \gtrsim \exp\Big(-c_{29}\log\frac{1}{\varepsilon_1\alpha}\cdot\varepsilon_1^{-1}\alpha^{1/3}\varphi\Big) \ge e^{-\varepsilon_1\varphi} \quad .$$

Consequently, by (4.15) we have

$$(4.20) \qquad |U(\mu)| \ge \frac{1}{2}e^{-\varepsilon_1\varphi} \quad .$$

Combining this with (4.11) - (4.12) we obtain the assertion of the Lemma.

REFERENCES

[1] Carlson, Über die Nullstellen der Dirichletschen Reihen und der Riemannschen ζ-Funktion, Arkiv för Mat. Astr. Och. Fysik 15(1920), No.20

[2] H.Cramér, Ein Mittelwertsatz in der Primzahltheorie, Math. Z. 12(1922), 147-153.

[3] A.E.Ingham, The distribution of prime numbers, University Press, Cambridge, 1932.

[4] J.Kaczorowski, On sign-changes in the remainder-term of the prime-number formula, I, Acta Arith., to appear

[5] S.Knapowski, On the mean values of certain functions in prime-number theory, Acta Math. Acad. Sci. Hungar. 10(1959), 375-390.

[6] S.Knapowski, On the sign changes in the remainder term in the prime number formula, Journ. Lond. Math. Soc. 36(1961), 451-460.

[7] S.Knapowski, On the sign changes of the difference $(\pi(x)-\text{li}x)$, Acta Arith. 7(1962), 107-120.

[8] S.Knapowski and P.Turán, On the sign changes of $(\pi(x)-\text{li}x)$ I., II., Topics in Number Theory, Coll. Math. Soc. János Bolyai 13., North-Holland P. C., Amsterdam-Oxford-New York, 1976, pp.153-169 and Monatshefte für Math. 82(1976), 163-175.

[9] J.E.Littlewood, Sur la distribution des nombres premiers, C.R. Acad. Sci. Paris 158(1914), 1869-1872.

[10] J.Pintz, On the remainder term of the prime number formula, I-VI, Acta Arith. 36(1979), 27-51 and 37(1980), 209-220, Studia Sci. Math. Hungar. 12(1977), 345-369, 13(1978), 29-42, 15(1980), 215-223 and 15(1980), 225-230.

[11] B.Riemann, Über die Anzahl der Primzahlen unter einer gegebenen Grösse, Monatsh. Preuss. Akad. Wiss., Berlin, 1859, pp.671-680.

[12] S.Skewes, On the difference $\pi(x)-\text{li}x$, II, Proc. Lond. Math. Soc. 5(1955), 48-70.

[13] Vera T.Sós and P.Turán, On some new theorems in the theory of diophantine approximation, Acta Math. Acad. Sci. Hungar. 6(1955), 241-255.

[14] W.Staś, Über die Umkehrung eines Satzes von Ingham, Acta Arith. 6(1961), 435-446.

[15] P.Turán, On the remainder-term of the prime-number formula, I-II, Acta Math. Acad. Sci. Hungar. 1(1950), 48-63 and 155-166.

[16] N.Wiener, A new method in Tauberian theorems, Journal of Math. and Phys., MIT, 7(1927-28), 161-184.

STRUCTURES GALOISIENNES

par

Jacques QUEYRUT

-:-:-:-

Introduction

1) Soit N une extension finie du corps des rationnels $\mathbb{Q}$, incluse dans $\mathbb{C}$. Soit Γ un groupe de $\mathbb{Q}$-automorphismes de N ; N est alors une extension galoisienne de $K = N^{\Gamma}$, de groupe de Galois Γ.

On note $\mathbb{Z}_N$ l'anneau des entiers de N, Cl(N) le groupe des classes d'idéaux de N, S_∞ l'ensemble des places infinies de N.

Soit S un ensemble de places de N stable par Γ et contenant S_∞. On note $Cl_S(N)$ le groupe des S-classes de N ; $Cl(N) = Cl_{S_\infty}(N)$ ($Cl_S(N)$ est le groupe des classes d'idéaux de l'anneau de Dedekind obtenu à partir de l'anneau $\mathbb{Z}_N$ en rendant inversibles tous les idéaux premiers appartenant à S).

On note U_S le groupe des S-unités, i.e. l'ensemble des éléments qui sont des unités à toutes les places de N qui ne sont pas dans S ; en particulier U_{S_∞} est le groupe des unités de l'anneau $\mathbb{Z}_N$.

Le groupe Γ opère sur $\mathbb{Z}_N$, Cl(N) et U_S. On appelle structure galoisienne de ces ensembles, leur structure en tant que $\mathbb{Z}[\Gamma]$-module.

2) Notons $\overline{\mathbb{Q}}$ la clôture algébrique de $\mathbb{Q}$ dans $\mathbb{C}$. Un corps de nombres est un sous-corps de $\overline{\mathbb{Q}}$ de degré fini sur $\mathbb{Q}$. Soit F un corps de nombres ; notons G_F le groupe de Galois de F sur $\overline{\mathbb{Q}}$. Ainsi avec cette notation, Γ est isomorphe à G_K/G_N. On note $R(G_F)$ le groupe de Grothendieck de la catégorie des G_F-modules topologiques de dimension finie sur $\mathbb{C}$.

Pour chaque place non archimédienne $\mathfrak{p}$ de F, on choisit une place $\mathfrak{P}$ de $\overline{\mathbb{Q}}$ au-dessus de $\mathfrak{p}$ et on note $I_{\mathfrak{P}}$ le groupe d'inertie de $\mathfrak{P}$ et $\sigma_{\mathfrak{P}}$ le Frobenius géométrique (voir [D 2], § 2-2-2). En suivant P. Deligne, on définit la fonction L d'Artin comme étant la fonction méromorphe sur $\mathbb{C}$ définie par

$$L_F(s, [V]) = \prod_{\mathfrak{p}} \det (1-\sigma_{\mathfrak{P}} N(\mathfrak{p})^{-s}, V^{I_{\mathfrak{P}}})^{-1}$$

pour toute classe $[V]$ de G_F-module topologique V de dimension finie sur $\mathbb{C}$, où $N(\mathfrak{p})$ est la norme de F sur $\mathbb{Q}$ de l'idéal $\mathfrak{p}$ et $V^{I_{\mathfrak{P}}}$ est le sous-espace de V stable par $I_{\mathfrak{P}}$. Cette fonction est prolongée à $R(G_F)$ par linéarité (attention, la fonction $L_F(s,\chi)$ définie ici est égale à la fonction L définie dans [Ta 2] et [M] prise en s et $\overline{\chi}$).

Si S est un ensemble fini de places de F, on pose

$$L_{F,S}(s,[V]) = \prod_{\mathfrak{p}\notin S} \det(1-\sigma_{\mathfrak{P}} N(\mathfrak{p})^{-s}, V^{I_{\mathfrak{P}}})^{-1} .$$

Pour tout $\chi \in R(G_F)$; la fonction L_F satisfait une équation fonctionnelle qui relie la valeur de $L_F(s,\chi)$ et la valeur de $L_F(1-s, \overline{\chi})$ (voir [M]).

Soit $C_{F,0}(\chi)$ (resp. $C_{F,1}(\chi)$ le premier coefficient non nul du développement de Laurent de $L_F(s,\chi)$ en $s=0$ (resp. $s=1$) et soit $r_{F,0}(\chi)$ (resp. $r_{F,1}(\chi)$) la multiplicité correspondante. On indexera par S les fonctions correspondant à $L_{F,S}(s,\chi)$.

On appelle somme de Gauss galoisienne le nombre complexe $\tau_F(\chi)$ défini par la relation suivante

$$(*) \qquad \tau_F(\chi)\, i^{r_{F,2}\dim\chi}\, d_F^{\dim\chi/2} = (-1)^{r_{F,1}(\chi)}\, 2^{b_{F,1}(\chi)}\, (2i\pi)^{b_{F,2}(\chi)}\, \frac{C_{F,0}(\chi)}{C_{F,1}(\overline{\chi})}$$

où d_F est la valeur absolue du discriminant de F sur $\mathbb{Q}$ et où $b_{F,1}$ et $b_{F,2}$ dépendent des places archimédiennes de F et sont définis de la façon suivante :

pour chaque place archimédienne v de F, on choisit une place w de $\overline{\mathbb{Q}}$ au-dessus de v et on note $G_{F,w}$ le groupe de décomposition de w ; soit $r_{F,2}$ le nombre de places complexes de F et posons pour tout G_F-module topologique V de dimension finie sur $\mathbb{C}$ de classe χ dans $R(G_F)$:

$$a_{F,1}(\chi) = \sum_{w \text{ réelle}} \dim V^{G_{F,w}} \qquad\qquad a_{F,2}(\chi) = \sum_{w \text{ réelle}} \operatorname{codim} V^{G_{F,w}}$$

on a alors

$$b_{F,2}(\chi) = a_{F,2}(\chi) + r_{F,2} \dim\chi$$

$$b_{F,1}(\chi) = a_{F,1}(\chi) - a_{F,2}(\chi)$$

([Ta 2], § 6, chapitre 0).

Enfin on pose $f_F(\chi) = \tau_F(\chi)\overline{\tau_F(\chi)}$ ($f_F(\chi)$ est la valeur absolue de la norme de F sur $\mathbb{Q}$ du conducteur d'Artin de χ).

Ainsi $W_F(\chi) = \dfrac{\tau_F(\chi)}{\sqrt{f_F(\chi)}\ i^{a_{F,2}(\chi)}}$ est un nombre complexe de module 1 .

Toutes les fonctions introduites dans ce paragraphe sont des homomorphismes sur le groupe $R(G_F)$.

3) De nombreux résultats obtenus ces dernières années, montrent qu'il existe une connexion étroite entre ces fonctions arithmétiques et les structures galoisiennes des modules décrits au paragraphe 1 :

* 1 → 2 - Les premiers résultats généraux dans cette direction sont dus à A. Fröhlich [F], puis à M. J. Taylor [Ty] qui ont montrés que les fonctions τ et W déterminent la structure galoisienne des anneaux d'entiers dans le cas où Γ ne contient pas d'automorphisme sauvagement ramifié. Ce résultat a été généralisé dans [Q 3] sous une forme légèrement différente sans l'hypothèse de ramification.

Stark [S] et Tate [Ta 2] conjecturent et montrent dans certains cas que la fonction C_0^S détermine la structure galoisienne de U_S. T. Chinburg a précisé et développé ces résultats et conjectures dans [C].

* 2 → 1 - Dans l'autre direction, on sait depuis les résultats de Langlands et Deligne (voir [D 2]) que W_F se décompose en produit de termes locaux $W_{F,\mathfrak{p}}$ sur l'ensemble des places $\mathfrak{p}$ de F . Il en est donc de même pour $\tau_{F,\mathfrak{p}}$. A. Fröhlich et M. J. Taylor [F - T] ont montré que la structure galoisienne des anneaux d'entie et certaines propriétés de congruences déterminent les fonctions $\tau_{F,\mathfrak{p}}$ dans le cas modéré par un procédé purement local. Ph. Cassou-Noguès et M. J. Taylor [C-N - T] ont montré que la structure galoisienne des anneaux d'entiers et le fait que la forme trace définisse une forme bilinéaire invariante par Γ déterminent dans le cas modéré $W_{F,\mathfrak{p}}(\chi)$ pour χ caractère symplectique. Enfin, P. Deligne [D1] a montré comment calculer les constantes $W_{F,\mathfrak{p}}(\chi)$ pour des caractères χ orthogonaux en utilisant la théorie du corps de classes et la classe de Stiefel-Withne de χ .

Le but de cet exposé est de donner un cadre algébrique dans lequel ces résultats puissent s'exprimer simplement et concrètement.

I. - Structures en tant que $\mathbb{Q}[\Gamma]$-modules

a) (comme additif) - Soit X le $\mathbb{Z}$-module libre de base les éléments de $Is_{\mathbb{Q}}(N, \mathbb{C})$, l'ensemble des $\mathbb{Q}$-isomorphismes de N dans $\mathbb{C}$.

En faisant opérer Γ sur $Is_{\mathbb{Q}}(N, \mathbb{C})$ par $(\gamma, v) \to v\gamma^{-1}$, on munit X d'une structure de $\mathbb{Z}[\Gamma]$-module de rang r, le degré de N^{Γ} sur $\mathbb{Q}$.

Soit $T_{N/\mathbb{Q}}$

$$\mathbb{C} \otimes_{\mathbb{Q}} N \longrightarrow \mathbb{C} \otimes_{\mathbb{Z}} X$$
$$\lambda \otimes z \longmapsto \lambda \sum_{v \in Is_{\mathbb{Q}}(N, \mathbb{C})} v(z) \otimes v .$$

L'application $T_{N/\mathbb{Q}}$ est un $\mathbb{C}[\Gamma]$-isomorphisme.

Les $\mathbb{Q}[\Gamma]$-modules N et $\mathbb{Q} \otimes_{\mathbb{Z}} X$ deviennent canoniquement isomorphes (en tant que $\mathbb{C}[\Gamma]$-modules) si on étend les scalaires de $\mathbb{Q}$ à $\mathbb{C}$.

m) (comme multiplicatif) - Soit Y_S le $\mathbb{Z}$-module libre de base les éléments de S et soit X_S le noyau de l'application $Y_S \longrightarrow \mathbb{Z}$ qui à $\sum_v n_v v$ associe $\sum_v n_v$. Comme précédemment Y_S et par suite X_S ont une structure de $\mathbb{Z}[\Gamma]$-module.

Soit $V_S = \mathbb{Q} \otimes_{\mathbb{Z}} U_S$; le noyau de l'homomorphisme canonique de U_S dans V_S, qui à u associe $1 \otimes u$, est le groupe des racines de l'unité contenues dans N.

Soit $Log_{S\, N/\mathbb{Q}}$:

$$\mathbb{C} \otimes_{\mathbb{Q}} V_S \longrightarrow \mathbb{C} \otimes_{\mathbb{Z}} X_S$$
$$\lambda \otimes z \longmapsto \lambda \sum_{v \in S} \log \|a\|_v \otimes v .$$

L'application $Log_{S, N/\mathbb{Q}}$ est un $\mathbb{C}[\Gamma]$-isomorphisme.

Les $\mathbb{Q}[\Gamma]$-modules V_S et $\mathbb{Q} \otimes_{\mathbb{Z}} X_S$ deviennent canoniquement isomorphes si on étend les scalaires de $\mathbb{Q}$ à $\mathbb{C}$.

On est donc amener à classifier les objets de la forme (V, α, W) où V et W sont des $\mathbb{Q}[\Gamma]$-modules et α est un $\mathbb{C}[\Gamma]$-isomorphisme de $\mathbb{C} \otimes_{\mathbb{Q}} V$ sur $\mathbb{C} \otimes_{\mathbb{Q}} W$. Cela se fait au moyen d'un groupe de Grothendieck, noté $\mathcal{K}_{o, \mathbb{C}/\mathbb{Q}}(\mathbb{Q}[\Gamma])$, (voir [Q 2], § 2) de la façon suivante :

soit $\mathcal{C}_{\mathbb{C}/\mathbb{Q}}(\mathbb{Q}[\Gamma])$ la catégorie suivante :

les objets sont les triplets (V, α, W) vérifiant les conditions précédentes,

un morphisme de (V, α, W) dans (V', α', W') est une paire (f, g) de $\mathbb{Q}[\Gamma]$-isomorphismes de V dans V' et de W dans W' respectivement telle que $\alpha' \circ f_{\mathbb{C}} = g_{\mathbb{C}} \circ \alpha$ où $f_{\mathbb{C}}$ et $g_{\mathbb{C}}$ proviennent de f et g par extensions des scalaires de $\mathbb{Q}$ à $\mathbb{C}$.

Le groupe $\mathcal{K}_{o,\mathbb{C}/\mathbb{Q}}(\mathbb{Q}[\Gamma])$ est le quotient du groupe abélien libre engendré par les classes d'isomorphismes, encore notées (V, α, W) des objets (V, α, W) de $\mathcal{C}_{\mathbb{C}/\mathbb{Q}}(\mathbb{Q}[\Gamma])$, par le sous-groupe engendré par les éléments de la forme

$$(V\oplus V', \alpha\oplus\alpha', W\oplus W') - (V, \alpha, W) - (V', \alpha', W')$$

$$(V, \beta\circ\alpha, W) - (V, \alpha, U) - (U, \beta, W).$$

On note $[V, \alpha, W]$ la classe de (U, α, W) dans $\mathcal{K}_{o,\mathbb{C}/\mathbb{Q}}(\mathbb{Q}[\Gamma])$.

On utilise la description de $\mathcal{K}_1(\mathbb{C}[\Gamma])$ qui le présente comme le groupe classifiant les objets de la forme (V,α) où V est un $\mathbb{C}[\Gamma]$-module de type fini et α est un $\mathbb{C}[\Gamma]$-automorphisme de V.

Soit $[V, \alpha]\in\mathcal{K}_1(\mathbb{C}[\Gamma])$; il existe un $\mathbb{C}[\Gamma]$-module W et un $\mathbb{Q}[\Gamma]$-module U tels que $V\oplus W \simeq \mathbb{C}\otimes_{\mathbb{Q}} U$. L'application qui à $[V, \alpha]$ associe $[U, \alpha\oplus 1_W, U]$ de $\mathcal{K}_{o,\mathbb{C}/\mathbb{Q}}(\mathbb{Q}[\Gamma])$ ne dépend pas du choix de W et de U (cf. [H]). Elle définit un homomorphisme, noté $\delta_{\mathbb{C}/\mathbb{Q}}$, rendant la suite suivante exacte :

$$0 \longrightarrow \mathcal{K}_1(\mathbb{Q}[\Gamma]) \longrightarrow \mathcal{K}_1(\mathbb{C}[\Gamma]) \xrightarrow{\delta_{\mathbb{C}/\mathbb{Q}}} \mathcal{K}_{o,\mathbb{C}/\mathbb{Q}}(\mathbb{Q}[\Gamma]) \longrightarrow 0 .$$

Il existe une dualité entre $\mathcal{K}_1(\mathbb{C}[\Gamma])$ et le groupe des caractères de Γ, identifié à $\mathcal{K}_0(\mathbb{C}[\Gamma])$ ([Q1], § 2). On obtient ainsi un isomorphisme, noté Det, de $\mathcal{K}_1(\mathbb{C}[\Gamma])$ dans $\mathrm{Hom}(R(\Gamma), \mathbb{C}^*)$.

Plus précisément, soit $[V, \alpha]\in\mathcal{K}_1(\mathbb{C}[\Gamma])$; $\mathrm{Det}([V,\alpha])$ est l'application qui à tout caractère χ de Γ d'un $\mathbb{C}[\Gamma]$-module W associe le déterminant sur $\mathrm{Hom}_{\mathbb{C}[\Gamma]}(W, V)$ du $\mathbb{C}$-isomorphisme $f \longmapsto \alpha\circ f$.

a) On note $\tau_{\mathbb{C}[\Gamma]}$ (resp. $W_{\mathbb{C}[\Gamma]}$) l'élément de $\mathcal{K}_1(\mathbb{C}[\Gamma])$ dont l'image par Det est l'homomorphisme de $R(\Gamma)$ dans $\mathbb{C}^*$ donné par le composé des homomorphismes suivants :

$$R(\Gamma) \xhookrightarrow{\mathrm{Inf}} R(G_K) \xrightarrow{\mathrm{Ind}_{G_K}^{G_{\mathbb{Q}}}} R(G_{\mathbb{Q}}) \xrightarrow{\tau_{\mathbb{Q}}\ (\text{resp. } W_{\mathbb{Q}})} \mathbb{C}^* .$$

On a $\mathrm{Det}(\tau_{\mathbb{C}[\Gamma]})(\chi) = i^{r_{K,2}\dim\chi}\, d_K^{\dim\chi/2}\, \tau_K(\chi)$, $\forall\chi\in R(\Gamma)$ (cf. [M], théorème 8.1. iii)).

m) On note $C^S_{0,\mathbb{C}[\Gamma]}$ l'élément de $\mathcal{K}_1(\mathbb{C}[\Gamma])$ dont l'image par Det est l'homomorphisme de $R(\Gamma)$ dans $\mathbb{C}^*$ donné par le composé des homomorphismes suivant

$$R(\Gamma) \xhookrightarrow{\mathrm{Inf}} R(G_K) \xrightarrow{C^s_{K,0}} \mathbb{C}^*$$

où s désigne ici l'ensemble des places de K au-dessous des places de S dans N

PROPOSITION I ([Q4], corollaire 2.2). - *Si* $W_{\mathbb{C}[\Gamma]}(\chi)=1$, *pour tout caractère symplectique* χ, *on a*

$$[N,\ T_{N/\mathbb{Q}},\ \mathbb{Q}\otimes_{\mathbb{Z}} X] = \delta_{\mathbb{C}/\mathbb{Q}}(\tau_{\mathbb{C}[\Gamma]}) .$$

CONJECTURE I' (Stark, voir [C]). - *Sous les hypothèses de la proposition I, on a*

$$[V_S,\ \mathrm{Log}_{S,N/\mathbb{Q}},\ \mathbb{Q}\otimes_{\mathbb{Z}} X_S] = \delta_{\mathbb{C}/\mathbb{Q}}(C^{s}_{0,\mathbb{C}[\Gamma]}) .$$

II. - *Structure en tant que* $\mathbb{Z}[\Gamma]$-*modules*

On est maintenant amené à classifier les objets du types (M, α, M') où M et M' sont des $\mathbb{Z}[\Gamma]$-modules de type fini et α un $\mathbb{C}[\Gamma]$-isomorphisme de $\mathbb{C}\otimes_{\mathbb{Z}} M$ sur $\mathbb{C}\otimes_{\mathbb{Z}} M'$.

Comme précédemment cela se fait au moyen d'un groupe de Grothendieck relatif, noté $\mathcal{G}^{\mathbb{C}}_{0,\,rel}(\mathbb{Z}[\Gamma])$, de la façon suivante ([Q2], § 2) :

Soit $\mathcal{C}_{\mathbb{C}/\mathbb{Z}}(\mathbb{Z}[\Gamma])$ la catégorie suivante :

les objets sont les triplets (M, α, M') vérifiant les conditions précédentes

un morphisme de (M, α, M') dans (P, α, P') est un couple (f, g) de $\mathbb{Z}[\Gamma]$-homomorphismes de M dans P et de N dans N' respectivement et tels que $\alpha'\circ f_{\mathbb{C}} = g_{\mathbb{C}}\circ\alpha$ où $f_{\mathbb{C}}$ et $g_{\mathbb{C}}$ proviennent de f et g par extensions des scalaires de $\mathbb{Z}$ à $\mathbb{C}$.

Une suite exacte (courte) d'objets et de morphismes de cette catégorie est donc donnée par un diagramme commutatif :

$$\begin{array}{ccccccccc}
0 & \longrightarrow & \mathbb{C}\otimes_{\mathbb{Z}} M & \xrightarrow{f'_{\mathbb{C}}} & \mathbb{C}\otimes_{\mathbb{Z}} P & \xrightarrow{f''_{\mathbb{C}}} & \mathbb{C}\otimes_{\mathbb{Z}} Q & \longrightarrow & 0 \\
 & & \downarrow{\scriptstyle\alpha} & & \downarrow{\scriptstyle\beta} & & \downarrow{\scriptstyle\gamma} & & \\
0 & \longrightarrow & \mathbb{C}\otimes_{\mathbb{Z}} M' & \xrightarrow{g'_{\mathbb{C}}} & \mathbb{C}\otimes_{\mathbb{Z}} P' & \xrightarrow{g''_{\mathbb{C}}} & \mathbb{C}\otimes_{\mathbb{Z}} Q' & \longrightarrow & 0 .
\end{array}$$

Le groupe $\mathcal{G}^{\mathbb{C}}_{0,\,rel}(\mathbb{Z}[\Gamma])$ est le quotient du groupe abélien libre engendré par les classes d'isomorphismes, encore notées (M, α, M') des objets (M, α, M') de $\mathcal{C}_{\mathbb{C}/\mathbb{Z}}(\mathbb{Z}[\Gamma])$, par le sous-groupe engendré par les éléments de la forme $(M, \beta\alpha, M') - (M, \alpha, M'') - (M'', \beta, M')$ et les éléments de la forme $(M, \alpha, M') - (P, \beta, P') - (Q, \gamma, Q')$ où ces trois objets sont liés par une suite exacte.

On obtient alors un diagramme

$$\begin{array}{ccccccc}
\mathcal{K}_1(\mathbb{C}[\Gamma]) & \xrightarrow{\ \delta\ } & \mathcal{G}^{\mathbb{C}}_{o,\,rel}(\mathbb{Z}[\Gamma]) & \longrightarrow & \mathcal{G}_o(\mathbb{Z}[\Gamma]) & \longrightarrow & R(\Gamma) \\
\wr\downarrow \text{Det} & \searrow^{\delta_{\mathbb{C}/\mathbb{Q}}} & \downarrow \text{Ext}^{\mathbb{Q}}_{\mathbb{Z}} & & & & \\
\text{Hom}(R(\Gamma)\,,\mathbb{C}^*) & & \mathcal{K}_{o,\,\mathbb{C}/\mathbb{Q}}(\mathbb{Q}[\Gamma]) & & & &
\end{array} \quad .$$

<u>Exemples</u>

1) Supposons que Γ est trivial, la suite exacte précédente devient

$$1 \longrightarrow \{\pm 1\} \longrightarrow \mathbb{C}^* \longrightarrow \mathcal{G}^{\mathbb{C}}_{0,\,rel}(\mathbb{Z}) \longrightarrow \mathbb{Z} \xrightarrow{\ \sim\ } \mathbb{Z}$$

et on obtient un diagramme commutatif si Γ n'est pas nécessairement trivial

$$\begin{array}{ccccccc}
 & \text{Hom}(R(\Gamma)\,,\mathbb{C}^*) & \longrightarrow & \mathcal{G}^{\mathbb{C}}_{0,\,rel}(\mathbb{Z}[\Gamma]) & \longrightarrow & \mathcal{G}_0(\mathbb{Z}[\Gamma]) & \longrightarrow R(\Gamma) \\
 & \downarrow f & & \downarrow \text{Res}^{\mathbb{Z}}_{\mathbb{Z}[\Gamma]} & & \downarrow g & \\
0 \longrightarrow & \mathbb{C}^*/\{\pm 1\} & \longrightarrow & \mathcal{G}^{\mathbb{C}}_{0,\,rel}(\mathbb{Z}) & \longrightarrow & \mathbb{Z} &
\end{array}$$

où l'application f est l'application $\varphi \longmapsto \varphi(r_\Gamma)$ où r_Γ est la représentation r
gulière de Γ et $g([M]) = \dim_{\mathbb{Q}}(\mathbb{Q}\otimes_{\mathbb{Z}} M)$.

THÉORÈME II ([Q 3]). - <u>Si</u> $W_{\mathbb{C}[\Gamma]}(\chi) = 1$ <u>pour tout caractère symplectique</u> χ , <u>on a</u>

$$[\mathbb{Z}_N\,,\ T_{N/\mathbb{Q}}\,, X] = \delta(\tau_{\mathbb{C}[\Gamma]})\,.$$

En écrivant $Cl_S(N)$ comme quotient d'un module libre

$$0 \longrightarrow M \longrightarrow \mathbb{Z}[\Gamma]^i \longrightarrow Cl_S(N) \longrightarrow 0$$

obtient un élément $[\,M\,,\,1\,,\,\mathbb{Z}[\Gamma]^i]$ de $\mathcal{G}^{\mathbb{C}}_{0,\,rel}(\mathbb{Z}[\Gamma])$ qui ne dépend pas du choix de i et de M , que l'on note par abus de notation $[Cl_S(N)]$.

On montre que l'élément

$$[\,U_S\,,\ Log_{S,\,N/\mathbb{Q}}\,,\ X_S\,] - \delta(C^s_{0,\,\mathbb{C}[\Gamma]}) - [\,Cl_S(N)\,]$$

ne dépend pas du choix de S ([C]).

CONJECTURE II'. - <u>Sous l'hypothèse du théorème II</u>, <u>on a</u> :

$$[U_S\,,\ Log_{S,\,N/\mathbb{Q}}\,,\ X_S] = \delta(C^s_{0,\,\mathbb{C}[\Gamma]}) + [\,Cl_S(N)\,] \quad .$$

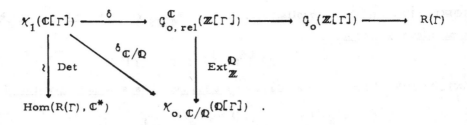

COROLLAIRE. - Si la conjecture est vraie, on a dans $\mathcal{C}_o(\mathbb{Z}[\Gamma])$

$$[U_S]=[X_S]+[Cl_S(N)]$$

et en particulier

$$[U_{S_\infty}]=[X_{S_\infty}]+[Cl(N)] \quad .$$

III. - Propriétés cohomologiques

a) On a les trois équivalences suivantes :

i) N/K est modérément ramifiée

ii) $\mathbb{Z}_N$ est cohomologiquement trivial

iii) $\mathbb{Z}_N$ est un $\mathbb{Z}[\Gamma]$-module projectif.

Ceci conduit au théorème de Fröhlich ([F], théorème II, voir [Q4]) et au théorème de Taylor ([Ty]).

m) On choisit S assez gros de telle sorte que S soit stable par Γ, S contienne S_∞ et les places ramifiées, et $Cl_S(N)$ soit trivial.

Soit J_S le groupe des S-idèles. On a donc deux suites exactes :

$$0 \longrightarrow U_S \longrightarrow J_S \longrightarrow J/K^* \longrightarrow 0$$

$$0 \longrightarrow X_S \longrightarrow Y_S \longrightarrow \mathbb{Z} \longrightarrow 0 \; .$$

THÉORÈME ([Ta 1]). - Il existe $\alpha \in H^2(\Gamma, \mathrm{Hom}(X_S, U_S))$ tel que le cup-produit par α donne un isomorphisme de

$$\hat{H}^r(\Gamma, X_S) \quad \text{sur} \quad \hat{H}^{r+2}(\Gamma, U_S) \; .$$

Cet élément α est construit à partir d'un élément α' de $H^2(\Gamma, \mathrm{Hom}(Y_S, J_S))$ donné par la théorie du corps de classe local et d'un élément α'' de $H^2(\Gamma, \mathrm{Hom}(\mathbb{Z}, J/K^*))$ donné par la théorie du corps de classe global. Ces trois éléments sont tels que par cup-produit on obtient des isomorphismes rendant le diagramme suivant commutatif

$$\begin{array}{ccccccc}
\longrightarrow & H^r(\Gamma, X_S) & \longrightarrow & H^r(\Gamma, Y_S) & \longrightarrow & H^r(\Gamma, \mathbb{Z}) & \longrightarrow \\
& \downarrow \alpha & & \downarrow \alpha' & & \downarrow \alpha'' & \\
\longrightarrow & H^{r+2}(\Gamma, U_S) & \longrightarrow & H^{r+2}(\Gamma, J_S) & \longrightarrow & H^{r+2}(\Gamma, J/K^*) & \longrightarrow
\end{array} \; .$$

Comme X_S est un $\mathbb{Z}$-module libre, $\mathrm{Ext}^2_\Gamma(X_S, U_S)$ s'identifie à $H^2(\Gamma, \mathrm{Hom}_{\mathbb{Z}}(X_S, U_S))$. On en déduit une suite exacte de $\mathbb{Z}[\Gamma]$-modules :

$$0 \longrightarrow U_S \longrightarrow A \longrightarrow B \longrightarrow X_S \longrightarrow 0$$

dont la classe dans $\mathrm{Ext}^2_\Gamma(X_S, U_S)$ est égale à α et où A et B sont cohomologiquement triviaux.

Soit $\mathcal{K}_0(\mathcal{C}\mathcal{T}(\mathbb{Z}[\Gamma]))$ le groupe de Grothendieck de la catégorie des $\mathbb{Z}[\Gamma]$-modules de type fini et cohomologiquement triviaux.

THÉORÈME ([C]). - L'élément $[A]-[B]$ de $\mathcal{K}_0(\mathcal{C}\mathcal{T}(\mathbb{Z}[\Gamma]))$ associé à X_S, U_S et α ne dépend pas du choix de S vérifiant les conditions précédentes ni du choix de A et B tels qu'il existe une suite exacte

$$0 \longrightarrow U_S \longrightarrow A \longrightarrow B \longrightarrow X_S \longrightarrow 0$$

dont la classe dans $\mathrm{Ext}^2_\Gamma(X_S, U_S)$ soit α et où A et B sont cohomologiquement triviaux.

On suppose que N est une extension modérée de K .

THÉORÈME III ([Ty]). - L'élément $[\mathbb{Z}_N]-[X]$ de $\mathcal{K}_0(\mathcal{C}\mathcal{T}(\mathbb{Z}[\Gamma]))$ est d'ordre 1 ou 2 . Il est d'ordre 1 si les constantes de l'équation fonctionnelle des séries L d'Artin pour les caractères symplectiques valent +1 .

CONJECTURE III' (Chinburg). - Si S est assez gros, l'élément $[A]-[B]$ de $\mathcal{K}_0(\mathcal{C}\mathcal{T}(\mathbb{Z}[\Gamma]))$ est d'ordre 1 ou 2 . Il est d'ordre 1 si les constantes de l'équation fonctionnelle des séries L d'Artin pour les caractères symplectiques valent +1 .

Une formulation plus précise du théorème III et de la conjecture III' peut être donnée en utilisant l'élément $W_{\mathbb{Z}[\Gamma]}$ défini dans [Q 5] .

IV. - Equation fonctionnelle "algébrique"

Si M est un $\mathbb{Z}[\Gamma]$-module, on note M_0 l'ensemble des éléments $m \in M$ tels que $\sum_{\gamma\in\Gamma} \gamma(m) = 0$. Soit $S_{\infty,\mathbb{R}}$ (resp. $S_{\infty,\mathbb{C}}$) l'ensemble des places archimédiennes réelles (resp. complexes) de N . Le groupe de Galois de $\mathbb{C}$ sur $\mathbb{R}$ opère sur $\mathrm{Is}_{\mathbb{Q}}(N, \mathbb{C})$ et $S_{\infty,\mathbb{R}} \cup S_{\infty,\mathbb{C}}$ s'identifie à un système de représentants des orbites de $\mathrm{Gal}(\mathbb{C}/\mathbb{R})$ dans $\mathrm{Is}_{\mathbb{Q}}(N, C)$. Ainsi $\mathrm{Is}_{\mathbb{Q}}(N, \mathbb{C})$ est égal à $S_{\infty,\mathbb{R}} \cup \{ w, \overline{w} / w \in S_{\infty,\mathbb{C}}\}$

où $\overline{w}$ est le composé de w et de la conjugaison complexe.

En utilisant les notations du paragraphe I, on a donc une suite exacte de $\mathbb{Z}[\Gamma]$-modules :

$$0 \longrightarrow Y_{S_\infty,\mathbb{C}} \xrightarrow{q} X_o \xrightarrow{p} X_{S_\infty} \longrightarrow 0$$

q est défini par $q(w) = w - \overline{w}$, $\forall w \in S_{\infty,\mathbb{C}}$.

Soit U^+ l'ensemble des unités de $\mathbb{Z}_N$ totalement positives aux places réelles de N. On plonge U^+ dans $\prod_{w\in S_\infty} N_w^* = (\mathbb{R}\otimes K)^*$ et on pose

$$\log_\infty U = \{ x = (x_w)_w \in \prod_{w\in S_\infty} N_w, (\exp(x_w))_w \in U^+ \} .$$

Si $\log_\infty 1$ désigne l'ensemble des $x = (x_w)_w \in \log_\infty U$ tels que $\exp(x_w) = 1$ pour tout $w \in S_\infty$, on a une suite exacte de $\mathbb{Z}[\Gamma]$-modules

$$0 \longrightarrow \log_\infty 1 \longrightarrow \log_\infty U \xrightarrow{\exp} U^+ \longrightarrow 0 .$$

Soit μ l'isomorphisme de $\mathbb{C}[\Gamma]$-module de $\mathbb{C}\otimes_{\mathbb{Z}} \log_\infty U$ dans $\mathbb{C}\otimes_{\mathbb{Z}} X_o$ qui à $x\otimes(y_w)_{w\in S_\infty}$ associe $x \sum_{w\in S_{\infty},\mathbb{R}} y_w \otimes v + x \sum_{w\in S_{\infty},\mathbb{C}} y_w\otimes w + \overline{y}_w \otimes \overline{w}$.

On obtient donc un diagramme commutatif dont les flèches horizontales sont rationnelles sur $\mathbb{Z}$:

$$\begin{array}{ccccccccc}
0 & \longrightarrow & \mathbb{C}\otimes_{\mathbb{Z}} \log_\infty 1 & \longrightarrow & \mathbb{C}\otimes_{\mathbb{Z}} \log_\infty U & \longrightarrow & \mathbb{C}\otimes_{\mathbb{Z}} U^+ & \longrightarrow & 0 \\
 & & \downarrow \mu_1 & & \downarrow \mu & & \downarrow \mathrm{Log}_{S_\infty, N/\mathbb{Q}} & & \\
0 & \longrightarrow & \mathbb{C}\otimes_{\mathbb{Z}} Y_{S_\infty,\mathbb{C}} & \longrightarrow & \mathbb{C}\otimes_{\mathbb{Z}} X_o & \longrightarrow & \mathbb{C}\otimes_{\mathbb{Z}} X_{S_\infty} & \longrightarrow & 0
\end{array}$$

où μ_1 désigne la restriction de μ à $\mathbb{C}\otimes_{\mathbb{Z}} \log_\infty 1$.

De la définition de $\mathcal{G}^{\mathbb{C}}_{o,\,rel}(\mathbb{Z}[\Gamma])$, on tire l'égalité

$$[\log_\infty U , \mu , X_o] = [U^+ , \mathrm{Log}_{S_\infty, N/\mathbb{Q}} , X_{S_\infty}] + [\log_\infty 1 , \mu_1 , Y_{S_\infty,\mathbb{C}}] .$$

Un calcul simple montre que

$$[\log_\infty 1 , \mu_1 , Y_{S_\infty,\mathbb{C}}] = \delta([\mathbb{C}\otimes_{\mathbb{Z}} Y_{S_\infty,\mathbb{C}} , \mu_1'])$$

où μ_1' est la multiplication par $2i\pi$ et que

$$\mathrm{Det}([\mathbb{C}\otimes_{\mathbb{Z}} Y_{S_\infty,\mathbb{C}} , \mu_1'])(\chi) = (2i\pi)^{b_{F,2}(\chi)} , \quad \forall \chi \in R(\Gamma) .$$

On obtient la relation suivante dans $\mathcal{G}^{\mathbb{C}}_{o,\,rel}(\mathbb{Z}[\Gamma])$

$$[\mathbb{Z}_N, T_{N/\mathbb{Q}}, X] = [\log_\infty 1, \mu_1, Y_{S_\infty,\mathbb{C}}] + [U^+_{S_\infty}, \mathrm{Log}_{S_\infty, N/\mathbb{Q}}, X_{S_\infty}] - [\log_\infty U, T^{-1}_{N/\mathbb{Q}} \circ \mu, (\mathbb{Z}_N)_o] .$$

D'où en appliquant $\mathrm{Ext}^{\mathbb{Q}}_{\mathbb{Z}}$, on a dans $\mathcal{K}_{o,\,\mathbb{C}/\mathbb{Q}}(\mathbb{Q}[\Gamma])$

$$[N, T_{N/\mathbb{Q}}, \mathbb{Q}\otimes_{\mathbb{Z}} X] = [\mathbb{Q}\otimes_{\mathbb{Z}} \log_\infty 1, \mu_1, \mathbb{Q}\otimes_{\mathbb{Z}} Y_{S_\infty,\mathbb{C}}] + [V_S, \mathrm{Log}_{S_\infty}, + [V_S, \mathrm{Log}_{S_\infty, N/\mathbb{Q}}, \mathbb{Q}\otimes_{\mathbb{Z}} X_{S_\infty}] - [\mathbb{Q}\otimes_{\mathbb{Z}} \log_\infty U, T^{-1}_{N/\mathbb{Q}} \circ \mu, N_o] .$$

Enfin la proposition I entraîne le théorème suivant :

THEOREME IV. - <u>La conjecture</u> I' <u>est équivalente à la conjecture suivante</u> :

<u>Si</u> $W_{\mathbb{C}[\Gamma]}(\chi) = 1$ <u>pour tout caractère symplectique</u> χ, <u>alors on a dans</u> $\mathcal{K}_{o,\,\mathbb{C}/\mathbb{Q}}(\mathbb{Q}[\Gamma])$ <u>l'égalité suivante</u> :

$$[\mathbb{Q}\otimes_{\mathbb{Z}} \log_\infty U, T^{-1}_{N/\mathbb{Q}} \circ \mu, N_o] = \delta(\overline{C}_{1,\mathbb{C}[\Gamma]})$$

<u>où</u> $\overline{C}_{1,\mathbb{C}[\Gamma]}$ <u>est le composé des homomorphismes suivants</u> :

$$R(\Gamma) \longrightarrow R(\Gamma) \xrightarrow{\mathrm{Inf}} R(G_K) \xrightarrow{C^{S_\infty}_{K,1}} \mathbb{C}^*$$

<u>le premier homomorphisme transformant un caractère en son conjugué</u>.

<u>Démonstration</u>. - Appliquer $\mathrm{Det}^{-1} \circ \delta_{\mathbb{C}/\mathbb{Q}}$ à la relation (*) du paragraphe 2 de l'introduction.

-:-:-:-

RÉFÉRENCES

[C] T. CHINBURG, <u>On the Galois structure of algebraic integers and units</u>, (à paraître).

[CN-T] Ph. CASSOU-NOGUÈS and M. J. TAYLOR, <u>Local Root Numbers and Hermitian - Galois Module Structure of Rings of Integers</u>, Math. Ann. 263 (1983), 251-261.

[D1] P. DELIGNE, <u>Les constantes locales de l'équation fonctionnelle de la fonctio L d'Artin d'une représentation orthogonale</u>, Invent. Math. 102 (1975), 391-325.

[D2] P. DELIGNE, <u>Les constantes des équations fonctionnelles des fonctions</u> L, Modular functions of one variable II, Lecture Notes in Math., 349, Springer-Verlag 1973, p. 501-597.

[F] A. FRÖHLICH, <u>Arithmetic and Galois module structure for tame extensions</u>, J. reine angew. Math., 286-287 (1976), 380-439.

[F-T] A. FRÖHLICH and M. J. TAYLOR, The arithmetic theory of local Galois Gauss sums for tame characters, Trans. Royal. Soc. A 298 (1980), 141-181.

[H] A. HELLER, Some exact sequences in algebraic K-theory, Topologie 3 (1969), 389-408.

[M] J. MARTINET, Algebraic number fields : L-functions and Galois properties, Proc. Sympos. Univ. Durham, Academic Press, London, 1977.

[Q1] J. QUEYRUT, S-groupes des classes d'un ordre arithmétique, J. of Algebra 76, I (1982), 234-260.

[Q2] J. QUEYRUT, Modules radicaux sur des ordres arithmétiques (à paraître dans Journal of Algebra).

[Q3] J. QUEYRUT, Structure galoisienne des anneaux d'entiers I, Ann. Inst. Fourier, Grenoble, 31, 3 (1981), 1-35.

[Q4] J. QUEYRUT, Anneaux d'entiers dans le même genre, (à paraître).

[Q5] J. QUEYRUT, Sommes de Gauss et structure galoisienne des anneaux d'entiers, Sém. de Th. des Nombres, Université de Bordeaux I, exp. n° 10, 1981-1982.

[S] H. M. STARK, L-functions at $s = 1$, I ; II ; III ; IV. Advances in Math. 7, (1971), 301-343 ; 17 (1973), 60-92 ; 22 (1976), 64-84 ; 35 (1980), 197-235.

[Ta 1] J. TATE, The cohomology groups of tori in finite Galois extensions of number fields, Nagoya Math. J. 27 (1966), 709-719.

[Ta 2] J. TATE, La conjecture de Stark, (à paraître).

[Ty] M. J. TAYLOR, On Fröhlich's conjecture for rings of integers for tame extensions, Invent. Math. 63 (1981), 41-79.

-:-:-:-

Jacques QUEYRUT
L.A. au C.N.R.S. n° 226
U.E.R. Mathématiques et Informatique
Université de Bordeaux I
351, cours de la Libération
F 33405 TALENCE CEDEX

NEW VERY LARGE AMICABLE PAIRS

Herman J.J. te Riele
Centrum voor Wiskunde en Informatica
Kruislaan 413
1098 SJ Amsterdam
The Netherlands

Abstract

Computations are described which led to the discovery of many very large amicable pairs, which are much larger than the largest amicable pair thus far known.

1. Introduction

About 200 years ago, to be more precise: on September 18, 1783, Leonhard Euler died. He left 59 new amicable pairs (APs) as a result of an extensive, systematic study ([7]). A pair of positive integers (m_1,m_2) is called amicable, if $m_1 \neq m_2$ and if each number is the sum of the proper divisors of the other, i.e., $\sigma(m_1) - m_1 = m_2$ and $\sigma(m_2) - m_2 = m_1$, where $\sigma(\cdot)$ denotes the sum of *all* the divisors - function. The following AP of Euler's will play a crucial rôle in this paper:

$$(1) \quad \left\{ \begin{array}{l} 11498355 \\ 12024045 \end{array} \right. = 3^4 5\cdot 11\cdot \left\{ \begin{array}{l} 29\cdot 89 \\ 2699 \end{array} \right. .$$

Prior to Euler, only three APs were known, namely,

$(220, 284) = (2^2 5\cdot 11, 2^2 71)$ (already known to the Pythagoreans [9, p. 97]),
$(17296, 18416) = (2^4 23\cdot 47, 2^4 1151)$ (Ibn Al-Bannā' [4]) and
$(9363584, 9437056) = (2^7 191\cdot 383, 2^7 73727)$ (Descartes [9, p. 99]).

After Euler, many more APs have been found (cf. [8] and [11]), most of them with the help of various variations of Euler's methods.*) A small minority of the known APs were found by systematic computer searches (i.e., by testing for *all* m in a given interval, whether $s(s(m)) = m$, where $s(m) = \sigma(m) - m$). It is generally believed, although unproved, that there are infinitely many APs. The largest AP thus far known consists of two 152-digit numbers ([10]).

In this short paper we describe computations by which we have found many very large APs, the largest pair consisting of two 282-digit numbers, and we indicate the rôle played by Euler's pair (1) in this work.

*) The author "collects" APs. He has a computer file of, currently, 7495 different APs. More than 45% of them were found by W. Borho and H. Hoffmann. Anyone who is really interested may send a request for a print-out, or a copy on tape.

2. A method for finding amicable pairs

Very recently, we have discovered methods for constructing APs from *given* APs, which turned out to be very "prolific": from a "mother" list of 1592 known APs, 2324 *new* APs were constructed ([11]). About half the number of these pairs were found by using the following lemma (which is a special case of Method 2 given in [11]; the proof of this lemma is left to the reader).

LEMMA 1 *Let (au,ap) be a given amicable pair with $gcd(a,u) = gcd(a,p) = 1$, where p is a prime. If a pair of prime numbers (r,s) with $r < s$ and $gcd(a,rs) = 1$ exists, satisfying the bilinear Diophantine equation*

$$(2) \qquad (r - p)(s - p) = \frac{\sigma(a)}{a}\left(\sigma(u)\right)^2 =: R,$$

and if a third prime q exists, with $gcd(au,q) = 1$ and

$$(3) \qquad q = r + s + u,$$

then (auq,ars) is also an amicable pair.

The right hand side, R, of (2) is a positive integer. If the factorization of R into primes is known, equation (2) can easily be solved by writing R in all possible ways as the product $R = A \cdot B$, with $2 \le A < B$, so that $r = p + A$ and $s = p + B$. For nearly all known APs of the form (au,ap), u is the product of 2, 3, 4 or 5 distinct prime numbers (compare the examples given in the introduction). As a consequence, R usually has *many* divisors, and this explains, at least heuristically, the large number of new APs found with Lemma 1.

Example For Borho's AP ([2]) mentioned in the "Note added in proof" in [11] we have $a = 2 \cdot 5^3 19 \cdot 67$, $u = 15959 \cdot 5346599$ and $p = 85331735999$, so that $R = 2^{17} 3^6 5^4 7^4 13 \cdot 17 \cdot 19^3 67$, a number with 44800 even divisors less than its square root, such that its co-divisor is also even. By testing all these cases, we found 145 new APs with Lemma 1. □

The program used in [11] could not handle cases with $R > 10^{25}$, so that we could not apply Lemma 1 to the largest known APs of the form (au,ap) (also the 152-digit AP mentioned above is of this form). Fortunately, my colleague D.T. Winter has recently developed a very fast package for multi-precision integer arithmetic. This package was used by A.K. Lenstra in his implementation of a primality proving program on a CDC Cyber 750 computer. The algorithm used in that program was based on ideas of Adleman, Pomerance and Rumeley ([1]) and of Cohen and H.W. Lenstra, Jr. ([5]). With this program it is possible now to prove primality of numbers of up to 200 decimal digits in a reasonable amount of computer time.

Winter's package and A.K. Lenstra's program enabled us to apply Lemma 1 to the largest known APs of the form (au,ap). In this way we found 3 new APs (with 123-,

127- and 141-digit members) from the 81-digit AP given in [10], and 11 new APs (with 231- up to 282-digit members) from the 152-digit AP given in [10]. Table 1 gives all the information needed to identify these 14 large APs.

Some details of our computations and some historical background which led to the 282-digit AP are given in the next section.

3. The 282-digit new amicable pair

In 1972, Borho ([3]) presented his so-called Thabit-rules, which were inspired by the following formula, due to the Arabian mathematician Thabit ibn Kurrah ([6]):

If $p = 3\cdot 2^{n-1} - 1$, $q = 3\cdot 2^{n} - 1$ *and* $r = 9\cdot 2^{2n-1} - 1$ *are primes and* $n \geq 2$, *then* $2^n pq$ *and* $2^n r$ *form an amicable pair.*

Examples are the three pre-Euler APs mentioned in the introduction. Many of Borho's Thabit-rules are *constructed* from given APs. In particular, Borho constructed the following Thabit-rule from Euler's AP (1):

If the two numbers $q_1 = 5281^n 2582 - 1$ *and* $q_2 = 5281^n 2582\cdot 2700 - 1$ *are primes, and* $n \geq 1$, *then* $3^4 5\cdot 11\cdot 29\cdot 89\cdot 5281^n q_1$ *and* $3^4 5\cdot 11\cdot 5281^n q_2$ *form an amicable pair.*

Lee ([3]) found that indeed q_1 and q_2 are both primes for $n = 1$, and te Riele ([10]) showed that $n = 19$ is the next value of n for which this rule is successful. Borho ([2]) found that these are the *only* successful cases for $n \leq 267$.

Application of Lemma 1 to the " $n = 19$ " - AP gave

$$R = 2^{11}3^4 5^4 11\cdot 19\cdot 41\cdot 139\cdot 311\cdot 1291^2 5281^{19} 6661\cdot 33331\cdot 13944481\cdot 75019421\cdot$$
$$\cdot 24027536081\cdot 92192755565941\cdot 155588291031361,$$

which is a 156-digit number with 30,720,000 even divisors less than its square root, with even co-divisor. Estimates of the running time of our program revealed that testing *all* these divisors would consume too much computer time. Therefore, we made a selection of about 700,000 divisors A of R such that $A \equiv R/A \equiv 0 \pmod{30}$. This enlarged the chance of finding primes r and s, since for these A and B ($= R/A$) none of the primes 2, 3 and 5 divides r and s. The divisor

$$A = 2\cdot 3\cdot 5\cdot 139\cdot 1291^2 5281^2 6661\cdot 33331$$

yielded the 282-digit amicable pair (the symbol " \ " means: continuation of the number on the next line):

$$\begin{cases} m_1 = 3^4 5\cdot 11\cdot 5281^{19} 29\cdot 89\cdot t\cdot q \\ m_2 = 3^4 5\cdot 11\cdot 5281^{19} r\cdot s \end{cases},$$

where

$t = $ (75 digits)
13917570188877597630885553289918626792708863255174423058328801872338 2689621,
$q = $ (130 digits)
6417976467106377983899074271257515017287808217523810874305651125787167909571 2287\
99804727940017355222105406135083828506969640009869,
$r = $ (78 digits)
3757743950996951360339099388278029234031393078897094656085740629703346 84854669,
$s = $ (130 digits)
64179764671063779838990742712575150172878082175238101393187694511612 370225051595\
59202047017062286716344081391043389211584233243399.

The decimal representation of the pair reads as follows (both members have 282 decimal digits) :

$m_1 = $
55361064940788699236737990872270382863 5124\
33844020585504589806863185485656131816319616559073822997568562257512746133163331\
90629669367246234005166063052241199078258371384155343551406293972127278208968662\
44929815949926004072029749149921701002222426683487022058391322136048764726481795 ,

$m_2 = $
579135510810265354278157982778496837394 357\
11498975428462763455455482685498887837085265040796911245187828896835018627257768\
19497946906167907515580359819567168202271836497015019139851462990225721043204383\
48107419189784957177823219886153751513747229819951505729702954824327231219438205 .

4. Historical comments

To the best of our knowledge, Lemma 1 has never been explicitly stated in the literature. Euler already gave two APs, viz.,

$$(4) \quad 2^4 \cdot \left\{ \begin{array}{l} 23 \cdot 47 \cdot 9767 \\ 1583 \cdot 7103 \end{array} \right. \quad \text{and} \quad 3^2 7 \cdot 13 \cdot \left\{ \begin{array}{l} 5 \cdot 17 \cdot 1187 \\ 131 \cdot 971 \end{array} \right. ,$$

which could have been found with Lemma 1 from the pairs (also known to Euler)

$$2^4 \cdot \left\{ \begin{array}{l} 23 \cdot 47 \\ 1151 \end{array} \right. \quad \text{and} \quad 3^2 7 \cdot 13 \cdot \left\{ \begin{array}{l} 5 \cdot 17 \\ 107 \end{array} \right. , \quad \text{respectively.}$$

Unfortunately, Euler did not explain how he found his pairs (4). Escott ([6]) gave at least 36 APs which could have been found with Lemma 1, so it is not unreasonable to assume that he was aware of it.

Acknowledgement

I wish to thank A.K. Lenstra for his help in proving all the pseudoprimes which I gave to him, to be prime.

Table 1

14 large APs of the form (auq, ars) found with Lemma 1 from the 81- and 152-digit APs given in [10]

$$r = p + A,\ s = p + R/A,\ q = r + s + u,\ R = \Pi_{i=1}^{\ell} v_i^{e_i},\ A = \Pi_{i=1}^{\ell} v_i^{f_i}$$

I. 3 APs with 123-, 127- and 141-digit members:

$a = 3^2 7^2 13 \cdot 19 \cdot 29 \cdot 14401^8$
$u = 41 \cdot 173 \cdot 13122977591352752053026748464924755893$
$p = 95902720237605912003519477781670116073351$
$\ell = 10$

i	v_i	e_i	f_i		
			digits: 123	127	141
1	2	10	4	6	8
2	3	6	0	1	5
3	5	2	1	0	2
4	7	2	1	1	1
5	29	1	0	0	1
6	79	1	1	0	1
7	3547	2	2	2	1
8	14401	8	1	5	2
9	125017	1	0	1	0
10	29732720450482311322203201	1	1	0	0

II. 11 APs with 231-, 232-, 233-, 235-, 239-, 246-, 248-, 249-, 250-, 263- and 282-digit members:

$a = 3^4 5 \cdot 11 \cdot 5281^{19}$
$u = 29 \cdot 89 \cdot$
$\cdot 1391757018887759763088555328991862679270886325517442305832880187233826896 21$
$p = 37577439509969513603390993882780292340313930788970942257487765055313326197939$
$\ell = 17$

i	v_i	e_i	f_i										
			digits: 231	232	233	235	239	246	248	249	250	263	282
1	2	11	1	1	1	1	1	5	1	1	1	5	1
2	3	4	2	2	1	1	1	1	1	1	1	2	1
3	5	4	1	1	3	2	1	3	3	1	2	3	1
4	11	1	0	0	1	1	0	1	0	1	0	1	0
5	19	1	0	1	0	0	1	1	0	1	1	0	0
6	41	1	1	0	0	0	0	1	1	1	0	0	0
7	139	1	0	0	0	0	1	0	1	0	0	0	1
8	311	1	0	0	0	0	0	1	0	0	1	1	0
9	1291	2	2	1	0	0	1	2	0	0	2	0	2
10	5281	19	17	5	11	10	11	12	0	1	2	10	2
11	6661	1	0	0	1	0	0	0	0	0	0	0	1
12	33331	1	1	1	1	1	1	0	0	1	0	0	1
13	13944481	1	0	0	0	0	1	0	1	1	0	0	0
14	75019421	1	0	1	1	0	1	0	1	0	0	0	0
15	24027536081	1	0	1	0	0	0	0	1	1	1	0	0
16	92192755565941	1	0	1	1	1	0	0	1	1	1	0	0
17	155588291031361	1	0	1	0	1	0	0	1	1	1	0	0

References

[1] Adleman, L.M., C. Pomerance & R.S. Rumely, *On distinguishing prime numbers from composite numbers*, Ann. Math. 117 (1983), 173-206.

[2] Borho, W., *Grosse Primzahlen und befreundete Zahlen: Über den Lucas-Test und Thabit-Regeln*, to appear in Mitt. Math. Gesells. Hamburg.

[3] Borho, W., *On Thabit ibn Kurrah's formula for amicable numbers*, Math. Comp. 26 (1972), 571-578.

[4] Borho, W., *Some large primes and amicable numbers*, Math. Comp. 36 (1981), 303-304.

[5] Cohen, H. & H.W. Lenstra, Jr., *Primality testing and Jacobi sums*, Math. Comp. 42 (1984), 297-330.

[6] Escott, E.B., *Amicable numbers*, Scripta Math. 12 (1946), 61-72.

[7] Euler, L., *De numeris amicabilibus*, *Leonhardi Euleri Opera Omnia*, Teubner, Leipzig and Berlin, Ser. I, vol. 2, 1915, 63-162.

[8] Lee, E.J. & J.S. Madachy, *The history and discovery of amicable numbers*, J. Recr. Math. 5 (1972), Part I: 77-93, Part II: 153-173, Part III: 231-249.

[9] Ore, O., *Number theory and its history*, McGraw-Hill Book Company, New York etc. 1948.

[10] Riele, H.J.J. te, *Four large amicable pairs*, Math. Comp. 28 (1974), 309-312.

[11] Riele, H.J.J. te, *On generating new amicable pairs from given amicable pairs*, Math. Comp. 42 (1984), 219-223.

RIGID-ANALYTIC L - TRANSFORMS

P. Schneider
Fakultät für Mathematik
Universitätsstr. 31
8400 Regensburg
Bundesrepublik Deutschland

In this talk I want to present a new method of defining p-adic L-functions for a certain class of elliptic curves. In the first section we shortly review the general philosophy of complex and p-adic L-functions and then explain the idea of the method which is based on the notion of a rigid-analytic automorphic form. The construction of a p-adic L-function associated with such an automorphic form is carried out in the second section.

I. THE STARTING POINT

Let $E_{/\mathbb{Q}}$ be an elliptic curve over the rationals. One of the most interesting invariants of E is its Hasse-Weil L-function

$$L(E,s) = \prod_{p \text{ good}} (1-t_p p^{-s}+p^{1-2s})^{-1} \cdot \prod_{p \text{ bad}} (1-t_p p^{-s})^{-1}$$

with $t_p := \begin{cases} p+1 - \#E(\mathbb{F}_p) & \text{if } E \text{ has good reduction at } p, \\ \pm 1 \text{ or } 0 & \text{otherwise,} \end{cases}$

It converges for $Re(s) > 3/2$ and apparently collects arithmetic information about E. But in order to study its properties one needs analytic methods. We therefore now assume that E is a Weil curve, i.e., there exists a nonconstant $\mathbb{Q}$-morphism

$$X_0(N) \xrightarrow{\pi} E$$

such that $\pi(i\infty) = 0$ and $\pi^*\omega = c_\omega \cdot f$ for any holomorphic differential form ω on E, where c_ω is a constant and f is a normalized newform of weight 2 for $\Gamma_0(N)$.

Commentary:

1) $\pi^*\omega$ always is a cuspform of weight 2 for $\Gamma_0(N)$ which is an eigenform for all Hecke operators T_p, $p \nmid N$. The requirement "$\pi^*\omega$ newform" means that N is the minimal possible number for which such a π exists

2) One of Weil's conjectures says that any $E_{/\mathbb{Q}}$ is a Weil curve.

The analytic properties of the Mellin transform

$$L(f,s) := \frac{(2\pi)^s}{\Gamma(s)} \cdot \int_0^\infty f(iy)y^{s-1}dy$$

of f are easy to obtain. But according to Eichler/Shimura, Igusa, and Deligne/Langlands we have

$$L(E,s) = L(f,s).$$

We therefore get analytic continuation and a functional equation also for $L(E,s)$.

On the other hand, at certain integer points $L(f,s)$ and its twists by Dirichlet characters have strong algebraicity and even integrality properties. Therefore there is a natural way to associate with $L(f,s)$ a p-adic analytic L-function $L_p(f,s)$ (p a prime number and s now a p-adic variable) such that the values of $L(f,s)$ and $L_p(f,s)$ at the "critical" integer points are closely related (Mazur/Swinnerton-Dyer, Manin, Amice-Velu, Visik). We emphasize that with this method $L_p(f,s)$ cannot be defined independently of $L(f,s)$. It also should be mentioned that there is a theory (Iwasawa,Mazur) how to define arithmetically a p-adic L-function $L_p(E,s)$; furthermore there is the "main conjecture" which relates $L_p(E,s)$ to $L_p(f,s)$.

Our idea to construct a p-adic L-function for E is to use directly Mumford's theory of p-adic uniformization. Let $\mathbb{C}_p$ denote the completion of an algebraic closure of $\mathbb{Q}_p$. The modular curve $X_0(N)_{/\mathbb{C}_p}$ itself is a Mumford curve if and only if $N = p$ (see [2]). Unfortunately, at present, no corresponding discrete group is known explicitly. But let us assume that N is square-free with an even number of prime divisors. Denote by D_N the quaternion algebra over $\mathbb{Q}$ which is ramified precisely at the prime divisors of N, and let Γ_N be the group of units of reduced norm 1 in a maximal order of D_N. If $S_{N/\mathbb{Q}}$ is the Shimura curve with $S_N(\mathbb{C}) = \Gamma_N \backslash \mathbb{H}$ then a result of Ribet ([8]) says that the Jacobian of S_N is $\mathbb{Q}$-isogenous to the new part of the Jacobian of $X_0(N)$:

$$J_0(N)^{new} \sim \text{Jac } S_N.$$

We now fix a prime divisor p of N and denote by D_N' the quaternion algebra over $\mathbb{Q}$ which is ramified precisely at ∞ and at the prime divisors of N different from p. The image Γ_N' in $PGL_2(\mathbb{Q}_p)$ of the group of p-units (with respect to a maximal order) in D_N' is a discrete and finitely generated subgroup of $PGL_2(\mathbb{Q}_p)$. According to Čerednik ([1]) one has a rigid-analytic isomorphism

$$S_N(\mathbb{C}_p) \cong \Gamma_N' \setminus (\mathbb{C}_p \setminus \mathbb{Q}_p) .$$

Thus any Weil curve E with an analytic conductor N which fulfills the above assumptions (and consequently has multiplicative reduction at p) has a p-adic analytic uniformization

$$\Gamma_N' \setminus (\mathbb{C}_p \setminus \mathbb{Q}_p) \xrightarrow{\psi} E(\mathbb{C}_p)$$

which is "defined over $\mathbb{Q}$". Furthermore the rigid-analytic automorphic form $\psi^*\omega$ of weight 2 for Γ_N' up to a constant only depends on E.

In the next section we shall construct a p-adic analogue $L_p(g,s)$ of the classical Mellin transform for any rigid-analytic automorphic form g of arbitrary weight. In particular, we view $L_p(\psi^*\omega,s)$ as the p-adic L-function of E; of course, one first has to normalize the constant correctly (using Hecke operators). But we will not discuss this problem here, neither the question whether $L_p(\psi^*\omega,s)$ and $L_p(f,s)$ agree.

II. THE L-TRANSFORM

Let $K \subseteq \mathbb{C}_p$ be a finite extension field of $\mathbb{Q}_p$, let $\Gamma \subseteq SL_2(K)$ be a finitely generated discrete subgroup, and denote by $\mathcal{L} \subseteq K \cup \{\infty\}$ its set of limit points. Γ then acts discontinuously (via fractional linear transformations) on the analytic set

$$H: = \mathbb{C}_p \cup \{\infty\} \setminus \mathcal{L}$$

and according to Mumford ([7] or [6]) $C: = \Gamma \setminus H$ has a natural structure of a smooth projective curve over $\mathbb{C}_p$. We always make the following assumptions:

a) $\mathcal{L}$ is infinite (and therefore compact and perfect);

b) $\infty \in \mathcal{L}$.

DEFINITION:

A rigid-analytic function $f: H \to \mathbb{C}_p$ is called an automorphic form of weight $n \in \mathbb{Z}$ for Γ if

$$f(\gamma x) = (cx+d)^n f(x) \text{ for all } \gamma = \begin{pmatrix} a & b \\ c & d \end{pmatrix} \in \Gamma \text{ and } x \in H.$$

Furthermore $M_n(\Gamma)$ denotes the $\mathbb{C}_p$-vector space of all automorphic forms of weight n for Γ.

In a completely analogous way as in the classical case of a co-compact Fuchsian group one can compute the dimension of the vector space $M_n(\Gamma)$

for $n \neq 1$. We state the result only for a Schottky group Γ.

PROPOSITION:

Suppose that Γ *is free of rank* $r > 1$. *Then*

$$\dim_{\mathbb{C}_p} M_n(\Gamma) = \begin{cases} 0 & \text{for } n < 0, \\ 1 & \text{for } n = 0, \\ r & \text{for } n = 2, \\ (n-1)(r-1) & \text{for } n \geq 3. \end{cases}$$

Proof: We have $M_0(\Gamma) = \mathbb{C}_p$ since C is projective. On the other hand r is equal to the genus of C. $M_2(\Gamma)$ which is isomorphic to the vector space of holomorphic differentials on C therefore has the dimension r. The considerations in §4 of [5] imply the existence of a nonvanishing meromorphic function f_0 on H such that

$$f_0(\gamma x) = (cx+d) f_0(x) \quad \text{for all} \quad \gamma = \begin{pmatrix} a & b \\ c & d \end{pmatrix} \in \Gamma \quad \text{and} \quad x \in H$$

and

$$\deg \operatorname{div}(f_0) = r - 1.$$

Consequently the map

$$\begin{aligned} \Gamma(C, \mathcal{O}(n \operatorname{div}(f_0))) &\longrightarrow M_n(\Gamma) \\ f &\longmapsto f \cdot f_0^n \end{aligned}$$

is an isomorphism. But the dimension on the left hand side for $n < 0$ or $n \geq 3$ is the required one by the Riemann-Roch theorem.

Γ not only acts on H but also on a certain tree T_Γ. Namely, let T_K be the Bruhat-Tits tree of $SL_2(K)$. The straight paths of T_K the ends of which correspond to the fixed points of a non-trivial hyperbolic element in Γ (i.e., the axes in T_K of the hyperbolic elements in Γ) form a subtree of T_K. The tree T_Γ is constructed from this subtree by neglecting all vertices P with the following two properties:

i. P has only two adjacent vertices P_1 and P_2;

ii. there is no nontrivial elliptic element in Γ which fixes P but not P_1 and P_2;

it only depends on Γ (not on the field K). The group Γ acts without inversion on T_Γ (use [9] II.1.3), and the quotient graph $S := \Gamma \backslash T_\Gamma$ is finite ([6] I.3.2.2). Furthermore, there is a canonical Γ-equivariant bijection

$$\mathcal{L} \longleftrightarrow \{\text{ends of } T_\Gamma\}$$

$$= \{\text{equivalence classes of halflines in } T_\Gamma\}$$

([6]I.2.5).

Notation: For any tree T we denote by Vert(T), resp. Edge(T), the set of vertices, resp. edges, of T. For any edge y of T, the vertices A(y) and E(y), resp. the edge $\bar{y}$, are defined to be the origin and the terminus, resp. the inverse edge, of y.

DEFINITION:

Let M be an abelian group. A harmonic cocycle on T_Γ with values in M is a map

$$c: \text{Edge}(T_\Gamma) \longrightarrow M$$

with the properties

i. $c(\bar{y}) = -c(y)$ for all $y \in \text{Edge}(T_\Gamma)$, and

ii. $\sum_{E(y)=P} c(y) = 0$ for all $P \in \text{Vert}(T_\Gamma)$.

Let $C_{har}(T_\Gamma, M)$ denote the abelian group of all M-valued harmonic cocycles on T_Γ.

Our first basic observation will be that by "integration" one can construct a map from vector-valued holomorphic differential forms on H to vector-valued harmonic cocycles on T_Γ. By "integration" we mean the theory of residues which we shortly recall in the following. (I am grateful to F. Herrlich for some clarifying discussion about this point Let

$$F = \mathbb{C}_p \cup \{\infty\} \setminus (D_0 \cup \cdots \cup D_m)$$

be a connected affinoid set where the D_i are pairwise disjoint open disks

$$D_0 = \{x : |x-a_0|_p > |b_0|_p\} \quad \text{and}$$

$$D_i = \{x : |x-a_i|_p < |b_i|_p\} \quad \text{for } 1 \le i \le m;$$

for simplicity we only consider the case that $m \geq 1$ and $\infty \notin F$. Furthermore we can assume that $a_0 \notin F$. Put

$$F_i := \mathbb{C}_p \cup \{\infty\} \setminus D_i$$

and

$$w_o(x) := \frac{x-a_0}{b_0}, \text{ resp. } w_i(x) := \frac{b_i}{x-a_i} \quad \text{for} \quad 1 \leq i \leq m .$$

These $w_i(x)$ obviously are invertible holomorphic functions on F. Any holomorphic differential form $\omega \in \Omega(F)$ on F has representations

$$\omega = f_i \, d \frac{1}{w_i} \qquad \text{with} \qquad f_i \in \mathcal{O}(F).$$

Let now

$$f_i = f_o^{(i)} + \ldots + f_m^{(i)} \qquad \text{with} \qquad f_j^{(i)} \in \mathcal{O}(F_j)$$

$$\text{and} \quad f_j^{(i)}(\infty) = 0 \qquad \text{for} \quad 1 \leq j \leq m$$

be the Mittag-Leffler decomposition of f_i ([6] p. 41), which is uniquely determined and fulfills the following condition on the norms

$$(*) \qquad \| f_i \|_F = \max_{0 \leq j \leq m} \| f_j^{(i)} \|_{F_j} .$$

The differential form

$$\omega_i := f_i^{(i)} \, d \frac{1}{w_i}$$

then is meromorphic on F_i with at most one pole at $x = a_o$ in case $i = 0$, resp. $x = \infty$ in case $1 \leq i \leq m$. If

$$\omega_i = \sum_{\nu \in \mathbb{Z}} c_\nu^{(i)} w_i^\nu \, dw_i$$

denotes its development into a Laurent series we define

$$\operatorname{res}_{D_i} \omega := c_{-1}^{(i)} .$$

This definition is independent of the particular representation of the disks D_i. Namely, for $1 \leq i \leq m$ already ω_i and therefore also $\operatorname{res}_{D_i} \omega$ (see [4] p. 21) is independent; the case $i = 0$ then follows from the subsequent theorem of residues. Although this result is well known we will include a proof for the convenience of the reader.

<u>PROPOSITION</u>:

$\sum_{i=0}^{m} \operatorname{res}_{D_i} \omega = 0$ *for any* $\omega \in \Omega(F)$.

<u>Proof</u>: If the assertion holds true for rational holomorphic differential forms on F then also for any holomorphic one by taking limits and using

(*). Let therefore

$$\omega = g(x)dx \in \Omega(F)$$

be a differential form where $g(x)$ is a rational function without poles in F. If we view ω as a meromorphic differential form on $\mathbb{C}_p \cup \{\infty\}$ then we of course have

$$\sum_{a\in\mathbb{C}_p\cup\{\infty\}} \mathrm{res}_a\omega = 0$$

where $\mathrm{res}_a\omega$ is defined in the usual way by

$$\mathrm{res}_a\omega := \alpha_{-1}, \quad \omega = \begin{cases} \sum_{\nu\in\mathbb{Z}} \alpha_\nu\cdot(x-a)^\nu d(x-a) & \text{for } a \neq \infty, \\ \sum_{\nu\in\mathbb{Z}} \alpha_\nu\cdot(\frac{1}{x})^\nu d\,\frac{1}{x} & \text{for } a = \infty. \end{cases}$$

Since ω is holomorphic on F we get

$$\sum_{i=0}^{m} r_i(\omega) = 0 \qquad \text{with} \quad r_i(\omega) := \sum_{a\in D_i} \mathrm{res}_a\omega .$$

The assertion then is proved if we show that one has

$$\mathrm{res}_{D_i}\omega = -r_i(\omega) .$$

Let us first consider the case $i \geq 1$. If

$$g = g_0+\ldots+g_m \quad \text{with rational} \quad g_j \in \mathcal{O}(F_j)$$
$$\text{and} \quad g_j(\infty) = 0 \text{ for } 1 \leq j \leq m$$

is the Mittag-Leffler decomposition of g, then $\left(\sum_{j\neq i} g_j(x)\right)dx$ is holomorphic on D_i which implies

$$r_i(\omega) = r_i(g_i(x)dx).$$

According to [4] p. 22 we have

$$r_i(\omega) = r_i(g_i(x)dx) = d_{-1}^{(i)}$$

where $g_i(x)dx = \sum_{\nu\in\mathbb{Z}} d_\nu^{(i)} (\frac{1}{w_i})^\nu d\,\frac{1}{w_i}$. On the other hand, from $\omega = g(x)dx = gb_i d\,\frac{1}{w_i}$ we derive $f_i = b_i g$ and therefore $\omega_i = b_i g_i d\,\frac{1}{w_i} = g_i(x)dx$. Together with $d\,\frac{1}{w_i} = -w_i^{-2}dw_i$ this implies

$$c_{\nu}^{(i)} = -d_{-\nu-2}^{(i)} .$$

In the case $i = 0$ the differential form $\left(\sum_{j=1}^{m} f_j^{(0)}\right) d\,\frac{1}{w_0}$ is holomorphic on D_0 and we get

$$r_0(\omega) = r_0(f_0^{(0)} d\,\frac{1}{w_0}) .$$

But $f_0^{(0)} d\,\frac{1}{w_0}$ is holomorphic on $\{x: |x-a_0|_p = |b_0|_p\}$. We thus have

$$r_0(\omega) = r_0(f_0^{(0)} d\frac{1}{w_0}) = - \sum_{|a-a_0|_p<|b_0|_p} \mathrm{res}_a\, f_0^{(0)}\, d\frac{1}{w_0}$$

which according to [4] p. 22 is equal to $-c_{-1}^{(0)}$. Q.E.D.

We have to list some further useful properties of the residues the proof of which is an easy exercise.

<u>Remark</u>:

i. $\mathrm{res}_{D_i}(\omega_1+\omega_2) = \mathrm{res}_{D_i}\omega_1 + \mathrm{res}_{D_i}\omega_2$;

ii. let $F' = \mathbb{C}_p \cup\{\infty\} \setminus (D_0 \cup \ldots \cup D_n \cup D'_{n+1} \cup \ldots) \supseteq F$ be an affinoid set containing F where the $D_0,\ldots,D_n,D'_{n+1},\ldots$ $(1 \leq n < m)$ are pairwise disjoint open disks; for $0 \leq i \leq n$ and any $\omega \in \Omega(F')$ we then have

$$\mathrm{res}_{D_i}\omega = \mathrm{res}_{D_i}\omega|F \quad ;$$

iii. for any $\gamma \in PGL_2(\mathbb{C}_p)$ with $\infty \notin \gamma(F)$ we have

$$\mathrm{res}_{\gamma(D_i)}{}^{\gamma}\omega = \mathrm{res}_{D_i}\omega \quad \text{with} \quad {}^{\gamma}\omega := \omega\circ\gamma^{-1}$$

The second ingredient which we need for the construction of a map from the holomorphic differential forms $\Omega(H)$ on H to $C_{har}(T_\Gamma,\mathbb{C}_p)$ is a certain natural family of affinoid subsets of H. Its definition relies on ideas of Drinfeld ([3], see also [4] Chap. V). We first put

$$U(y) := \{a \in \mathcal{L}: \text{a halfline in } T_\Gamma \text{ corresponding to } a \text{ passes through } y\}$$

for any $y \in \mathrm{Edge}(T_\Gamma)$. The $U(y)$ are compact and open in $\mathcal{L}$ and form a basis of the topology of $\mathcal{L}$.

Remark:

i. $\mathcal{L} = U(y) \cup U(\bar{y})$ and $\mathcal{L} = \bigcup_{E(y)=P} U(\bar{y})$ where the union is disjoint in eac

case;

ii. $U(\gamma(y)) = \gamma(U(y))$ for any $\gamma \in \Gamma$.

Let now

$$R: \mathbb{C}_p \cup \{\infty\} \longrightarrow \overline{\mathbb{F}}_p \cup \{\infty\}$$

$$a \longmapsto \begin{cases} a \bmod \mathfrak{m} & \text{if } |a|_p \leq 1, \\ \infty & \text{otherwise} \end{cases}$$

be the usual reduction map where $\mathfrak{m}$, resp. $\overline{\mathbb{F}}_p$, denotes the maximal ideal resp. the residue class field, of $\mathbb{C}_p$; we set $R_\sigma := R \circ \sigma^{-1}$ for $\sigma \in PGL_2(K$ Furthermore, we denote by P_o that vertex of T_K which is defined by the lattice $\mathfrak{o}_K \oplus \mathfrak{o}_K$ where $\mathfrak{o}_K$ is the ring of integers in K.

LEMMA:

For any $y \in Edge(T_\Gamma)$, *the set*

$$D_y := R_\sigma^{-1}(R_\sigma(U(\bar{y}))) \subseteq \mathbb{C}_p \cup \{\infty\}$$

where $\sigma \in PGL_2(K)$ *is such that* $E(y) = \sigma(P_o)$ *is an open disk and does not depend on the special choice of* σ.

Proof: The fibres of R_σ are open disks. So, it remains to show that $R_\sigma(U(\bar{y}))$ is a one-point set. We obviously can assume that $T_\Gamma = T_K$ and $\sigma = 1$ in which case that property is easily checked by explicit computation.

Thus, for any $P \in Vert(T_\Gamma)$,

$$F(P) := \mathbb{C}_p \cup \{\infty\} \setminus \bigcup_{E(y)=P} D_y$$

is a connected affinoid subset of H, and we have

$$F(\gamma(P)) = \gamma(F(P)) \qquad \text{for } \gamma \in \Gamma.$$

We now associate with a holomorphic differential form $\omega \in \Omega(H)$ the ma

$$c_\omega: \; Edge(T_\Gamma) \longrightarrow \mathbb{C}_p$$

$$y \longmapsto res_{D_y}(\omega | F(E(y))) .$$

LEMMA:

c_ω is a harmonic cocycle on T_Γ.

Proof: The above proposition immediately implies $\sum_{E(y)=P} c_\omega(y) = 0$. Fix now an edge y of T_Γ and put $Q: = A(y)$ and $P: = E(y)$. The open disks D_z with $E(z) = P$, $z \neq y$ or $E(z) = Q$, $z \neq \bar{y}$ then are pairwise disjoint such that

$$F(y): = \mathbb{C}_p \cup \{\infty\} \setminus \bigcup_z D_z \supseteq F(Q) \cup F(P) \ ;$$

this follows from the general fact that, for any two edges y_1, y_2 of T_Γ with $E(y_2) = A(y_1)$ and $y_2 \neq \bar{y}_1$, we have

$$D_{y_2} \subseteq D_{y_1}$$

(reduce to the case $T_\Gamma = T_K$ and apply [6] I§2). Using again the above proposition we compute

$$c_\omega(y) = \operatorname{res}_{D_y}(\omega|F(P)) = -\sum_{\substack{E(z)=P \\ z\neq y}} \operatorname{res}_{D_z}(\omega|F(P))$$

$$= -\sum_{\substack{E(z)=P \\ z\neq y}} \operatorname{res}_{D_z}(\omega|F(y)) = \sum_{\substack{E(z)=Q \\ z\neq \bar{y}}} \operatorname{res}_{D_z}(\omega|F(y))$$

$$= \sum_{\substack{E(z)=Q \\ z\neq \bar{y}}} \operatorname{res}_{D_z}(\omega|F(Q)) = -\operatorname{res}_{D_{\bar{y}}}(\omega|F(Q))$$

$$= -c_\omega(\bar{y}) .$$

Q.E.D.

We therefore get the Γ-equivariant homomorphism

$$I: \quad \Omega(H) \longrightarrow C_{har}(T_\Gamma, \mathbb{C}_p)$$

$$\omega \longmapsto I(\omega): = c_\omega \quad .$$

In order to derive from it maps from the automorphic forms to the harmonic cocycles we introduce the symmetric powers

$$W^n: = \operatorname{Sym}^n W \qquad (n \geq 0)$$

of the natural representation of $\Gamma \subseteq SL_2(K)$ on the $\mathbb{C}_p$-vector space

$W = \mathbb{C}_p \oplus \mathbb{C}_p$. We then have the homomorphisms

$$I_n: \quad M_{n+2}(\Gamma) \longrightarrow H^0(\Gamma,\Omega(H)\otimes W^n) \longrightarrow H^0(\Gamma,C_{har}(T_\Gamma,W^n))$$
$$f \longmapsto \omega_f \longmapsto c_f := (I\otimes id_{W^n})(\omega_f)$$

where $\omega_f := \sum_{i=0}^{n} x^i f(x)dx \otimes (1,0)^i \cdot (0,1)^{n-i}$.

Remark:

There is a canonical map $\varepsilon_n: H^0(\Gamma,C_{har}(T_\Gamma,W^n)) \longrightarrow H^1(\Gamma,W^n)$ (see [9] II.2.8). We will show in another paper that

$$\varepsilon_n \circ I_n : M_{n+2}(\Gamma) \xrightarrow{\cong} H^1(\Gamma,W^n)$$

is an isomorphism (which can be viewed as an analogue of the Shimura isomorphism in the classical theory of automorphic forms).

The next basic observation is that harmonic cocycles on T_Γ are nothing else than certain distributions on the set of limit points $\mathcal{L}$.

DEFINITION:

For any abelian group M and any locally compact and totally disconnected space X let D(X,M) denote the abelian group of all M-valued finitely additive functions on the family of compact open subsets of X ("distributions on X"). In case X is compact put
$D_o(X,M) := \{\mu \in D(X,M): \mu(X) = 0\}$.

The following result due to Drinfeld ([3]) now is easy to prove.

LEMMA:

The map $D_o(\mathcal{L},M) \longrightarrow C_{har}(T_\Gamma,M)$ *is an isomorphism.*

$$\mu \longmapsto c_\mu(y) := \mu(U(y)).$$

Furthermore, if we set $\mathcal{L}_o := \mathcal{L}\setminus\{\infty\}$ then restriction of distributions induces an isomorphism $D_o(\mathcal{L},M) \xrightarrow{\cong} D(\mathcal{L}_o,M)$. Altogether we thus have constructed homomorphisms

$$M_{n+2}(\Gamma) \longrightarrow C_{har}(T_\Gamma,W^n) \cong D_o(\mathcal{L},W^n) \cong D(\mathcal{L}_o,W^n)$$
$$f \longmapsto c_f \longmapsto \mu_f .$$

We consider μ_f as the <u>p-adic L-transform</u> of the automorphic form f. If f has weight 2 then μ_f even is a $\mathbb{C}_p$-valued measure (i.e., a bounded distribution) on $\mathcal{L}_o$. Namely, because of its Γ-invariance and the finiteness of the quotient graph S the cocycle c_f takes on only a finite number of different values. In general μ_f will not be a measure but we can describe its growth rather precisely. Let f always be an automorphic form of weight n+2 for Γ.

<u>Notation</u>: For any $\omega \in \Omega(H)$ and any $y \in \mathrm{Edge}(T_\Gamma)$ we put

$$\mathrm{res}_y \omega := \mathrm{res}_{D_y} (\omega | F(E(y))).$$

<u>LEMMA</u>:

For $0 \leq i \leq n$, $y \in \mathrm{Edge}(T_\Gamma)$, $\gamma = \begin{pmatrix} a & b \\ c & d \end{pmatrix} \in \Gamma$, *and* $e \in \mathbb{C}_p$ *such that* $\gamma(e) \neq \infty$ *we have*

$$\mathrm{res}_{\gamma(y)} (x-\gamma e)^i f(x)dx = (ce+d)^{n-2i} \cdot \sum_{j=0}^{n-i} \binom{n-i}{j} \left(e+\frac{d}{c}\right)^{-j} \cdot \mathrm{res}_y (x-e)^{i+j} f(x)dx .$$

<u>Proof</u>: Using $(-cx+a) = (a-c\gamma(e)) - c(x-\gamma(e))$ and $(ce+d)(a-c\gamma(e)) = 1$ we compute

$$\mathrm{res}_y (x-e)^i f(x)dx = \mathrm{res}_{\gamma(y)} {}^{\gamma}((x-e)^i f(x)dx)$$

$$= \mathrm{res}_{\gamma(y)} \left(\frac{dx-b}{-cx+a} - e\right)^i (-cx+a)^{n+2} f(x) (-cx+a)^{-2} dx$$

$$= \mathrm{res}_{\gamma(y)} (ce+d)^i (x-\gamma e)^i (-cx+a)^{n-i} f(x)dx$$

$$= \sum_{j=0}^{n-i} \binom{n-i}{j} \mathrm{res}_{\gamma(y)} (ce+d)^i (x-\gamma e)^i (a-c\gamma(e))^j (-c)^{n-i-j} (x-\gamma e)^{n-i-j} f(x)dx$$

$$= \sum_{j=0}^{n-i} \binom{n-i}{j} (ce+d)^{i-j} (-c)^{n-i-j} \mathrm{res}_{\gamma(y)} (x-\gamma e)^{n-j} f(x)dx.$$

In particular, our assertion holds true if i = n. The general case then follows by an inductive argument using identities like

$$\sum_{j=i}^{m-1} (-1)^j \binom{m}{j} \binom{j}{i} = (-1)^{m+1} \binom{m}{i} \qquad \text{for } i < m. \qquad \text{Q.E.D.}$$

<u>PROPOSITION</u>:

There exists a constant $C > 0$ *such that we have*

$$\rho_y^{n/2-i} \cdot |\mathrm{res}_y (x-e)^i f(x)\,dx|_p < C$$

for all $0 \leq i \leq n$, $y \in \mathrm{Edge}(T_\Gamma)$ *with* $U(\bar{y}) \subseteq \mathcal{L}_o$, *and* $e \in U(\bar{y})$ *where*

$$\rho_y := \sup\{|u-v|_p : u,v \in D_y\} .$$

Proof: Since the quotient graph S is finitely we can choose finitely man edges $y_1, \ldots, y_m$ of T_Γ such that $\infty \notin U(\bar{y}_1) \cup \ldots \cup U(\bar{y}_m)$ and such that any $y \in \mathrm{Edge}(T_\Gamma)$ with $\infty \notin U(\bar{y})$ is Γ-equivalent to one of the y_ν, say $y = \gamma(y_\nu)$ with $\gamma = \begin{pmatrix} a & b \\ c & d \end{pmatrix} \in \Gamma$. Using

$$|\gamma^{-1}(e) + \frac{d}{c}|_p = |\gamma^{-1}(e) - \gamma^{-1}(\infty)|_p \geq \rho_{y_\nu}$$

and

$$\rho_y = |c\gamma^{-1}(e)+d|_p^{-2} \cdot \rho_{y_\nu}$$

we derive from the above lemma

$$\rho_y^{n/2-i} \cdot |\mathrm{res}_y (x-e)^i f(x)\,dx|_p$$

$$\leq \rho_{y_\nu}^{n/2-i} \cdot \max_{0\leq j\leq n-i} |\gamma^{-1}(e)+\frac{d}{c}|_p^{-j} \cdot |\mathrm{res}_{y_\nu} (x-\gamma^{-1}(e))^{i+j} f(x)\,dx|_p$$

$$\leq \max_{0\leq j\leq n-i} \rho_{y_\nu}^{n/2-i-j} \cdot |\mathrm{res}_{y_\nu} (x-\gamma^{-1}(e))^{i+j} f(x)\,dx|_p .$$

But the last term obviously is bounded independently of $\gamma^{-1}(e) \in U(\bar{y}_\nu)$.

Q.E.D.

Let us define the $\mathbb{C}_p$-valued distributions $\mu_f^{(o)}, \ldots, \mu_f^{(n)}$ on $\mathcal{L}_0$ by

$$\mu_f = \sum_{i=0}^{n} \mu_f^{(i)} \cdot (1,0)^i \cdot (0,1)^{n-i} .$$

Putting

$$\int_U x^i d\mu_f := \mu_f^{(i)}(U)$$

for $0 \leq i \leq n$ and any compact open subset $U \subseteq \mathcal{L}_0$ then induces a $\mathbb{C}_p$-linear map

$$\int \cdot\, d\mu_f : \mathcal{C}^n(\mathcal{L}_0) \longrightarrow \mathbb{C}_p$$

on the space $\mathcal{C}^n(\mathcal{L}_0)$ of all functions with compact support on $\mathcal{L}_o$ which are locally a polynomial in x of degree $\leq n$. The above proposition shows that this map satisfies a certain growth condition; we namely have

$$\int_{U(\overline{y})} (x-e)^i d\mu_f = - \operatorname{res}_y (x-e)^i f(x) dx$$

(under the appropriate assumptions). That property allows us to extend $\int . d\mu_f$ to a map on all functions with compact support on $\mathcal{L}_o$ which satisfy a certain condition of Lipschitz type. In order to be more specific let us make the following assumption which from an arithmetic point of view seems to be a natural one:

$$\Gamma \text{ is cocompact in } SL_2(\mathbb{Q}_p).$$

Then $T_\Gamma = T_{\mathbb{Q}_p}$ (use [9] II.1.5.5) and μ_f is a distribution on $\mathcal{L}_o = \mathbb{Q}_p$. In fact, the above proposition shows that $\int . d\mu_f$ induces an "admissible measure" on $\mathbb{Z}_p^x$ in the sense of Visik ([10]). The function

$$L_p(f,\chi) := \int_{\mathbb{Z}_p^x} \chi \, d\mu_f$$

therefore is well-defined and analytic in $\chi \in \mathrm{Hom}_{cont}(\mathbb{Z}_p^x, \mathbb{C}_p^x)$ (see [10]). In particular, if $\kappa: \mathbb{Z}_p^x \to 1+p\mathbb{Z}_p \subseteq \mathbb{C}_p^x$ denotes the canonical projection map then

$$L_p(f,s) := L_p(f,\kappa^{1-s})$$

is an analytic function on the open disk $\{s \in \mathbb{C}_p : |s|_p < qp^{-1/(p-1)}\}$ where $q = 4$ for $p = 2$ and $q = p$ otherwise.

REFERENCES

[1] Cerednik, I.V.: Uniformization of algebraic curves by discrete arithmetic subgroups of $PGL_2(k_w)$ with compact quotients. Math. USSR Sbornik 29, 55-78 (1976)

[2] Deligne, P., Rapoport, M.: Les schémas de modules de courbes elliptiques. In Modular Functions of One Variable II. Lecture Notes in Math. vol. 349, 143-316. Berlin-Heidelberg-New York: Springer 1973

[3] Drinfeld, V.G.: Coverings of p-adic symmetric regions. Funct. Anal. Appl. 10, 107-115 (1976)

[4] Fresnel, J., van der Put, M.: Géométrie Analytique Rigide et Applications. Boston-Basel-Stuttgart: Birkhauser 1981

[5] Gerritzen, L.: Zur nichtarchimedischen Uniformisierung von Kurven. Math. Ann. 210, 321-337 (1974)

[6] Gerritzen, L., van der Put, M.: Schottky Groups and Mumford Curves. Lecture Notes in Math. vol. 817. Berlin-Heidelberg-New York: Springer 1980

[7] Mumford, D.: An Analytic Construction of Degenerating Curves over Complete Rings. Compositio Math. 24, 129-174 (1972)

[8] Ribet, K.: Sur les variétés abéliennes à multiplications reelles. C.R. Acad. Sc. Paris 291, 121-123 (1980)

[9] Serre, J.-P.: Trees. Berlin-Heidelberg-New York: Springer 1980

[10] Visik, M.M.: Non-archimedean measures connected with Dirichlet series. Math. USSR Sbornik 28, 216-228 (1976)

UN RAPPORT SUR DE RECENTS TRAVAUX EN THEORIE ANALYTIQUE DES NOMBRES

G. Tenenbaum

Nous nous proposons ici de décrire succinctement un certain nombre de résultats récents qui constituent un progrès significatif dans l'étude de la structure multiplicative des entiers.

Le problème consiste, étant donné un entier n, à décrire le plus exactement et le plus précisément possible la suite de ses diviseurs. Il est clair qu'il ne faut attendre aucun renseignement universel non trivial. L'étude probabiliste donne un éclairage plus pertinent : on cherche à déterminer, en particulier par leur densité naturelle, des suites pour lesquelles on peut exhiber un comportement spécifique.

Toute l'information concernant les diviseurs d'un entier est en principe contenue dans ses facteurs premiers. Cependant, alors que la connaissance probabiliste des facteurs premiers paraît assez avancée (voir par exemple [1], [6], [7], [14], [15]), la description subséquente des diviseurs n'est pas encore satisfaisante.

On peut interpréter la situation de la manière suivante. Etant donné un entier n, l'étude de ses diviseurs se confond avec celle de la variable aléatoire D_n prenant les valeurs $(\log d)/(\log n)$, lorsque d parcourt les diviseurs de n, avec probabilité uniforme $1/\tau(n)$. A la factorisation canonique de n correspond la décomposition de D_n en une somme de variables aléatoires indépendantes

$$D_n = \sum_{p|n} X_p$$

où, pour chaque nombre premier p, X_p prend les valeurs $(j \log p)/(\log n)$, $0 \leqslant j \leqslant v_p(n)$, avec probabilité uniforme $1/(v_p(n)+1)$, la notation v_p désignant la valuation p-adique. Cette décomposition est traitée directement par des techniques probabilistes dans [10], [27], [34]. Le mauvais rendement qui est constaté lors de la transition de l'information des X_p aux D_n reflète non seulement le traditionnel défaut d'indépendance des nombres premiers dans le modèle de la Théorie Probabiliste des Nombres mais également certaines difficultés de nature spécifiquement probabiliste - comme par exemple le problème de la fonction de concentration d'une somme de variables indépendantes, qui est sous-jacent à la question arithmétique traitée dans [28] .

La situation est bien illustrée par l'étude en loi des D_n . La régularité en moyenne de la distribution des facteurs premiers suffit à impliquer l'existence d'une valeur moyenne [2]

$$x^{-1} \sum_{n \leqslant x} \operatorname{Prob}(D_n \leqslant u) = (2/\pi) \operatorname{Arc} \sin \sqrt{u} + O((\log x)^{-1/2})$$

uniformément pour $x \geqslant 2$ et $0 \leqslant u \leqslant 1$. En revanche, la violence des turbulences locales induit une sorte de principe d'incertitude : toute suite d'entiers $\mathcal{A}$ pour laquelle la suite $\{D_n : n \in \mathcal{A}\}$ possède une loi limite est nécessairement de densité nulle [30] .

Historiquement, la première information précise sur la répartition des facteurs premiers est le Théorème de Hardy et Ramanujan [24] :

$$\omega(n) = (1 + o(1)) \log\log n \,, \qquad \text{(p.p.)} \,,$$

où $\omega(n)$ désigne le nombre des facteurs premiers distincts de n , et la notation p.p. (presque partout) indique que la relation mentionnée a lieu pour une suite d'entiers de densité unité.

Connu depuis 1917, ce résultat n'a pas été véritablement utilisé dans l'étude probabiliste des nombres avant les années 30 et les travaux d'Erdös. Une généralisation très utile en est fournie par la proposition que la somme

$$\sum_{n \leqslant x} y^{\omega(n)}$$

est dominée, pour y borné, par les entiers n satisfaisant à

$$\omega(n) = y \log\log x + O(\xi(x) \sqrt{\log\log x}) \,,$$

où $\xi(x)$ désigne une fonction tendant vers l'infini arbitrairement lentement. Pour innocent qu'il paraisse, ce lemme n'en constitue pas moins un résultat précieux. Il permet de pondérer une somme arithmétique par un coefficient multiplicatif dont les variations épousent celles de la fonction caractéristique de l'ensemble des entiers ayant une quantité prescrite de facteurs premiers [29] .

Le fameux principe d'Erdös énonçant que le j-ième facteur premier d'un entier "normal" est de l'ordre de exp exp j peut être utilisé de manière similaire. La version quantitative la plus précise de ce résultat est :

$$\omega(n,t) := \sum_{p|n, p\leqslant t} 1 = \log\log t + O(\sqrt{\log\log t \, \log\log\log\log t})$$

pour presque tout n et uniformément pour $\xi(n) < t \leqslant n$, [4] . Comme dans le cas du Théorème de Hardy et Ramanujan, on peut construire des poids multiplicatifs adaptés aux fonctions $\omega(n,t)$ et établir des lemmes idoines. Ces résultats sont les prototypes d'une famille d'assertions élémentaires fondamentales qui sous-tendent des étapes cruciales dans la plupart des démonstrations sur ce sujet. La manière dont il faut les employer est cependant rarement immédiate et peut requérir une certaine sophistication.

Par nécessité de concision, nous nous limiterons ici aux travaux concernant les ordres moyens et normaux des fonctions arithmétiques liées aux diviseurs. Nous avons écarté, bien que des progrès significatifs aient également été réalisés dans ces domaines, les questions de fonctions de répartition, d'ensembles de multiples, et, de propriétés extrêmales.

Selon une idée couramment répandue, l'ordre moyen est l'information la plus facile à obtenir concernant une fonction arithmétique. C'est effectivement le cas lorsque la fonction envisagée peut être décomposée en une somme d'une manière qui rend pertinente une interversion de sommations. La plupart des fonctions classiques, comme $\omega(n)$, $\varphi(n)$ [l'indicateur d'Euler] , $\tau(n)$ [nombre des diviseurs], $\sigma_k(n)$ [somme des puissances k-ièmes des diviseurs], relèvent d'un tel traitement. Toutefois, la théorie fine des diviseurs fourmille d'exemples dans lesquels une manipulation semblable est inopérante. Les quatre cas suivants sont typiques :

$$\rho(n) := \sum_{1\leqslant i<\tau(n)} d_i/d_{i+1} ,$$

$$g(n) := \text{card } \{i,\ 1\leqslant i<\tau(n) : d_i | d_{i+1}\}$$

$$f(n) := \text{card } \{i,\ 1\leqslant i<\tau(n) : (d_i, d_{i+1}) = 1\}$$

$$\Delta(n) := \max_u \text{card } \{d : d|n ,\ u<d\leqslant eu\} ;$$

ici et dans la suite nous désignons par

$$1 = d_1 < d_2 < \ldots < d_{\tau(n)} = n$$

la suite ordonnée des diviseurs de n . Bien que les trois premières fonctions soient définies sous forme de sommes, le calcul de la moyenne par interversion est, en l'état actuel des choses, impossible : il suppose une estimation de

$$A(x;u,v) := \text{card }\{n \leqslant x : u|n,\ v|n;\ w\nmid n,\ u < w < v\} \quad .$$

Cette quantité dépend intimement des propriétés arithmétiques de u et v ; on n'en connaît pas d'évaluation satisfaisante.

Des quatre fonctions introduites plus haut, seule la première possède un ordre moyen connu. Il découle du résultat général suivant

THEOREME 1. ([13]). *Soit* θ *une application de classe* C^2 *de* $[0,1]$ *dans* $\mathbb{R}$. *On a*

$$(1) \quad \sum_{n \leqslant x} \ \sum_{1 \leqslant i < \tau(n)} \theta(d_i/d_{i+1}) = x \log x \left\{\theta(1) + O\left(\frac{\log\log\log x}{(\log x)^{\delta}\sqrt{\log\log x}}\right)\right\}$$

où l'exposant δ *défini par*

$$(2) \quad \delta = 1 - (\log(e \log 2))/(\log 2) = 0{,}086071 \ldots ,$$

est optimal dès que $t\,\theta'(t)$ *est monotone sur* $[0,1]$.

On peut se demander comment les difficultés décrites précédemment ont été contournées. Le point de départ de la démonstration consiste à remarquer que, pour $n < x$, la somme intérieure de (1) vaut

$$\theta(1)(\tau(n)-2) - \int_1^{\sqrt{x}} \int_1^{\sqrt{x}} \varphi(y/z)\ \varepsilon(n;y,z) z^{-2}\, dy dz + O(\log(2x/n))$$

où l'on a posé $\varphi(t) = (t\,\theta'(t))'$ et où $\varepsilon(n;y,z)$ prend la valeur 1 si n possède un diviseur dans $[y,z[$ et la valeur 0 dans le cas contraire. Après permutation de la sommation en n et de l'intégrale double, on constate que le problème se ramène à l'estimation de

$$H(x,y,z) := \sum_{n \leqslant x} \varepsilon(n;y,z) \quad .$$

Cette quantité correspond au cas le plus simple de crible par des nombres composés : le crible par tous les entiers d'un intervalle. Elle intervient naturellement dans de nombreux problèmes de Théorie des Nombres (une liste de références est donnée dans [31]). Dans [33], nous dégageons quatre comportements asymptotiques possibles pour $H(x,y,z)$.

Nous citons seulement le résultat suivant, qui implique assez facilement le Théorème 1.

THEOREME 2. ([33]). *Sous l'hypothèse*

$$(3) \qquad 1 \leqslant 2y \leqslant z \leqslant \min(y^{3/2},\ x^{1/2})$$

et, en définissant u *par la relation* $y^{1+u} = z$ *, on a*

$$(4) \qquad x\, u^{\delta}\, L_1(1/u) < H(x,y,z) < x\, u^{\delta}\, L_2(1/u)$$

où la constante positive δ *est définie par (2) et où* L_1 *et* L_2 *sont des fonctions à variation lente, tendant vers* 0 *à l'infini, dont un choix possible est*

$$L_1(v) = \exp\{-c_1 \sqrt{\log v \cdot \log\log 2v}\},\quad L_2(v) = c_2(\log v)^{-1/2} \log\log 2v\ .$$

Si l'on remplace, dans (3), $2y$ *par* $(1+\varepsilon)y$ *, la formule (4) reste valable à condition de considérer* c_1 *et* c_2 *comme des fonctions de* $\varepsilon > 0$ *. De plus, dans le cas où* $z = O(y)$ *, on peut omettre, quitte à modifier* c_2 *, le facteur* $\log\log 2v$ *dans* $L_2(v)$ *.*

Ainsi le Théorème 1 a finalement été prouvé à l'aide d'une interversion de sommations, mais d'un type un peu inhabituel.

Un autre exemple d'une telle situation apparaît dans la théorie de la fonction $\Delta(n)$. Hooley a montré [25] que la valeur moyenne de cette fonction a des applications dans diverses branches de l'Arithmétique, comme l'Approximation Diophantienne, ou le Problème de Waring. L'analyse de Fourier classique fournit facilement l'encadrement suivant

$$(1/\tau(n))\int_{-1}^{1} |\tau(n,\theta)|^2\, d\theta \ll \Delta(n) \ll \int_{-1}^{1} |\tau(n,\theta)|\, d\theta\ ,\quad (n \geqslant 1)\ ,$$

où l'on a posé

$$\tau(n,\theta) := \sum_{d|n} d^{i\theta}\ ,\quad (\theta \in \mathbb{R})\ .$$

La fonction multiplicative $\tau(n,\theta)$ est un instrument essentiel dans l'étude des diviseurs. Hall a déterminé en 1975, [17], la valeur moyenne de $|\tau(n,\theta)|^u$ pour chaque valeur du paramètre positif u . L'application de ses résultats pour $u = 1$ et $u = 2$ fournit l'estimation suivante.

THEOREME 3. *Posons* $S(x) = \sum_{n \leq x} \Delta(n)$. *On a*

$$x \log\log x \ll S(x) \ll x(\log x)^{(4/\pi)-1} .$$

La borne supérieure a été démontrée par Hooley en 1979 par une méthode voisine. La borne inférieure, qui améliore un résultat d'Erdös, figure dans un travail en collaboration avec Hall [21] . Dans ce même article nous affinons la majoration grâce à une utilisation récursive de l'inégalité élémentaire

$$\Delta(n)^2 \leq \sum_{\substack{dd' \mid n \\ |\log d'/d| \leq 1}} \Delta(n/dd') .$$

Les deux méthodes peuvent être combinées pour améliorer encore l'exposant.

THEOREME 4.([23]). *Soit* α *l'infimum des réels* ξ *tels que*

$$S(x) \ll x(\log x)^{\xi}$$

On a

$$\alpha < 0,2197 .$$

A titre de comparaison, $(4/\pi) - 1 = 0,2732 \ldots$.

Il serait trop long de décrire ici toutes les méthodes directes ou indirectes qui ont été employées pour estimer des ordres moyens de fonctions liées à la répartition des diviseurs. Nous nous contenterons de signaler que le Théorème 2 est souvent utile dans ce type de problèmes. Il est appliqué dans [13] pour les fonctions $f(n)$ et $g(n)$ introduites précédemment.

Nous portons maintenant notre attention sur la question des ordres normaux. Un problème célèbre dans ce domaine est la conjecture d'Erdös [5]

$$(5) \qquad \Delta(n) \geq 2 , \qquad \text{(p.p.)} ,$$

que l'on peut encore énoncer

$$E(n) := \min_{1 \leq i < \tau(n)} d_{i+1}/d_i < e \ , \qquad \text{(p.p.)} .$$

Le nombre des rapports distincts d'/d , $d|n$, $d'|n$, vaut $\prod_{p^\nu \| n} (2\nu + 1)$ et s'approche donc de

$$3^{\omega(n)} = (\log n)^{\log 3 + o(1)} \ , \qquad \text{(p.p.)} .$$

En supposant l'équirépartition asymptotique, on est donc amené à la conjecture plus forte

$$E(n) = 1 + (\log n)^{1 - \log 3 + o(1)} \ , \qquad \text{(p.p.)} .$$

La minoration contenue dans cette formule a été prouvée par Erdös et Hall en 1979, [9] . Leur résultat est même plus précis : on a

$$E(n) > 1 + (\log n)^{1-\log 3} \exp \{- c \sqrt{\log\log n \cdot \log\log\log\log n}\}, \qquad \text{(p.p.)} ,$$

pour une constante positive convenable c .

La majoration a été récemment établie par Maier. Sa méthode est indirecte : il établit d'abord un *théorème de comparaison* permettant de traduire le problème en des termes purement probabilistes ; dans un deuxième temps, il développe un argument récursif très ingénieux pour prouver le résultat probabiliste requis. Sa démonstration est très longue et difficile. En retenant l'idée de base de la seconde étape, nous avons construit une preuve arithmétique fondée sur l'utilisation de poids multiplicatifs du type décrit au début de cet article. Cette démonstration, qui figure dans un article en commun [28] , est considérablement plus courte et plus simple. Elle fournit aussi une légère amélioration du résultat final : on a

$$E(n) < 1 + (\log n)^{1-\log 3} \exp \{\xi(n) \sqrt{\log\log n}\} \ , \qquad \text{(p.p.)} ,$$

pour toute fonction $\xi(n)$ tendant vers l'infini. Cette borne est probablement optimale.

La même méthode permet encore une minoration de l'ordre normal de $\Delta(n)$, [28] :

$$\Delta(n) > (\log\log n)^{\gamma} \ , \qquad \text{(p.p.)} ,$$

pour toute constante $\gamma < (-\log 2)/(\log(1-1/\log 3)) = 0,28754\ldots$.
Maier a annoncé une majoration du type

$$\Delta(n) < (\log\log n)^{\beta} \quad , \qquad \text{(p.p.)}$$

pour une constante convenable β . Nous ne savons pas encore si une approche directe est susceptible de fournir un tel résultat.

Un autre ensemble de théorèmes liés aux ordres normaux trouve son origine dans le concept d'*équirépartition sur les diviseurs*, essentiellement dû à Kátai [26] . On dit qu'une fonction arithmétique est équirépartie sur les diviseurs (en abrégé *erd*) si l'on a pour tout z , $0 \leqslant z \leqslant 1$,

$$\text{card}\,\{d : d|n,\ f(d) \leqslant z (\text{mod}\, 1)\} = (z + o(1))\ \tau(n) \quad , \qquad \text{(p.p.)}\ .$$

Kátai a montré qu'une condition nécessaire et suffisante pour qu'une fonction additive f soit *erd* est que

$$\sum_{p} \|\nu f(p)\|^2/p = +\infty \;, \qquad (\nu = 1, 2, \ldots) \;,$$

où $\|x\|$ désigne la distance de x à l'ensemble des entiers. La fonction logarithme a été étudiée plus en détail par Hall [16], [17], et Erdös et Hall [8] . Le cas général est plus complexe [3], [18], [19], [20], [22], [32] .

Le théorème suivant fournit un analogue du critère classique de Weyl.

THEOREME 5. ([32]). *Une condition nécessaire et suffisante pour qu'une fonction* f *soit* *erd* *est que l'on ait pour* x *infini*

$$\sum_{k<x} \Big| \sum_{\substack{n<x \\ n \equiv 0 (\text{mod}\, k)}} \frac{e(\nu f(n))}{n\ 4^{\Omega(n)}} \Big| = o(\sqrt{\log x}) \;, \quad (\nu = 1, 2, \ldots) \;,$$

où $\Omega(n)$ *désigne le nombre des facteurs premiers de* n *, comptés avec leur ordre de multiplicité, et où l'on a posé* $e(x) = \exp(2i\pi x)$.

Comme la croissance des diviseurs est exponentielle p.p. on pourrait croire, par analogie avec l'équirépartition classique, que le cas des fonctions à croissance plus rapide qu'une puissance pose un problème hors de portée. Il n'en est rien. La restriction à une suite de densité 1 est une condition suffisamment souple pour permettre une étude assez avancée avec les outils usuels. Nous donnons ci-dessous un tableau synoptique des principaux résultats typiques connus.

La fonction	*est erd si et seulement si*	*références*
$(\log\log d)^{\alpha}$	$\alpha > 1$	[19]
$(\log d)^{\beta}$	$\beta > 0$	[32]
$\exp\{(\log d)^{\gamma}\}$, $(0 < \gamma < 3/5)$	-	[22]
θd	$\theta \in \mathbb{R} \setminus \mathbb{Q}$	[3]
d^{α}	$\alpha \in \mathbb{R}_{+} \setminus \mathbb{N}$	[22]

Nous concluons ce rapport par un bref aperçu de la démonstration de la conjecture d'Erdös (5) donnée dans [28] . Nous nous restreignons pour simplifier l'exposé à la version initiale de la conjecture.

Posons pour chaque entier n

$$n_k = \prod_{\substack{p \mid n \\ p < \exp\exp k}} p \quad .$$

Alors n_k possède $(1 + o(1))\, k$ facteurs premiers p.p. dès que $k \geqslant k_0(n) \to \infty$. Nous voulons montrer que, si k est assez grand en fonction de n , on a

$$\Delta(n_k) \geqslant 2 \quad , \qquad\qquad \text{(p.p.)} .$$

Intuitivement, l'idée de la démonstration consiste à établir que la "probabilité conditionnelle" pour que $\Delta(n_{k+\ell}) = 1$ sachant que $\Delta(n_k) = 1$ ne dépasse pas $1 - \varepsilon$, où $\varepsilon = \varepsilon(n) \to 0$ assez lentement, pour des valeurs relativement petites de ℓ . Par itération, on obtient que l'évènement $\Delta(n_{r\ell}) = 1$ se produit avec une "probabilité" au plus égale à $(1-\varepsilon)^r$. Si l'on peut alors choisir r de sorte que $\varepsilon r \to \infty$, cette majoration tend vers 0, d'où la conclusion.

Pour atteindre ce résultat, on montre que la mesure $\lambda(n_k)$ de l'ensemble

$$\bigcup_{dd'|n_k} (\log d'/d + [-1,1])$$

est suffisamment grande p.p. . Cela permet de minorer efficacement la probabilité qu'un diviseur premier de $n_{k+\ell}/n_k$ soit proche d'un rapport d'/d, donc la probabilité que l'on puisse exhiber deux diviseurs proches de $n_{k+\ell}$. L'analyse de Fourier fournit la minoration

$$\lambda(n_k) \geqslant 3^{2\omega(n_k)} \Big/ \left(2\pi \int_{-1}^{1} \prod_{p|n_k} (1+2\cos(\theta \log p))^2 \, d\theta \right) .$$

On obtient une borne supérieure p.p. pour le dénominateur en utilisant des moyennes pondérées par des coefficients multiplicatifs du type décrit au début de cet article.

Bibliographie

[1] Bovey, J.D., On the size of prime factors of integers, *Acta Arith.* 33 (1977), 65-80.

[2] Deshouillers, J.-M., Dress, F., Tenenbaum, G., Lois de répartitio[n] des diviseurs, 1, *Acta Arith.* 34 (1979), 273-285.

[3] Dupain, Y., Hall, R.R., Tenenbaum, G., Sur l'équirépartition modulo 1 de certaines fonctions de diviseurs, *J. London Math. Soc.* (2 26 (1982), 397-411.

[4] Erdös, P., On the distribution function of additive functions, *Ann. of Math.* 47 (1946), 1-20.

[5] Erdös, P., On the density of some sequences of integers, *Bull. Amer. Math. Soc.* 54 (1948), 685-692.

[6] Erdös, P., Some remarks on prime factors of integers, *Canadian J. Math.* 11 (1959), 161-167.

[7] Erdös, P., On the distribution of prime divisors, *Aequationes Math.* 2 (1969), 177-183.

[8] Erdös, P., Hall, R.R., Some distribution problems concerning the divisors of integers, *Acta Arith.* 26 (1974), 175-188.

[9] Erdös, P., Hall, R.R., The propinquity of divisors, *Bull. London Math. Soc.* 11 (1979), 304-307.

[10] Erdös, P., Nicolas, J.-L., Méthodes probabilistes et combinatoires en théorie des nombres, *Bull. Sc. Math.* (2° série), 100 (1976), 301-320.

[11] Erdös, P., Nicolas, J.-L., Propriétés probabilistes des diviseurs d'un nombre, *Astérisque* 41-42 (1977), 203-214.

[12] Erdös, P., Tenenbaum, G., Sur la structure de la suite des diviseurs d'un entier, *Ann. Inst. Fourier* 31 (1981), 17-37.

[13] Erdös, P., Tenenbaum, G., Sur les diviseurs consécutifs d'un entier, *Bull. Soc. Math. de France* 111, fasc. 2 (1983), à paraître.

[14] Friedlander, J.B., Integers free from large and small primes, *Proc. London Math. Soc.* (3) 33 (1976), 565-576.

[15] Galambos, J., The sequences of prime divisors of integers, *Acta Arith.* 31 (1976), 213-218.

[16] Hall, R.R., The divisors of integers I, *Acta Arith.* 26 (1974), 41-46.

[17] Hall, R.R., Sums of imaginary powers of the divisors of integers, *J. London Math. Soc.* (2) 9 (1975), 571-580.

[18] Hall, R.R., The distribution of $f(d) \pmod 1$, *Acta Arith.* 31 (1976), 91-97.

[19] Hall, R.R., A new definition of the density of an integer sequence, *J. Austral. Math. Soc.* Ser. A, 26 (1978), 487-500.

[20] Hall, R.R., The divisor density of integers sequences, *J. London Math. Soc.* (2) 24 (1981), 41-53.

[21] Hall, R.R., Tenenbaum, G., On the average and normal orders of Hooley's Δ-function, *J. London Math. Soc.* (2) 25 (1982), 392-406.

[22] Hall, R.R., Tenenbaum, G., Les ensembles de multiples et la densité divisorielle, *Ann. of Math.*, à paraître.

[23] Hall, R.R., Tenenbaum, G., The average orders of Hooley's Δ_r-functions, soumis à *Mathematika*.

[24] Hardy, G.H., Ramanujan, S., The normal number of prime factors of a number n, *Quart. J. Math.* 48 (1917), 76-92.

[25] Hooley, C., On a new technique and its application to the theory of numbers, *Proc. London Math. Soc.* (3) 38 (1979), 115-151.

[26] Kátai, I., Distributions (mod *1)* of additive functions on the set of divisors, *Acta Arith.* 30 (1976), 209-212.

[27] Kátai, I., The distribution of additive functions on the set of divisors, *Publicationes Math.* 24 (1-2) (1977), 91-96.

[28] Maier, H., Tenenbaum, G., On the set of divisors of an integer, soumis à *Inventiones Math.*

[29] Tenenbaum, G., Sur une technique en théorie analytique des nombres *Sém. Théorie des Nombres,* Bordeaux (1979/80).

[30] Tenenbaum, G., Lois de répartition des diviseurs, 2, *Acta Arith.* 38 (1980), 1-36.

[31] Tenenbaum, G., Sur la probabilité qu'un entier possède un diviseur dans un intervalle donné, *Sém. Théorie des Nombres,* Bordeaux (1981/82).

[32] Tenenbaum, G., Sur la densité divisorielle d'une suite d'entiers, *J. Number Theory* 15 (1982), 331-346.

[33] Tenenbaum, G., Sur la probabilité qu'un entier possède un diviseur dans un intervalle donné, *Compositio Math.* (1983), à paraître.

[34] Vose, M.D., *Limit theorems for sequences of divisor distributions,* Ph. D., Austin at Texas (1981).

Gérald Tenenbaum
Université de Nancy I
UER Sciences Mathématiques
Boîte Postale 239
54506 Vandoeuvre les Nancy Cedex
France

SUR LA RÉSOLUTION D'UN PROBLÈME DE PLONGEMENT

N. Vila
Secció de Matemàtiques,
Universitat Autònoma de Barcelona,
Bellaterra, Barcelona, Espanya.

Le problème inverse de la théorie de Galois demande si, donné un groupe fini G, il existe un corps de nombres K galoisien sur Q avec groupe de Galois $G(K/Q)$ isomorphe à G.

Comme il est bien connu, Šafarevič (1954) a prouvé que ce problème a une reponse affirmative si G est un groupe résoluble. Le problème pour un groupe quelconque reste sans une solution générale. On sait seulement que certaines familles de groupes non-résolubles se réalisent comme groupe de Galois sur Q.

Dans ce contexte, avec P. Bayer, nous avons mis en question si les extensions centrales du groupe alterné A_n sont groupe de Galois sur Q. On a trouvé une réponse affirmative au problème pour presque la moitié des valeurs de n.

Nous abordons cette question comme un problème de plongement d'une extension à groupe de Galois A_n dans une extension ayant comme groupe de Galois le groupe $\hat{A}_n$ des représentations de A_n. Pour ceci, on construit d'abord des familles d'équations à groupe de Galois A_n sur $Q(T)$. En utilisant un résultat récent de Serre, on prouve que l'obstruction au plongement est nulle pour presque la moitié des valeurs de n.

1. Le problème de plongement

Soit $\hat{A}_n$ le groupe des représentations (Darstellungsgruppe) de A_n ([2], 23.4). Rappelons que Schur [4] a montré que $\hat{A}_n$ est la seule extension non triviale de A_n avec noyeau $Z/2$, pour $n \neq 6$ et 7, ce que nous supposerons par la suite.

D'autre part, on peut voir que le groupe des représentations de A_n (ou plus généralement celui d'un groupe parfait quelconque) est caractérisé par la propieté universelle suivante:

Pour toute extension centrale E de A_n, il existe un homomorphisme de groupes unique $h : \hat{A}_n \longrightarrow E$ tel que le diagramme

$$\begin{array}{ccccccccc} 1 & \longrightarrow & Z/2 & \longrightarrow & \hat{A}_n & \xrightarrow{\pi} & A_n & \longrightarrow & 1 \\ & & & & h \downarrow & & \| \, id & & \\ 1 & \longrightarrow & \ker j & \longrightarrow & E & \xrightarrow{j} & A_n & \longrightarrow & 1 \end{array}$$

est commutatif ([3], th 5.7).

On a alors le lemme de réduction suivant

<u>LEMME 1.1.</u> *Soit* K *un corps de nombres et* N/K *une extension gal*

sienne à groupe de Galois isomorphe à A_n *. Si* N/K *admet un plongement dans une extension à groupe de Galois* $\hat{A}_n$*,* N/K *admet aussi un plongement dans une extension ayant comme groupe de Galois une extension centrale de* A_n *quelconque.*

Ainsi, notre problème de réaliser toutes les extensions centrales du groupe alterné se réduit à réaliser seulement une, $\hat{A}_n$.

Soit $a_n \in H^2(A_n, Z/2)$ l'élément associé à la suite exacte

$$1 \longrightarrow Z/2 \longrightarrow \hat{A}_n \longrightarrow A_n \longrightarrow 1.$$

Soit K un corps commutatif de caractéristique $\neq 2$. Soit L/K une extension séparable de degré n et N/K sa clôture normale. Supposons que son groupe de Galois G(N/K) est isomorphe à A_n, on a donc un épimorphisme $G_K \longrightarrow A_n$. Considérons l'homomorphisme d'inflation associé à cet épimorphisme

$$\inf : H^2(A_n, Z/2) \longrightarrow H^2(G_K, Z/2).$$

Il est bien connu que $\inf(a_n)$ mesure l'obstruction au plongement de N/K dans une extension à groupe de Galois $\hat{A}_n$ (cf. [1]).

Un résultat récent de Serre nous donne, dans notre cas, la valeur de l'obstruction au plongement dans $\hat{A}_n$ en termes d'un invariant de Hasse-Witt. On a

<u>THÉORÈME 1.2</u> (Serre, [5]).

$$\inf (a_n) = w(L/K),$$

où $w(L/K)$ *désigne l'invariant de Hasse-Witt de la forme quadratique* $Tr_{L/K}(X^2)$.

D'après ce résultat le problème de réaliser $\hat{A}_n$ comme groupe de Galois sur Q se concrète à résoudre les questions suivantes:

a) Trouver des polynômes $f(X) \in Q[X]$ irréductibles de degré n tels que son groupe de Galois soit isomorphe à A_n.

b) Calculer $w(L/Q)$, où $L = Q(\theta)$, étant θ une racine de $f(X)$.

c) Donner des conditions sur $f(X)$ à fin que $w(L/Q) = 1$.

2. Construction d'équations avec groupe de Galois A_n

Soit G un groupe fini. Soient $t_1,\ldots,t_r$ éléments de G, nous dirons que $(t_1,\ldots,t_r)$ est une r - *présentation de Hurwitz* de G si

a) $t_1,\ldots,t_r$ génèrent G,

b) $t_1\cdot\ldots\cdot t_r = 1$.

Nous désignerons par $\mathcal{H}_r(G)$ l'ensemble des r-présentations de Hurwitz de G.

Si $(t_1,\ldots,t_r) \in \mathcal{H}_r(G)$, soit $\mathcal{H}(t_1,\ldots,t_r)$ l'ensemble des $(s_1,\ldots,s_r) \in \mathcal{H}_r(G)$ tels que les subgroupes $\langle s_i \rangle$ et $\langle t_i \rangle$, $1 \leq i \leq r$, sont conjugués à G. Nous appelons *nombre de Hurwitz* de $(t_1,\ldots,t_r)$, $h(t_1,\ldots,t_r)$, au cardinal

$$h(t_1,\ldots,t_r) = \#(\mathcal{H}(t_1,\ldots,t_r) / \mathrm{Aut}(G))$$

Soient $F = \overline{Q}(T)$ et $k_o \subset \overline{Q}$ un subcorps, où $\overline{Q}$ est une clôture algébraïque de Q. On dit qu'une extension de Galois N/F est *galoisienne* k_o-*définie* s'il existe une extension de Galois N_o/F_o, $F_o = k_o(T)$, telle que

$$N_o\overline{Q} = N \quad \text{et} \quad G(N_o/F_o) \cong G(N/F)$$

En général il est un problème difficile de trouver les subcorps de définition d'une extension galoisienne de F. Tout de suite, nous donnons un critère qui garantit que certains corps sont corps de définition dans le cas où le groupe de Galois soit un groupe complet (c'est-à-dire qu'il s'agit d'un groupe dans lequel le centre est trivial et tout automorphisme est interne).

THÉORÈME 2.1. *Soit* G *un groupe fini complet et soit* $(t_1,\ldots,t_r) \in \mathcal{H}_r(G)$ *telle que* $h(t_1,\ldots,t_r) = 1$. *Si* $k_o \subset \overline{Q}$ *est un subcorps et* $S = \{\wp_1,\ldots,\wp_r\}$ *est une famille de premiers de* F

k_o*- définis, alors il existe une extension galoisienne* N/F *avec groupe de Galois isomorphe à* G *qui est galoisienne* k_o*-définie.*

Démonstration. Soit F^S l'extension galoisienne de F maximale non-ramifiée en dehors de S, et G^S son groupe de Galois sur F. Soient $u_1,\dots,u_r$ générateurs du groupe profini G^S satisfaisant $u_1\cdot\ldots\cdot u_r = 1$. Considérons l'épimorphisme $\pi : G^S \longrightarrow G$ défini par $\pi(u_i) = t_i$. Soit $N = (F^S)^{\ker(\pi)}$ le corps fixe par ker(π) on a $G(N/F) \cong G$. En utilisant que $h(t_1,\dots,t_r) = 1$ et que les premiers φ_i sont k_o-définis, on peut prouver que l'extension N/F_o est normale. Soit $E = G(N/F_o)$ et H le centralisateur de G dans E. E opère sur G par conjugation et étant G complet on peut définir un épimorphisme de groupes $E \longrightarrow G$ tel que la suite

$$1 \longrightarrow H \longrightarrow E \longrightarrow G \longrightarrow 1,$$

est exacte et décompose. Alors $N_o = N^H$ est l'extension galoisienne de F_o que l'on cherche.

Remarque. Toute groupe fini complet admettant une présentation de Hurwitz avec nombre de Hurwitz 1 est groupe de Galois sur Q(T) et, d'après le théorème d'irréductibilité de Hilbert, sur Q.

Comme une conséquence de ce dernier théorème et d'une étude de la ramification des extensions finies de $\bar{Q}(T)$ ([6]), nous avons obtenu un résultat qui nous permettra de construir des polynômes sur Q(T) avec groupe de Galois certains groupes complets.

THÉORÈME 2.2. *Soit* G *un groupe fini complet,* π *une représentation par permutations de* G *fidèle et transitive de degré* n. *Soit* $(t_1,\ldots,t_r) \in \mathcal{H}_r(G)$ *avec* $h(t_1,\ldots,t_r) = 1$. *Soient* $\mathfrak{P}_1,\ldots,\mathfrak{P}_r$ *premiers de* F *Q-définis. On suppose qu'il y a une seule extension* K/F_0, *à moins de* F_0*-isomorphismes, de degré* n *avec ramification seulement aux* $\mathfrak{P}_i$, $1 \leq i \leq r$, *et du type:*

$$\mathfrak{P}_i = \mathfrak{A}_{i,1}^{e_{i,1}} \cdots \mathfrak{A}_{i,k}^{e_{i,k}} \mathfrak{A}_{i,0},$$

où les $\mathfrak{A}_{i,j}$, $0 \leq j \leq k$, *sont diviseurs de* K *non-ramifiés;* $e_{i,0} = 1$, $e_{i,j}$, $0 < j \leq k$, *sont les différents longueurs des cicles disjoints dans* $\pi(t_i)$ *plus grands que* 1 *et* $\deg(\mathfrak{A}_{i,j})$ *est le nombre de cicles disjoints dans* $\pi(t_i)$ *de longueur* $e_{i,j}$, $0 \leq j \leq k$. *Alors le groupe de Galois de la clôture galoisienne de* K/F_0 *est isomorphe à* G.

En utilisant ce résultat, nous construirons de nouvelles familles de polynômes à groupe de Galois S_n, $n \geq 5$.

PROPOSITION 2.3. *Soient* n,k *des entiers positifs, premiers entre eux,* $k \leq n$. *Considérons les permutations*

$$s_1 = (n\ n-1\ \ldots\ 3\ 2\ 1),$$
$$s_2 = (1\ 2\ \ldots\ k)\ (k+1\ \ldots\ n),$$
$$s_3 = (1\ k).$$

Alors $(s_1, s_2, s_3) \in \mathcal{H}_3(S_n)$ *et* $h(s_1, s_2, s_3) = 1$.

Démonstration. Soit G le groupe engendré par s_1, s_2; $s_3 = s_1 s_2$. On a $(1\ 2 \ldots k)$, $(k+1\ k+2 \ldots n) \in G$, car $s_2^{n-k} = (1\ 2 \ldots k)^{n-k}$ et $(k, n-k) = 1$. D'autre part

$$(1\ 2) = s_1\ (k+1 \ldots n)\ s_3\ (1\ 2 \ldots k)\ s_3 \in G,$$

en conséquence, $G = S_n$. On doit montrer maintenant que $h(s_1, s_2, s_3) = 1$. Soit $(t_1, t_2, t_3) \in \mathcal{H}(s_1, s_2, s_3)$, t_1 est un n-cicle, alors il existe $b \in S_n$ tel que $s_1 = b\ t_1\ b^{-1}$. Si $t_i' = b\ t_i\ b^{-1}$, $i = 2,3$, (t_1, t_2, t_3) et (s_1, t_2', t_3') sont equivalent module $\mathrm{Aut}(S_n) = S_n$. D'autre part, il existe un entier r tel qu $s_1^{-r}\ t_3'\ s_1^{r} = (1\ j)$, ou $j \leqslant (n+1)/2$. Il doit être $j = k+1$, donc $s_1^{-r}\ t_2'\ s_1^{r} = s_1^{-1}\ (1j)$ engendre un subgroupe conjugué de $<s_2$

THÉORÈME 2.4. *Soient* n,k *des entiers positifs, premiers entre eux,* $k \leqslant n$, $n \geqslant 5$. *Le polynôme*

$$G(X,T) = X^{n-k}\left(X - \frac{n}{n-k}\right)^k - \left(\frac{-k}{n-k}\right)^k T \qquad (1)$$

a groupe de Galois sur $Q(T)$ *isomorphe à* S_n.

Démonstration. Considérons $\wp_0, \wp_1, \wp_\infty$, les premiers de $\overline{Q}(T)$ définis par

$$\mathrm{div}\ (T) = \wp_0\ \wp_\infty^{-1}\ ,\ \mathrm{div}\ (T-1) = \wp_1\ \wp_\infty^{-1}.$$

Supposons que $K/Q(T)$ soit une extension de degré n, qui ramifie seulement à $\mathfrak{P}_\infty, \mathfrak{P}_0, \mathfrak{P}_1$, et de la façon:

$$\mathfrak{P}_\infty = \mathfrak{p}_\infty^n, \quad \mathfrak{P}_0 = \mathfrak{p}_{00}^{n-k}\, \mathfrak{p}_{01}^k, \quad \mathfrak{P}_1 = \mathfrak{p}_1^2 \mathfrak{a},$$

où $\mathfrak{p}_\infty, \mathfrak{p}_{00}, \mathfrak{p}_{01}, \mathfrak{p}_1$, sont premiers de K de degré 1 et $\mathfrak{a}$ est un diviseur non-ramifié de degré $n - 2$. Le corps K a genre zero et premiers de degré 1, alors K est un corps de fonctions rationelles à une variable sur Q. Choisissons la variable x telle que $K = Q(x)$ et

$$\operatorname{div}(x) = \mathfrak{p}_{00}\, \mathfrak{p}_\infty^{-1}, \qquad \operatorname{div}(x-1) = \mathfrak{p}_1\, \mathfrak{p}_\infty^{-1}$$
$$\operatorname{div}(x-a) = \mathfrak{p}_{01}\, \mathfrak{p}_\infty^{-1}, \quad \operatorname{div}(x^{n-2} + a_{n-3}\, x^{n-3} + \dots + a_0) = \mathfrak{a}\,\mathfrak{p}_\infty^{-1}$$

où $a, a_0, \dots, a_{n-3} \in Q$. On peut voir, alors, que x est une racine du polynôme (1), et d'après le théorème 2.3, on obtient le résultat.

Soit N le corps de décomposition du polynôme (1). Soit $L = N^{A_n}$ le corps fixe par $A_n \subset S_n$. L est aussi un corps de fonctions rationelles à une variable sur Q. On peut montrer

THÉORÈME 2.5. *Soient* n, k *des entiers positifs, premiers entre eux,* $k \leq n/2$, $n \geq 5$. *Le polynôme*

$$F_{n,k}(X,T) = \begin{cases} X^n - A(nX-k(n-k))^k, & \textit{si } n \textit{ est impair} \\ X^n + k^{n-2k}\, B^{n-k-1}\, (nX+(n-k)kB)^k, & \textit{si } n \textit{ est pair}, \end{cases}$$

où $A = k^{n-2k}(1-(-1)^{(n-1)/2}nT^2)$, $B = ((-1)^{n/2}k(n-k)T^2+1)$, *a*
groupe de Galois A_n *sur* $Q(T)$.

COROLLAIRE 2.6. *Le groupe de Galois de* $F_{n,k}(X,T)$ *sur* $Q(i,T)$ *est* A_n.

3. Calcul de l'invariant de Hasse-Witt

Soit $E = Q(T,\theta)$, étant θ une racine du polynôme $F_{n,k}(X,$
Pour calculer l'invariant de Hasse-Witt de la forme quadratique $Tr_{E/Q(T)}(X^2)$ il faut d'abord diagonaliser cette forme quadratiqu

THÉORÈME 3.1. *Si* k *est impair et* $k \leqslant (n+1)/3$, *l'espace quadratique* E *décompose*

$$E = \begin{cases} <1> \perp <\theta^{(n-k)/2}> \perp ((n-3)/2)H \perp <v>, & si\ n\ est\ impair \\ <1> \perp ((n-2)/2)H \perp <v>, & si\ n\ est\ pair, \end{cases}$$

où H *est un plan hyperbolique.*

Pour montrer ce résultat, on doit distinguer selons la parité de n. Les démonstrations dans les deux cas sont analogues, nous aborderons ici seulement le cas n impair.

Soit n impair, les valeurs $Tr(\theta^i)$, $0 \leq i \leq 2n-2$, déterminent la forme quadratique $Tr_{E/Q(T)}(X^2)$, car $1,\theta,\ldots,\theta^{n-1}$ est une base du $Q(T)$-espace vectoriel E. On peut calculer sans dificulté:

$Tr(1) = n,$

$Tr(\theta^i) = 0, \qquad 1 \leq i \leq n-k-1$

$Tr(\theta^i) = i\,A\binom{k}{n-i}\, n^{n-i}\,(-k(n-k))^{k-(n-i)}, \quad n-k \leq i \leq n$

$Tr(\theta^{n+i}) = 0, \qquad 1 \leq i \leq n-2k-1$

$$Tr(\theta^{n+i}) = \sum_{j=\max\{n-k-i,0\}}^{\min\{k,n-i\}} (i+j)A^2\binom{k}{j}\binom{k}{n-i-j} n^{n-i}(-k(n-k))^{i-(n-2k)}, \quad n-2k \leq i \leq n-2.$$

Alors, les vecteurs $1,\theta,\ldots,\theta^{(n-k)/2}$ sont deux à deux ortogonaux et $\theta,\ldots,\theta^{(n-k-2)/2}$ sont vecteurs isotropes. En conséquence, nous avons une première décomposition de l'espace quadratique E,

$$E = \langle 1\rangle \perp \langle\theta^{(n-k)/2}\rangle \perp ((n-k-2)/2)H \perp E',$$

où H est un plan hyperbolique et E' est un subespace de dimension k, contenu dans un suplémentaire de $\langle\theta,\ldots,\theta^{(n-k-2)/2}\rangle$ dans $\langle 1,\theta,\ldots,\theta^{(n-k)/2}\rangle^{\perp}$.

Pour finir l'étude de l'espace quadratique E, nous devons classifier l'espace quadratique E'. Pour faire ceci, nous calculons (cf.[6]) la valeur de la forme quadratique restringée au subespace $\langle 1,\theta,\ldots,\theta^{m}\rangle^{\perp}$, où $m = (n-k)/2$.

LEMME 3.2. *Soient* $e, v \in \langle 1, \theta, \ldots, \theta^m \rangle^{\perp}$,

$$e = \sum_{i=0}^{n-1} \lambda_i \theta^i, \quad v = \sum_{i=0}^{n-1} \mu_i \theta^i.$$

Alors

$$Tr_{E/Q(T)}(ev) =$$

$$=(n-k)nA[(-\lambda_m \mu_m n^{k-1} + \sum_{j=1}^{k-1} \lambda_{m+j} (\sum_{i=1}^{k-j} (n-k+i+j) \binom{k}{k-i-j} n^{k-i-j-1} (-k)^{i+j} (n-k)^{i+j-1} \mu_{m+i}))$$

$$+kA(\sum_{i=1}^{k} \lambda_{n-i} (k(n-k))^{k-i} (-1)^i n^{i-1} \binom{k-1}{i-1}) (\sum_{i=1}^{k} \mu_{n-i} (k(n-k))^{k-i} (-1)^i n^{i-1} \binom{k-1}{i-1})].$$

D'après ce calcul nous pouvons finir la classification de l'espace quadratique E' en construisant effectivement suffisanment de vecteurs isotropes dans l'espace E'.

LEMME 3.3. *L'espace* E' *décompose*

$$E' = ((k-1)/2)H \perp \langle v \rangle,$$

où H *est un plan hyperbolique et* v *est un vecteur avec* $Tr\, v^2 = (-1)^{(n-3)/2} A(n-k) \in Q(T)^*/Q(T)^{*2}$.

Démonstration. Soit $e = \sum_{i=1}^{n-1} \lambda_i \theta^i$, $e \in E'$. On a que les scalaires $\lambda_{m+k}, \ldots, \lambda_{n-1}, \lambda_o \in Q(T)$ dépenent linéairement de $\lambda_m, \ldots, \lambda_{m+k-1} \in Q(T)$, et ces derniers sont libres. En conséquence, ils existent des constantes $A_o, \ldots, A_{k-1} \in Q(T)$ telles que

$$\sum_{i=1}^{k}\lambda_{n-i}(k(n-k))^{k-i}\ (-1)^{i}\ n^{i-1}\ \binom{k-1}{i-1} = A_0\lambda_m + \ldots + A_{k-1}\ \lambda_{m+k-1}.$$

Si pour tout $i > (k-1)/2$, $A_i = 0$, alors les vecteurs

$$e_i = \sum_{j=0}^{n-1} \lambda_j^i\ \theta^j, \quad 1 \leq i \leq (k-1)/2,$$

où $\lambda_j^i = 0$ pour tout $j \neq m + i + (k-1)/2$, $m \leq j \leq m + k - 1$, et $\lambda^i_{m+i+(k-1)/2} \neq 0$,

sont linéairement indépendants et d'après le lemme 3.2, on a

$$Tr(e_i^2) = Tr(e_i\ e_j) = 0, \quad 1 \leq i,j \leq (k-1)/2.$$

En conséquence, $E' = (k-1)/2\ H \perp \langle v \rangle$.

Si par contre, el existe un r, $(k+1)/2 \leq r \leq k-1$, tel que $A_r \neq 0$, nous considérons les vecteurs

$$e_i = \sum_{j=0}^{n-1} \lambda_j^i\ \theta^j, \quad (k+1)/2 \leq i \leq k-1,\ i \neq r,$$

où $\lambda_j^i = 0$, $j \neq m+r, m+i$, $0 \leq j \leq n-1$ et $\lambda^i_{m+i} \neq 0$, λ^i_{m+r} tels que $A_0\ \lambda_m + \ldots + A_{k-1}\ \lambda_{m+k-1} = 0$. D'après le lemme 3.2, on a

$$Tr\ (e_i^{\ 2}) = Tr\ (e_i e_j) = 0, \quad (k+1)/2 \leq i,j \leq k-1, \quad i,j \neq r.$$

D'autre part, les vecteurs $\{e_i, \theta^j\}$, $0 \leq j \leq m$, $(k+1)/2 \leq i \leq k-1$ sont linéairement indépendants. Nous avons donc $(k-3)/2$ vecteurs dans E'.

Soit $v = \sum_{i=0}^{n-1} \mu_i \theta^i$,

où $\mu_m \neq 0$, $\mu_{m+i} = 0$, $1 \leqslant i \leqslant k-1$, $i \neq r$, et $\mu_{m+r} \neq 0$ tel que $A_o \mu_m + \ldots + A_{k-1} \mu_{m+k-r} \neq 0$. On peut voir alors que

$$E' = ((k-3)/2)\, H \perp \langle v \rangle \perp E",$$

où $\dim E" = 2$ et le discriminat de $E"$ est -1. C'est-à-dire E est un plan hyperbolique.

Une fois diagonalisée la forme quadratique $Tr_{E/Q(T)}(X^2)$ on peut calculer facilement son invariant de Hasse-Witt.

THÉORÈME 3.4. *L'invariant de Hasse-Witt de l'espace quadratique* E *est*

$$w(E/Q(T)) = \begin{cases} (-(n-k)k,\ (-1)^{(n-1)/2}n) \otimes (-1,-1)^{(n+1)(n-1)/8}, & \textit{si } n \textit{ est impair} \\ (n, (-1)^{n/2}) \otimes (-1,-1)^{n(n-2)/8}, & \textit{si } n \textit{ est pair.} \end{cases}$$

4. Solutions au problème

Définition. On dit qu'un entier n $(n \neq 4\,m,\ n \neq 8\,m+7)$ la propiété (*N*) s'il existe une décomposition en somme de trois carrés

$$n = k_1^2 + k_2^2 + k_3^2$$

telle que $(k_1,n) = 1$ et $k_1^2 \leqslant (n+1)/3$.

D'après les théorèmes 1.2, 3.4, on a

THÉORÈME 4.1. *Soient* n,k *entiers positifs, premiers entre eux,* k *impair et* $k \leqslant (n+1)/3$. *Soit* $N_{n,k}$ *le corps de décomposition du polynôme* $F_{n,k}(X,T)$ *du théoreme* 2.5. *L'extension* $N_{n,k}/Q(T)$ *admet un plongement dans une extension à groupe de Galois* $\hat{A}_n$ *dans les cas suivants:*

$n \equiv 0 \pmod{8}$, $k > 0$,

$n \equiv 1 \pmod{8}$, k *un carré*,

$n \equiv 2 \pmod{8}$ *et somme de deux carrés*, $k>0$,

$n \equiv 3 \pmod{8}$ *et satisfaisant* (N), $k = k_1$.

Si $n \equiv 4,5,6$, ou $7 \pmod{8}$, $N_{n,k}/Q(T)$ *n'admet pas un plongement dans* $\hat{A}_n$ *pour aucunne valeur de* k.

Remarque. Moyennant l'ordinateur nous avons vu que tout entier $n \leqslant 600.000$ et $n \equiv 3 \pmod{8}$ a la propiété (N). Il paraît donc que la propiété (N) n'est pas restrictive.

COROLLAIRE 4.2. *Toute extension centrale de* A_n *est groupe de Galois sur* Q *pour les valeurs de* n *suivantes:*

$n \equiv 0,1 \pmod{8}$,

$n \equiv 2 \pmod{8}$ *et somme de deux carrés,*

$n \equiv 3 \pmod{8}$ *et satisfaisant* (N).

On peut observer finalment que sur $Q(i)$, toute extension centrale du groupe alterné se réalise comme groupe de Galois pour toute valeur de n, $n \neq 6,7$.

THÉORÈME 4.3. *Soient* n,k *entiers positifs, premiers entre eux,* k *impair,* $n \neq 6,7$, *et* $k \leq (n+1)/3$. *Soit* $N_{n,k}$ *le corps de décomposition du polynôme* $F_{n,k}(X,T)$ *du théorème 2.5. L'extension* $N_{n,k}/Q(i,T)$ *admet un plongement dans une extension à groupe de Galois* $\hat{A}_n$ *pour toute valeur de* k, *si* n *est pair et pour toute valeur de* k *carré, si* n *est impair.*

COROLLAIRE 4.4. *Toute extension centrale de* A_n *est groupe de Galois sur* $Q(i)$ *pour toute valeur de* n, $n \neq 6,7$.

Bibliographie

1. K. Hoechsmann; Zum Einbettungsproblem, J. Reine Angew. Math. 229 (1968), 81-106.

2. B. Huppert; Endliche Gruppen I, Die Grund. der Math. Wiss 134, Springer, 1967.

3. J. Milnor; Introduction to algebraic K-theory, Princenton University Press, 1971.

4. I. Schur; Über die Darstellungen der symmetrischen und alternierenden Gruppen durch gebrochene lineare Substitutionen, J. Math. 139 (1911), 155-250.

5. J.P. Serre; lettre à Martinet, 10 février 1982.

6. N. Vila; Sobre la realització de les extensions centrals del grup alternat com a grup de Galois sobre el cos dels racionals. (Thèse doctorale), Publ. Mat. U.A.B., Vol 27, Nº 3 (1983).

MIXING PROPERTIES OF THE LINEAR PERMUTATION GROUP

G. Wagner
University of Stuttgart
Pfaffenwaldring 57
D-7000 Stuttgart 80 / FRG

1. Introduction. For an arbitrary natural number $N \geq 1$ let $X = \{1,2,\dots,N\}$ be the set of residue classes mod N, with the usual multiplicative structure imposed on it. For any $a \in X$ satisfying $(a,N) = 1$ consider the mapping $\tau = \tau_a : X \to X$ defined by $\tau_a(x) = ax \pmod N$, $x \in X$. The mappings τ_a are one-to-one and form a group of order $\varphi(N)$, which we call the linear group Γ on X. (Translations are not considered here, since they do not affect the results in any respect.)

Let $A \subset X$ be a subset. What can be said about the behaviour of the sets τA, $\tau \in \Gamma$, with respect to uniform distribution in X? More precisely: For any $B \subset X$ denote by $\Delta(B)$ the discrepancy of B in X, defined as usual by ($\#$ denotes the number of elements of a set)

$$\Delta(B) = \max_{1 \leq k \leq N} \left| \#(B \cap \{1,2,\dots,k\}) - \frac{\#B}{N} \cdot k \right| .$$

$\Delta(B)$ always satisfies the inequality $0 \leq \Delta(B) \leq \frac{N}{4}$, and is a current measure for the deviation from uniform distribution.
Going back to the original subset $A \subset X$: Does there always exist a linear permutation $\tau \in \Gamma$ such that $\Delta(\tau A)$ is "small"? Do there exist subsets $A \subset X$ such that $\Delta(\tau A)$ is small for all $\tau \in \Gamma$?
We may formulate these questions in a similar way for sequences instead of sets. Let $A \subset X$, $\#A = r$, be arranged in arbitrary order as a sequence $\omega = (a_1, a_2, \dots, a_r)$. For a linear permutation $\tau \in \Gamma$ consider the transformed sequence $\tau\omega = (\tau a_1, \tau a_2, \dots, \tau a_r)$. (For A = X, this is simply a rearrangement of the original sequence.) Can we always find a $\tau \in \Gamma$ making $\Delta(\tau\omega) := \max_{1 \leq \rho \leq r} \Delta(\{\tau a_1, \tau a_2, \dots, \tau a_\rho\})$ small? Do there exist sequences ω with $\Delta(\tau\omega)$ small for all $\tau \in \Gamma$? These questions will be answer[ed] by essentially best possible estimates.
The main purpose of this paper, however, is to give a new interpretati[on] of certain well-known discrepancy estimates for the set of quadratic residues mod p, p prime, and the sequence of powers $g, g^2, \dots$ of a primitive root g (mod p^ℓ), $\ell \geq 1$. These estimates will appear as a natural consequence of certain invariance properties of the set (sequence) with respect to the linear group Γ.
We restrict ourselves to prime powers $N = p^\ell$, $p = 2,3,5, \dots$, althoug[h]

similar results (in a more complicated form) can be proved for composite N.

2. Upper estimates for subsets. Let $X = \{1,2, \dots ,p^\ell\}$, p prime, $\ell \geq 1$; let $A \subset X$. We give an upper estimate for $\min_{\tau\in\Gamma} \Delta(\tau A)$ by showing that $\sum_{\tau\in\Gamma} \Delta(\tau A)$ is small. Our result can be written in a more suggestive form by introducing the notion of "density" $\delta = \delta(A) = \frac{\# A}{p^\ell}$ of the subset A in the set X.

THEOREM 1. *Let* $A \subset X = \{1,2, \dots ,N\}$, $N = p^\ell$, *be a subset with density* $\delta = \delta(A)$. *The following inequality holds:*

$$\min_{\tau\in\Gamma} \Delta(\tau A) \leq 30 \sqrt{\delta(1-\delta)}\, N^{1/2} \log N .$$

Proof. We use the Erdös-Turan inequality as proved in |1|. We have

$$\sum_{(\tau,p^\ell)=1} \Delta(\tau A) \leq \# A\cdot \frac{6}{m+1}\varphi(p^\ell) + \frac{4}{\pi}\sum_{h=1}^{m}\frac{1}{h} \sum_{(\tau,p^\ell)=1} \left|\sum_{a\in A} e^{2\pi i h\tau a/p^\ell}\right| .$$

Here $1 \leq m < p^\ell$ is an arbitrary integer to be chosen later on.
Using the Cauchy-Schwarz inequality, we get

$$\sum_{(\tau,p^\ell)=1} \left|\sum_{a\in A} e^{2\pi i h\tau a/p^\ell}\right| \leq \varphi(p^\ell)^{1/2} \left[\sum_{a,b\in A}\ \sum_{(\tau,p^\ell)=1} e^{2\pi i h\tau\frac{a-b}{p^\ell}}\right]^{1/2} . \tag{1}$$

Our first aim is to give an estimate for the sum in square brackets.
Let $(h,p^\ell) = p^\mu$ $(0 \leq \mu < \ell)$ be the greatest common divisor.
For $a,b \in A$ fixed and $h^* = h\cdot p^{-\mu}$ we have

$$\sum_{(\tau,p^\ell)=1} e^{2\pi i h\tau\frac{a-b}{p^\ell}} = \sum_{(\tau,p^\ell)=1} e^{2\pi i h^*\tau\frac{a-b}{p^{\ell-\mu}}} =$$

$$= \sum_{\tau=1}^{p^\ell} e^{2\pi i h^*\tau\frac{a-b}{p^{\ell-\mu}}} - \sum_{\tau^*=1}^{p^{\ell-1}} e^{2\pi i h^*\tau^*\frac{a-b}{p^{\ell-\mu-1}}} . \tag{2}$$

The first sum in (2) is nonzero if and only if $p^{\ell-\mu} \mid (a-b)$, the second if and only if $p^{\ell-\mu-1} \mid (a-b)$. Denote by $R_\nu(A)$ $(0 \leq \nu \leq \ell)$ the number of pairs $a,b \in A$ satisfying the congruence relation $a \equiv b \mod p^\nu$. Then

$$\sum_{a,b\in A}\ \sum_{(\tau,p^\ell)=1} e^{2\pi i h\tau\frac{a-b}{p^\ell}} = p^\ell\cdot R_{\ell-\mu}(A) - p^{\ell-1}\cdot R_{\ell-\mu-1}(A) \leq$$

$\leq (p^\ell - p^{\ell-1})\, R_{\ell-\mu}(A)$, since the lefthand side is nonnegative and $R_{\ell-\mu-1}(A) \geq R_{\ell-\mu}(A)$. For each $a \in A$ there exist at most p^μ elements $b \in A$ such that $p^{\ell-\mu} \mid (a-b)$, hence $R_{\ell-\mu}(A) \leq \# A\cdot p^\mu$. This gives for each h,

$1 \leq h < p^\ell$, satisfying $(h,p^\ell) = p^\mu$:

$$\sum_{a,b \in A} \sum_{\tau} e^{2\pi i h\tau \frac{a-b}{p^\ell}} \leq \varphi(p^\ell)\, \#A \cdot p^\mu \quad . \tag{3}$$

Substituting (3) into (1) yields

$$\varphi(p^\ell)^{-1} \sum_{(\tau,p^\ell)=1} \Delta(\tau A) \leq \#A \cdot \frac{6}{m+1} + \frac{4}{\pi} \sum_{\mu=0}^{\ell-1} \sum_{\substack{h=1 \\ (h,p^\ell)=p^\mu}}^{m} \frac{1}{h} (\#A)^{1/2} p^{\mu/2}$$

$$\leq \#A \cdot \frac{6}{m+1} + \frac{4}{\pi} (\#A)^{1/2} \left(\sum_{h=1}^{m} \frac{1}{h}\right) \left(\sum_{\mu=0}^{\ell-1} p^{-\mu/2}\right)$$

$$\leq \#A \cdot \frac{6}{m+1} + \frac{4}{\pi} (\#A)^{1/2} (\log m + 1) \cdot 4$$

$$\leq 6 \left[\frac{\#A}{m+1} + (\#A)^{1/2} (\log m + 1)\right] .$$

Now choose $m = [\#A]^{1/2}$. This implies

$$\min_{\tau \in \Gamma} \Delta(\tau A) \leq 6(\#A)^{1/2} + 6(\#A)^{1/2} \left(1 + \frac{1}{2} \log(\#A)\right) \leq 21(\#A)^{1/2} \log p^\ell .$$

Here we used the inequality $2 < 3 \log 2 \leq 3 \log p$.

Now replace $\#A$ by δN, $N = p^\ell$. This gives

$$\min_{\tau \in \Gamma} \Delta(\tau A) \leq 21\sqrt{\delta}\, N^{1/2} \log N . \tag{4}$$

For $\delta \leq \frac{1}{2}$ we may add the factor $\sqrt{2(1-\delta)}$ on the righthand side of (4) for $\delta > \frac{1}{2}$ we do the same calculations for the complementary set $X \setminus A$. since complementary sets are mapped onto complementary sets by each $\tau \in \Gamma$ and since they have equal discrepancy, we get our final result

$$\min_{\tau \in \Gamma} \Delta(\tau A) \leq 30 \sqrt{\delta(1-\delta)}\, N^{1/2} \log N .$$

This proves Theorem 1.

3. Lower estimates for subsets. To get a lower bound for $\max_{\tau \in \Gamma} \Delta(\tau A)$ we shall use a method due to K.F.Roth |3|. Our aim is to prove the relatic $\max_{\tau \in \Gamma} \Delta(\tau A) \geq c \cdot N^{1/2}$ (5) , but the constant c cannot be expected to depend only on the number $\sqrt{\delta(1-\delta)}$ for the following simple reason: for $N = p^\ell$, $\ell > 1$, there exist low discrepancy subsets A which are invariant with respect to all $\tau \in \Gamma$. For example, take A to be the set of residues $\equiv 0 \bmod p$ in $X = \{1,2, \ldots ,p^\ell\}$. This set has discrepancy $\Delta(A) \leq 1$, density $\delta = 1/p$, and is invariant with respect to all $\tau \in \Gamma$. So in this case, we have $\max_{\tau \in \Gamma} \Delta(\tau A) \leq 1$. The constant c in (4) not only depends on the density δ, but also on the "densities" with respect to

residue classes $\equiv\lambda \pmod{p^\mu}$ in X ($1\leq\lambda\leq p^\mu; 0\leq\mu\leq\ell$). For any such residue class, let $\delta_\lambda(A,p^\mu) = \dfrac{\#\{A\cap\lambda \pmod{p^\mu}\}}{p^{\ell-\mu}}$. Define the "mean quadratic density" $\bar\delta^2(A,p^\mu)$ by $\bar\delta^2(A,p^\mu) = p^{-\mu}\sum_{\lambda=1}^{p^\mu}\delta_\lambda^2(A,p^\mu)$. Note that, in case $N=p$ is a prime, we have $\bar\delta^2(A,p^0) = \delta^2$ and $\bar\delta^2(A,p^1) = \delta$, where δ is the usual density of A in X as explained in 2.

The differences $\theta_\mu(A) := \bar\delta^2(A,p^{\ell-\mu}) - \bar\delta^2(A,p^{\ell-\mu-1})$ $(0\leq\mu\leq\ell-1)$ play the essential role in the next theorem. Note that $0\leq\theta_\mu(A)\leq 1$ (always) and, for $N=p$ prime, $\theta_0(A)$ coincides with our previous expression $\delta(1-\delta)$.

We first prove a lemma on the numbers $\theta_\mu(C)$ for intervals $C = \{1,2,\ldots,m\}$ $\subset X = \{1,2,\ldots,p^\ell\}$.

<u>LEMMA 1</u>. *There exists an interval* $C = \{1,2,\ldots,m\}\subset X = \{1,2,\ldots,p^\ell\}$ *such that* $\theta_\mu(C)\geq\frac{1}{8}p^{-2\mu}$ *for* $\mu = 0,1,\ldots,\ell-1$.

<u>Proof</u>. For $C = \{1,2,\ldots,m\}$, denote by ρ_μ the remainder $m - p^\mu\left[\frac{m}{p^\mu}\right]$, $\mu = 0,1,\ldots,\ell$. An easy but lengthy calculation shows that

$$\theta_\mu(C) = p^{-\ell-\mu}\left(\rho_{\ell-\mu}\left(1+2\left[\frac{m}{p^{\ell-\mu}}\right]\right) - p^{-1}\rho_{\ell-\mu-1}\left(1+2\left[\frac{m}{p^{\ell-\mu-1}}\right]\right)\right) +$$

$$+ p^{-2\mu}\left[\frac{m}{p^{\ell-\mu}}\right]^2 - p^{-2\mu-2}\left[\frac{m}{p^{\ell-\mu-1}}\right]^2 \qquad (0\leq\mu\leq\ell-1). \tag{6}$$

For p odd, choose $m = \frac{p-1}{2}(1+p+\ldots+p^{\ell-1}) < p^\ell$.

We have $[m/p^{\ell-\mu}] = (1/2)(p^\mu-1)$ and $\rho_{\ell-\mu} = (1/2)(p^{\ell-\mu}-1)$, $\mu = 0,\ldots,\ell$.

Substituting these relations into (6) yields

$$\theta_\mu(C) = \frac{1}{4}p^{-2\mu}\left(1-\frac{1}{p^2}\right)\geq\frac{2}{9}p^{-2\mu}.$$

For $N=2^\ell$, choose $m = 2^{\ell-1}+2^{\ell-3}+\ldots$. There are two cases:

If $\rho_{\ell-\mu} = \rho_{\ell-\mu-1}$ then $\left[\frac{m}{2^{\ell-\mu-1}}\right] = 2\left[\frac{m}{2^{\ell-\mu}}\right]$ and $\rho_{\ell-\mu}\geq 2^{\ell-\mu-2}$.

We obtain $\theta_\mu(C) = \frac{1}{2}\rho_{\ell-\mu}\cdot 2^{-\ell-\mu}\geq\frac{1}{8}\cdot 2^{-2\mu}$.

If $\rho_{\ell-\mu-1} < \rho_{\ell-\mu}$, then $\left[\frac{m}{2^{\ell-\mu-1}}\right] = 2\left[\frac{m}{2^{\ell-\mu}}\right]+1$, $\rho_{\ell-\mu} = \rho_{\ell-\mu-1}+2^{\ell-\mu-1}$ and $\rho_{\ell-\mu-1} < \frac{4}{3}\cdot 2^{\ell-\mu-3}$.

We obtain $\theta_\mu(C) = (-\frac{1}{2})\cdot 2^{-\ell-\mu}\rho_{\ell-\mu-1} + 2^{-2\mu-1} - 2^{-2\mu-2}\geq\frac{1}{6}\cdot 2^{-2\mu}$.

This proves Lemma 1.

THEOREM 2. *Let* $A \subset X = \{1,2, \dots ,N\}$, $N = p^{\ell}$, *be a subset with characteristic numbers* $\theta_\mu(A)$, $\mu = 0,1, \dots ,\ell-1$.
The following inequality holds:

$$\max_{\tau\in\Gamma} \Delta(\tau A) \geq \frac{1}{6}\sqrt{\sum_{\mu=0}^{\ell-1} p^{-\mu}\cdot\theta_\mu(A)} \cdot N^{1/2} .$$

Proof. For $\xi = 1,2, \dots ,p^{\ell}$; $A \subset X$ and $C = \{1, \dots ,m\} \subset X$ chosen in accordance with Lemma 1, define the Fourier transforms $\phi_A(\xi)$, $\phi_{\tau C}(\xi)$ $(\tau \in \Gamma)$ by $\phi_A(\xi) = \sum_{x=1}^{p^\ell} (\chi_A(x) - \delta)\, e^{2\pi i x\xi/p^\ell}$ ($\chi_A(x)$ denotes the indicator function of the set A) and $\phi_{\tau C}(\xi) = \sum_{x=1}^{m} e^{2\pi i\tau x\xi/p^\ell}$,$(\tau,p^\ell) = 1$. First we give a lower estimate for the sum $\sum_\xi |\phi_A(\xi)|^2 \sum_\tau |\phi_{\tau C}(\xi)|^2$ and then show that this sum can be rewritten in terms of discrepancies $\Delta(\tau A)$.

Let $(\xi,p^\ell) = p^\mu$, $0 \leq \mu < \ell$; put $\xi^* = p^{-\mu}\cdot\xi$. (The case $\xi=p^\ell$ is of no interest, since $\phi_A(p^\ell) = 0$.) We have

$$\sum_\tau |\phi_{\tau C}(\xi)|^2 = \sum_{(\tau,p^\ell)=1} \sum_{x,y=1}^{m} e^{2\pi i\xi\tau\frac{x-y}{p^\ell}} = \sum_{x,y} \sum_{\tau=1}^{p^\ell} e^{2\pi i\xi^*\tau\frac{x-y}{p^{\ell-\mu}}} - \sum_{x,y} \sum_{\tau^*=1}^{p^{\ell-1}} e^{2\pi i\xi^*\tau^*\frac{x-y}{p^{\ell-\mu-1}}} .$$

The first sum has value p^ℓ if $p^{\ell-\mu} \mid (x-y)$ and vanishes otherwise. Similarly, the second sum has value $p^{\ell-1}$ if $p^{\ell-\mu-1} \mid (x-y)$ and vanishes otherwise. According to our definition of the numbers $\theta_\mu(C)$ and the special choice of the interval C we get

$$\begin{aligned}\sum_\tau |\phi_{\tau C}(\xi)|^2 &= p^\ell \# \{x,y\in C:\ p^{\ell-\mu} | (x-y)\} - p^{\ell-1} \# \{x,y\in C:\ p^{\ell-\mu-1} | (x-y)\} \\ &= p^\ell\cdot p^{2\mu} \sum_\lambda \delta_\lambda^2(C,p^{\ell-\mu}) - p^{\ell-1}\cdot p^{2\mu+2} \sum_\lambda \delta_\lambda^2(C,p^{\ell-\mu-1}) \\ &= p^{2\ell+\mu}\cdot\theta_\mu(C) \geq \frac{1}{8}\, p^{-2\mu}\cdot p^{2\ell+\mu} = \frac{1}{8}\, p^{2\ell-\mu} \text{ by Lemma 1.}\end{aligned}$$

Now we evaluate $\sum |\phi_A(\xi)|^2$, summing over all ξ with $(\xi,p^\ell) = p^\mu$.

$$\sum_{(\xi,p^\ell)=p^\mu} |\phi_A(\xi)|^2 = \sum_{x,y\in A} \sum_{(\xi^*,p^{\ell-\mu})=1} e^{2\pi i\xi^*\frac{x-y}{p^{\ell-\mu}}} = p^{2\ell}\cdot\theta_\mu(A). \quad (8)$$

Combining (7) and (8) yields

$$\sum_{\xi=1}^{p^\ell} |\phi_A(\xi)|^2 \sum_\tau |\phi_{\tau C}(\xi)|^2 \geq \sum_{\mu=0}^{\ell-1} \frac{1}{8}\, p^{2\ell-\mu}\cdot p^{2\ell}\cdot\theta_\mu(A) = \frac{1}{8}\, p^{4\ell} \sum_{\mu=0}^{\ell-1} \theta_\mu(A)\cdot p^{-\mu} \quad (9)$$

By a transformation of indices we have, on the other hand, for each ξ:

$$\phi_A(\xi)\ \phi_{\tau C}(\xi) = \sum_{y=1}^{m} \sum_{x=1}^{p^\ell} e^{2\pi i\tau y\xi/p^\ell} . e^{2\pi i x\xi/p^\ell} (\chi_A(x) - \delta) =$$

$$= \sum_{a=1}^{p^\ell} e^{2\pi i a\xi/p^\ell} \sum_{y=1}^{m} (\chi_A(a-\tau y) - \delta) = \sum_{a=1}^{p^\ell} g_\tau(a)\ e^{2\pi i a\xi/p^\ell} .$$

Each number $g_\tau(a) := \sum_{y=1}^{m} (\chi_A(a-\tau y) - \delta)$ is equal to the "discrepancy" of the set A in the arithmetic progression $a-\tau, a-2\tau, \dots, a-m\tau \pmod{p^\ell}$. This, however, coincides with the "discrepancy" of the set $\tau^{-1}A := \tau^*A$ in the progression $\tau^*a-1, \tau^*a-2, \dots, \tau^*a-m \pmod{p^\ell}$. This latter progression is the union of at most two intervals in X, hence

$$\max_\tau \Delta(\tau A) \geq \frac{1}{2} \max_{a,\tau} |g_\tau(a)| . \tag{10}$$

We express the sum $\sum_\xi |\phi_A(\xi)|^2 \sum_\tau |\phi_{\tau C}(\xi)|^2$ in terms of the numbers $g_\tau(a)$:

$$\sum_\xi |\phi_A(\xi)|^2 \sum_\tau |\phi_{\tau C}(\xi)|^2 = \sum_{\xi=1}^{p^\ell} \sum_{a,b=1}^{p^\ell} \sum_{(\tau,p^\ell)=1} g_\tau(a)\ g_\tau(b)\ e^{2\pi i\xi(a-b)/p^\ell}$$

$$= p^\ell \sum_{a=1}^{p^\ell} \sum_{(\tau,p^\ell)=1} g_\tau^2(a) . \tag{11}$$

Comparing (11) and (9) yields

$$\max_{a,\tau} g_\tau^2(a) \geq p^{-2\ell} \varphi(p^\ell)^{-1} \cdot \frac{1}{8} p^{4\ell} \sum_{\mu=0}^{\ell-1} p^{-\mu}\ \theta_\mu(A) \geq \frac{1}{8} p^\ell \sum_{\mu=0}^{\ell-1} p^{-\mu}\ \theta_\mu(A).$$

Together with (10) this gives $\max_\tau \Delta(\tau A) \geq \frac{1}{6}\sqrt{\sum_{\mu=0}^{\ell-1} p^{-\mu}\ \theta_\mu(A)} \cdot N^{1/2}, N = p^\ell$, and proves Theorem 2.

Note: For N=p prime Theorem 2 reduces to $\max_\tau \Delta(\tau A) \geq \frac{1}{6} \sqrt{\delta(1-\delta)}\ N^{1/2}$.

4. Upper estimates for sequences. Let $A \subset X = \{1,2, \dots, p^\ell\}$, $\#A = r$, be a subset arranged in arbitrary order as a sequence $\omega = (a_1, a_2, \dots, a_r)$. To give an estimate for $\min_\tau \Delta(\tau\omega)$ we use a well-known device. Instead of the sequence $\tau\omega$ we consider the following subset $G_{\tau\omega}$ of the two-dimensional torus: $G_{\tau\omega} = \{ (\frac{1}{r}, \frac{\tau a_1}{p^\ell}), (\frac{2}{r}, \frac{\tau a_2}{p^\ell}), \dots, (\frac{r}{r}, \frac{\tau a_r}{p^\ell})\}$. It is known that the one-dimensional discrepancy $\Delta(\tau\omega)$ of the sequence $\tau\omega$ and the two-dimensional discrepancy $\Delta(G_{\tau\omega})$ of the subset $G_{\tau\omega}$ are of the same order, moreover $\Delta(\tau\omega) \leq \Delta(G_{\tau\omega})$.

THEOREM 3. *Let $A \subset X = \{1,2, \dots, p^\ell\}$ be a subset with density δ, $\#A = r$. Let $\omega = (a_1, a_2, \dots, a_r)$ be an arbitrary arrangement of A.*

There exists a numerical constant $c > 0$ such that the followin inequality holds:

$$\min_{\tau\in\Gamma} \Delta(\tau\omega) \leq c\cdot\sqrt{\delta}\, N^{1/2} \log^2 N \, , \quad N = p^{\ell} \, .$$

Proof. We use the two-dimensional Erdös-Turan inequality as stated in |1 Denote by $h = (h_1,h_2) \neq (0,0)$ an arbitrary lattice point in $\mathbb{R}^2$. Define $\|h\| = \max(|h_1|,|h_2|)$ and $\rho(h) = \max(|h_1|,1)\cdot\max(|h_2|,1)$. Then, for any positive integer m and positive constant $C = 216$, we have

$$\sum_{\tau} \Delta(G_{\tau\omega}) \leqslant C\cdot \sum_{\tau} \left\{\frac{r}{m} + \sum_{0<\|h\|\leqslant m} \frac{1}{\rho(h)} \left| \sum_{j=1}^{r} e^{2\pi i(h_1\cdot\frac{j}{r} + h_2\cdot\frac{\tau a_j}{p^\ell})} \right| \right\} \leqslant$$

$$\leqslant C\cdot\frac{r}{m}\,\varphi(p^\ell) + C\cdot \sum_{h} \frac{1}{\rho(h)}\,\varphi(p^\ell)^{1/2} \left(\sum_{\tau} \sum_{j,k=1}^{r} e^{2\pi i h_1\frac{j-k}{r}}\cdot e^{2\pi i h_2 \tau\frac{a_j-a_k}{p^\ell}}\right.$$

We first sum over all lattice points $h = (h_1,h_2)$, $0<\|h\|\leqslant m$, with $h_2=0$.

We have $\displaystyle\sum_{h_2=0} \frac{1}{\rho(h)} \left(\sum_{\tau} \sum_{j,k=1}^{r} e^{2\pi i h_1\frac{j-k}{r}} \right)^{1/2} \leqslant c_1\cdot\varphi(p^\ell)^{1/2} r^{1/2}(\log m +$

For $h_2\neq 0$ we get $\displaystyle\sum_{h_2\neq 0} \frac{1}{\rho(h)} \left(\sum_{\tau} \sum_{j,k=1}^{r} e^{2\pi i h_1\frac{j-k}{r}}\cdot e^{2\pi i h_2\tau\frac{a_j-a_k}{p^\ell}} \right)^{1/2} \leqslant$

$$\leqslant c_2\cdot r^{1/2}\,\varphi(p^\ell)^{1/2}\,(\log m + 1) \sum_{h_2=1}^{m} \frac{1}{h_2} \sum_{\mu=0}^{\infty} p^{-\mu/2}$$

$$\leqslant c_3\cdot r^{1/2}\,\varphi(p^\ell)^{1/2}\cdot(\log m + 1)^2 \, .$$

Here c_1,c_2,c_3 denote numerical constants. Both estimates together yield

$$\sum_{\tau} \Delta(\tau\omega) \leqslant \sum_{\tau} \Delta(G_{\tau\omega}) \leqslant c_4\cdot\left(\frac{r}{m}\,\varphi(p^\ell) + \varphi(p^\ell)\, r^{1/2}(\log m + 1)^2\right). \qquad (12)$$

Choosing $m = \left[r^{1/2}\right]$ leads to our final result

$$\min_{\tau} \Delta(\tau\omega) \leqslant c_5\cdot r^{1/2}(\log r^{1/2} + 1)^2 \leqslant c\sqrt{\delta}\,N^{1/2}\log^2 N,$$

where $c > 0$ is a numerical constant not depending on $N=p^\ell$. This proves the assertion.

5. Lower estimates for sequences. For sequences one would expect an improvement of the lower bound in Theorem 2 by a factor log N, but we have not been able to obtain this result. So, for any subset $A\subset X$ arranged as a sequence $\omega = (a_1,a_2, \ldots ,a_r)$, we have to apply Theorem 2 to each section $(a_1,a_2, \ldots ,a_\rho) \subset (a_1,a_2, \ldots ,a_r)$. This gives $\max_{\tau} \Delta(\tau\omega) \geqslant \frac{1}{6}\cdot c\cdot N^{1/2}$, where c is the maximum over all numbers $\sqrt{\sum_{\mu=0}^{\ell-1} \theta_\mu(\{a_1, \ldots ,a_\rho\})\cdot p^{-\mu}}$ $(1 \leqslant \rho \leqslant r)$. As in the case of subsets, ther

exist sequences ω for which $\max_\tau \Delta(\tau\omega)$ is extremely small. To give a striking example, let $A = X = \{1,2, \ldots ,2^\ell\}$ be arranged in the order ω of the van der Corput-sequence. Each section of ω is the union of at most ℓ complete arithmetic progressions mod 2^ν $(\nu=1,2,\ldots,\ell)$ in X. When applying a linear permutation $\tau \in \Gamma$, arithmetic progressions of this type are mapped onto progressions of the same difference. Hence, in this case, we have $\max_\tau \Delta(\tau\omega) \ll \log 2^\ell / \log 2$, far below the expected bound. Theorem 2 applied to this example, by the way, yields the trivial estimate $\max_\tau \Delta(\tau\omega) \gg c\cdot\sqrt{\log 2^\ell}$.

6. Examples. We will show that, apart from logarithmic factors, our results are best possible. This fact is due to the existence of subsets (sequences) which, in a certain sense, are "almost" invariant with respect to Γ. Let us give two examples.

For N=p, p prime, consider the subset $Q \subset X = \{1,2, \ldots ,p\}$ of quadratic residues mod p. Application of a linear permutation $\tau \in \Gamma$ either maps Q onto itself or onto the complementary set (with the point p excluded). But complementary sets have equal discrepancy, hence by Theorem 1 and Theorem 2: $p^{1/2} \ll \Delta(Q) \ll p^{1/2} \log p$. This estimate now appears as a stringent consequence of the invariance property of Q.

For $N=p^\ell$, $p \geq 3$ prime, let g be a primitive root mod p^ℓ. Consider the sequence $\omega = (g,g^2, \ldots ,g^r)$ with $r := \varphi(p^\ell)$. Let $\tau \in \Gamma$ be a linear permutation. Since $(\tau,p^\ell) = 1$, τ can be represented in the form $\tau = g^t$. Hence τ maps ω onto the sequence $\tau\omega = (g^{t+1}, \ldots ,g^r=1,g, \ldots ,g^t)$, which is simply a cyclical rearrangement of the original sequence ω. Discrepancy of such a rearrangement differs by a factor between $\frac{1}{2}$ and 2 only, hence by Theorem 3: $\Delta(\omega) \ll N^{1/2} \log^2 N$. To apply Theorem 2, let us estimate $p^{-\mu}\, \theta_\mu(A)$ for $A = \{g,g^2,\ldots,g^{r/2}\}$ with $r = \varphi(p^\ell)$. For $\delta_0^2(A)$ we get $\delta_0^2(A) = \frac{1}{2}(1-\frac{1}{p})$. Note that, since g is also a primitive root mod $p^{\ell-1}$, each residue class mod $p^{\ell-1}$ in X contains at most $\frac{p+1}{2}$ elements of A. The class $\equiv 0$ mod p does not contain any points of A, so $\delta_1^2(A) \leq$ $(1-\frac{1}{p})(\frac{p+1}{2p})^2$. This shows $\theta_0(A) \geq \frac{1}{4}(1-\frac{1}{p})(2-(1+\frac{1}{p})^2) \geq \frac{1}{27}$. Theorem 2 now yields $\Delta(\omega) \gg \frac{1}{33} N^{1/2}$. Again these results are a consequence of the invariance property of ω with respect to the linear group Γ.

References.

|1| Kuipers-Niederreiter Uniform distribution of sequences. Wiley 1974

|2| Niederreiter On the distribution of pseudo-random numbers II. Math.Comp. 28 (1974), 1117-1132.

|3| Roth Remark concerning integer sequences. Acta Arithmetica 9 (1964), 257-260.

Indépendance algébrique et exponentielles en plusieurs variables

Michel WALDSCHMIDT

Résumé. On montre que certains corps engendrés sur $\mathbb{Q}$ par des nombres de la forme $\exp\langle x,y\rangle$, avec $x\in\mathbb{C}^n$ et $y\in\mathbb{C}^n$, ont un grand degré de transcendance.

§1. Introduction.

Soit n un entier ≥ 1. Quand $X=\mathbb{Z}x_1+\ldots+\mathbb{Z}x_d$ est un sous-group de type fini de $\mathbb{C}^n$, on note

$$\mu(X)=\min_W(\mathrm{rang}_{\mathbb{Z}}\, X/X\cap W)/\dim_{\mathbb{C}}\,\mathbb{C}^n/W$$

quand W décrit les sous-espaces vectoriels de $\mathbb{C}^n$ sur $\mathbb{C}$ avec $W\neq\mathbb{C}^n$. Ainsi $\mu(X)\leq d/n$, et pour $n=1$ on a $\mu(X)=\mathrm{rang}_{\mathbb{Z}}X$.

On considère deux sous-groupes $X=\mathbb{Z}x_1+\ldots+\mathbb{Z}x_d$ et $Y=\mathbb{Z}y_1+\ldots+\mathbb{Z}y_\ell$ de $\mathbb{C}^n$. On note $\langle\ ,\ \rangle$ le produit scalaire usuel dan $\mathbb{C}^n$, on désigne par K le corps obtenu en adjoignant à $\mathbb{Q}$ les ℓd nombres

$$\exp\langle x_i,y_j\rangle, \qquad (1\leq i\leq d,\ 1\leq j\leq \ell),$$

et par t le degré de transcendance de K sur $\mathbb{Q}$.

On sait déjà [Wal 1] :

<u>si</u> $\mu(X)\mu(Y)>\mu(X)+\mu(Y)$, <u>alors</u> $t\geq 1$

<u>si</u> $\mu(X)\mu(Y)\geq 2(\mu(X)+\mu(Y))$, <u>alors</u> $t\geq 2$.

Il semble raisonnable d'espérer démontrer, dans un avenir assez proche, que l'hypothèse $\mu(X)\mu(Y)>\mu(X)+\mu(Y)$ implique

$$t+1>\mu(X)\mu(Y)/(\mu(X)+\mu(Y)).$$

Il y a encore deux obstacles pour y arriver. Le premier vient du critèr d'indépendance algébrique de [W-Z]: au lieu d'avoir l'exposant $t+1$

qu'on attend, on a seulement un exposant qui croît exponentiellement avec t. La solution de ce problème pourrait venir de la voie ouverte par Philippon [P2].

Le deuxième obstacle est de nature technique. Curieusement, dans tous les développements actuels de la méthode de Gel'fond (voir un aperçu historique dans [Wal 2]), pour obtenir de grands degrés de transcendance on est conduit à imposer une hypothèse d'approximation diophantienne. Pour simplifier nous demanderons une inégalité légèrement plus restrictive que celle dont nous avons vraiment besoin.

1.1. <u>Pour tout</u> $\varepsilon > 0$, <u>il existe</u> $H_0(\varepsilon) > 0$ <u>tel que si</u> $\lambda_1,\dots,\lambda_d$, $h_1,\dots,h_\ell$ <u>sont des entiers rationnels vérifiant</u>

$$\max(|\lambda_1|,\dots,|\lambda_d|,|h_1|,\dots,|h_\ell|) = H > H_0(\varepsilon)$$

<u>et</u>

$$\langle \sum_{i=1}^{d} \lambda_i x_i \, , \, \sum_{j=1}^{\ell} h_j \, y_j \rangle = \xi \neq 0 ,$$

<u>alors</u>

$$|\xi| > \exp(-H^{\varepsilon}) .$$

Sous cette hypothèse nous démontrerons :

THÉORÈME 1.2. <u>Si</u> $\mu(X)+\mu(Y) \neq 0$, <u>on a</u>

$$2^t \geq \mu(X)\mu(Y)/(\mu(X)+\mu(Y)) .$$

En utilisant des travaux récents de Zhu Yao Chen, on peut remplacer 2^t par $2^{t-2}(2+\sqrt{3})$ quand $t \geq 2$. En particulier quand $n=1$ cela améliore les résultats antérieurs de Chudnovsky [C], Warkentin [War], Philippon-Reyssat [P1,R], Endell [E1,E2] et Nesterenko [N].

Voici un autre corollaire qui fait intervenir plusieurs variables. Soit K un sous-corps de $\mathbb{C}$ de degré de transcendance $t \geq 0$ sur $\mathbb{Q}$, et soient α_{ij}, $(1 \leq i,j \leq m)$ m^2 éléments multiplicativement indépendants de K^*. On suppose que pour tout $\varepsilon > 0$ il existe $H_0(\varepsilon) > 0$ tel que pour tout $h_{ij} \in \mathbb{Z}$, $(1 \leq i, j \leq m)$ vérifiant

$$\max_{1 \leq i,j \leq m} |h_{ij}| = H \geq H_0(\varepsilon)$$

on ait

$$\left|1 - \prod_{i=1}^{m} \prod_{j=1}^{m} \alpha_{ij}^{h_{ij}}\right| > \exp(-H^{\varepsilon}) .$$

Pour $1 \leq i, j \leq m$ on choisit une détermination $\log \alpha_{ij}$ du logarithme de α_{ij}.

COROLLAIRE 1.3. <u>Le rang</u> r <u>de la matrice</u> $(\log \alpha_{ij})_{1\leqslant i,j\leqslant m}$ <u>vérifie</u>

$$r \geqslant m/2^{t+1} .$$

Dans le cas $t=0$ on retrouve un résultat de [Wal 1] §7. Le cas général se démontre de la même manière.

Il n'y a pas de difficulté à donner des analogues p-adiques de ces résultats.

Voici le plan de cet exposé. Au §2 on énonce un résultat plus précis que le théorème 1.2, en y remplaçant 2^t par un nombre $J(K)$ (vérifiant $J(K) \leqslant 2^t$ d'après le critère d'indépendance algébrique de [W-Z]). La définition de $J(K)$ nous amène à choisir une base de transcendance $\theta_1,\ldots,\theta_t$ de K sur $\mathbb{Q}$, et à lui faire subir des petites perturbations (§3). On montre au §4 comment intervient l'hypothèse technique (1.1). On introduit ensuite une fonction auxiliaire (§5) pour terminer la démonstration au §6.

Quelques mots sur la démonstration. On ne dispose pas, actuellement de "lemme de petites valeurs" (analogue au théorème de Tijdeman) pour les polynômes exponentiels en plusieurs variables. Pour cette raison les méthodes développées dans [C,E1,E2,N,P1,R,War] ne s'appliquent pas immédiatement ici. On pourrait en revanche utiliser la méthode de Masser et Wüstholz dans [M-W], mais le résultat serait légèrement moins précis que notre théorème 1.2.

Nous utiliserons ici un mélange de ces différentes méthodes, faisant intervenir à la fois un critère d'indépendance algébrique, et un lemme de zéros sous une forme raffinée due à Masser et Wüstholz [M-W]. Notre démonstration permet de travailler avec un groupe algébrique commutatif G quelconque (au lieu de $\mathbb{G}_m^d$). Nous indiquerons (sans démonstration) un résultat dans cette direction au §7.

§2. <u>Le critère d'indépendance algébrique et le coefficient</u> J .

Quand $P \in \mathbb{C}[X_1,\ldots,X_n]$ est un polynôme non nul en n variables à coefficients complexes, on note $H(P)$ sa hauteur (maximum des modules de ses coefficients), $d(P)$ le maximum de ses degrés partiels, et $t(P)$ la taille de P :

$$t(P) = \max(\log H(P), 1+d(P)) .$$

Soient $\theta_1,\ldots,\theta_t$ des nombres complexes, $t \geqslant 1$. On désigne par $A(\theta_1,\ldots,\theta_t)$ l'ensemble des nombres réels $\eta \geqslant 1$ ayant la propriété suivante : il existe une constante $T_0 > 0$ telle que pour tout $T \geqslant T_0$

et tout $(\tilde{\theta}_1,\dots,\tilde{\theta}_t)\in\mathbb{C}^t$ vérifiant

$$\max_{1\leq i\leq t} |\theta_i-\tilde{\theta}_i| < \exp(-2T^{\eta}) ,$$

il existe un polynôme $P\in\mathbb{Z}[X_1,\dots,X_t]$ de taille $\leq T$ avec

$$0 < |P(\tilde{\theta}_1,\dots,\tilde{\theta}_t)| < \exp(-T^{\eta}) .$$

Si l'ensemble $A(\theta_1,\dots,\theta_t)$ est vide, on pose $J(\theta_1,\dots,\theta_t) = 1$. S'il n'est pas vide, on désigne par $J(\theta_1,\dots,\theta_t)$ sa borne supérieure.

Quand K est un sous-corps de $\mathbb{C}$ de degré de transcendance $t\geq 1$ sur $\mathbb{Q}$, on note $J(K)$ la borne supérieure des nombres $J(\theta_1,\dots,\theta_t)$, quand $(\theta_1,\dots,\theta_t)$ décrit les bases de transcendance de K sur $\mathbb{Q}$. Du théorème de [W-Z] on déduit

$$J(K) \leq 2^t .$$

Le théorème 1.2 est donc une conséquence de l'énoncé suivant

THÉORÈME 2.1. Soient $x_1,\dots,x_d,y_1,\dots,y_\ell$ des éléments de $\mathbb{C}^n$ vérifiant l'hypothèse (1.1), et tels que les sous-groupes $X = \mathbb{Z}x_1+\dots+\mathbb{Z}x_d$ et $Y = \mathbb{Z}y_1+\dots+\mathbb{Z}y_\ell$ satisfassent $\mu(X)+\mu(Y) \neq 0$. Soit K un sous-corps de $\mathbb{C}$, de degré de transcendance fini sur $\mathbb{Q}$, contenant les ℓd nombres

$$\exp\langle x_i,y_j\rangle , \qquad (1\leq i\leq d ,\ 1\leq j\leq \ell).$$

Alors

$$J(K) \geq \mu(X)\mu(Y)/(\mu(X)+\mu(Y)) .$$

Les méthodes développées par R. Endell dans [E2] pourraient conduire à une inégalité stricte, au moins dans le cas $n=1$.

Le théorème 2.1 contient plus d'information que le théorème 1.2. Par exemple quand on suppose $\mu(X)\mu(Y) > \mu(X)+\mu(Y)$ l'inégalité $J(K) > 1$ que l'on déduit du théorème 2.1 se traduit par un résultat d'approximation simultanée des ℓd nombres $\exp\langle x_i,y_j\rangle$ par des nombres algébriques (cf. [W-Z] lemme 4.1).

§3. Petites perturbations.

On considère une extension K de $\mathbb{Q}$ de type fini, et une base de transcendance $\theta_1,\dots,\theta_t$ de K sur $\mathbb{Q}$. Soit $\theta_{t+1}\in K$ tel que $K = \mathbb{Q}(\theta_1,\dots,\theta_{t+1})$, et soit $B\in\mathbb{Q}(\theta_1,\dots,\theta_t)[X]$ le polynôme irréductible unitaire de θ_{t+1} sur $\mathbb{Q}(\theta_1,\dots,\theta_t)$. Quitte à multiplier θ_{t+1} par un élément non nul de l'anneau $A_o = \mathbb{Z}[\theta_1,\ldots,\theta_t]$ (un "dénominateur" commun des coefficients de B), on peut supposer de plus $B \in A_o[X]$. Pour utiliser la définition de J (§2), on est amené à considérer des éléments $\tilde{\theta}_1,\ldots,\tilde{\theta}_t$ de $\mathbb{C}$, proches respectivement de $\theta_1,\ldots,\theta_t$:

$$\max_{1\leqslant i\leqslant t} |\theta_i-\tilde{\theta}_i| < \varepsilon .$$

Pour $\varepsilon > 0$ suffisamment petit (dépendant de $\theta_1,\ldots,\theta_{t+1}$), le polynôme unitaire $B(\tilde{\theta}_1,\ldots,\tilde{\theta}_t,X) \in \mathbb{C}[X]$ a exactement une racine $\tilde{\theta}_{t+1}$ à distance minimale de θ_{t+1}, cette racine est simple, et

$$|\theta_{t+1} - \tilde{\theta}_{t+1}| \leqslant c\varepsilon$$

où c ne dépend que de $\theta_1,\ldots,\theta_{t+1}$. On peut le voir par exemple en considérant le semi-résultant du polynôme $B(\tilde{\theta}_1,\ldots,\tilde{\theta}_t,X)$ avec lui-même (cf. [R] lemme 3.7). On note $\tilde{K}$ le corps $\mathbb{Q}(\tilde{\theta}_1,\ldots,\tilde{\theta}_{t+1})$.

Prenons maintenant $x_1,\ldots,x_d$, $y_1,\ldots,y_\ell$ dans $\mathbb{C}^n$ tels que les nombres $\gamma_{ij} = \exp\langle x_i,y_j\rangle$, $(1\leqslant i\leqslant d$, $1\leqslant j\leqslant \ell)$ appartiennent tous à K^*. On écrit ces nombres comme des fractions rationnelles en $\theta_1,\ldots,\theta_{t+1}$:

$$\gamma_{ij} = \frac{a_{ij}}{b_{ij}}(\theta_1,\ldots,\theta_{t+1}) , \qquad (1\leqslant i\leqslant d ,\ 1\leqslant j\leqslant \ell)$$

où a_{ij} et b_{ij} sont des éléments de $\mathbb{Z}[X_1,\ldots,X_{t+1}]$, non nuls au point $\theta_1,\ldots,\theta_{t+1}$. Pour ε suffisamment petit (dépendant de $\theta_1,\ldots,\theta_{t+1}$, et des a_{ij} et b_{ij}) les nombres $a_{ij}(\tilde{\theta}_1,\ldots,\tilde{\theta}_{t+1})$ et $b_{ij}(\tilde{\theta}_1,\ldots,\tilde{\theta}_{t+1})$ sont tous différents de 0, et on peut définir

$$\tilde{\gamma}_{ij} = \frac{a_{ij}}{b_{ij}}(\tilde{\theta}_1,\ldots,\tilde{\theta}_{t+1}) \in \tilde{K}^* , \qquad (1\leqslant i\leqslant d ,\ 1\leqslant j\leqslant \ell)$$

On pose ensuite $\tilde{\gamma}_j = (\tilde{\gamma}_{1j},\ldots,\tilde{\gamma}_{dj}) \in \tilde{K}^{*d}$, $(1\leqslant j\leqslant \ell)$.

Quand E est un sous-ensemble fini de $\mathbb{C}^{*d}$, on note $\omega(E)$ le plus petit des degrés des polynômes non nuls de $\mathbb{C}[X_1,\ldots,X_d]$ qui s'annulent sur E (il s'agit du degré total).

Au paragraphe 6 on démontrera le résultat suivant.

PROPOSITION 3.1. <u>On suppose qu'il existe</u> $S_o > 0$ <u>tel que pour tout</u> $S \geqslant S_o$ <u>et pour tout</u> $(\tilde{\theta}_1,\ldots,\tilde{\theta}_t) \in \mathbb{C}^t$ <u>vérifiant</u>

$$\max_{1\leqslant i\leqslant t} |\theta_i-\tilde{\theta}_i| \leqslant \exp(-S) ,$$

<u>le sous-ensemble</u> $\tilde{\Gamma}(S)$ <u>de</u> $\mathbb{C}^{*d}$ <u>formé des éléments</u>

$$\tilde{\gamma}_1^{h_1}\ldots\tilde{\gamma}_\ell^{h_\ell} = \Big(\prod_{j=1}^{\ell}\tilde{\gamma}_{1j}^{h_j},\ldots,\prod_{j=1}^{\ell}\tilde{\gamma}_{dj}^{h_j}\Big)\ ,\ (h_1,\ldots,h_\ell)\in\mathbb{Z}^\ell\ ,\ 0\leqslant h_j\leqslant S$$

<u>vérifie</u>

(3.2) $$\omega(\tilde{\Gamma}(S)) \geqslant (S/d)^{\ell/d}\ .$$

<u>Si</u> $X = \mathbb{Z}x_1+\ldots+\mathbb{Z}x_d$ <u>et</u> $Y = \mathbb{Z}y_1+\ldots+\mathbb{Z}y_\ell$ <u>satisfont</u> $\mu(X) = d/n$ <u>et</u> $\mu(Y) = \ell/n$, <u>alors</u>

$$J(\theta_1,\ldots,\theta_t) \geqslant \ell d/n(\ell+d)\ .$$

Au paragraphe suivant on montre que la proposition 3.1 implique le théorème 2.1.

§4. <u>Utilisation de l'hypothèse technique</u> (1.1).

Montrons d'abord qu'il n'y a pas de restriction, pour le théorème 2.1, à supposer $\mu(X) = d/n$ et $\mu(Y) = \ell/n$. On se ramène déjà au cas où $\mathrm{rang}_{\mathbb{Z}}X = d$ et $\mathrm{rang}_{\mathbb{Z}}Y = \ell$, de manière évidente. On utilise ensuite le lemme 5.2 de [Wal 1] : il existe $n' \geqslant 1$, $X' \subseteq \mathbb{C}^{n'}$ et $Y' \subseteq \mathbb{C}^{n'}$, avec $\mathrm{rang}_{\mathbb{Z}}X' = d'$, $\mathrm{rang}_{\mathbb{Z}}Y' = \ell'$, $\mu(X') = d'/n' \geqslant \mu(X)$, $\mu(Y') = \ell'/n' \geqslant \mu(Y)$, et $\langle X',Y'\rangle \subseteq \langle X,Y\rangle$. Alors

$$\ell'd'/n'(\ell'+d') \geqslant \mu(X)\mu(Y)/(\mu(X)+\mu(Y))\ ,$$

et l'hypothèse (1.1) pour X et Y implique la même hypothèse pour X' et Y'. Ainsi, quitte à remplacer X et Y par X' et Y', on peut supposer $\mu(X) = d/n$ et $\mu(Y) = \ell/n$.

Pour déduire le théorème 2.1 de la proposition 3.1, il ne reste plus qu'à vérifier l'hypothèse (3.2) en utilisant (1.1). C'est l'objet du lemme suivant.

LEMME 4.1. <u>Soient</u> $x_1,\ldots,x_d$, $y_1,\ldots,y_\ell$ <u>des éléments de</u> $\mathbb{C}^n$ <u>vérifiant la condition</u> (1.1), <u>avec</u> $\ell d \geqslant n(\ell+d)$ <u>et tels que</u> $X = \mathbb{Z}x_1+\ldots+\mathbb{Z}x_d$ <u>et</u> $Y = \mathbb{Z}y_1+\ldots+\mathbb{Z}y_\ell$ <u>satisfassent</u> $\mu(X) = d/n$ <u>et</u> $\mu(Y) = \ell/n$. <u>Soit</u> $\varepsilon > 0$. <u>Il existe</u> $S_0(\varepsilon) > 0$ <u>tel que pour tout</u> $S \geqslant S_0(\varepsilon)$ <u>et pour tout</u> $z_{ij}\in\mathbb{C}$, $(1\leqslant i\leqslant d\ ,\ 1\leqslant j\leqslant \ell)$ <u>avec</u>

$$\max_{i,j}|z_{ij} - \langle x_i,y_j\rangle| < \exp(-S^\varepsilon)\ ,$$

<u>l'ensemble</u>

$$E = \Big\{\Big(\prod_{j=1}^{\ell} e^{h_j z_{1j}},\ldots,\prod_{j=1}^{\ell} e^{h_j z_{dj}}\Big)\in\mathbb{C}^{\times d}\ ;\ (h_1,\ldots,h_\ell)\in\mathbb{Z}^\ell\ ,\ 0\leqslant h_j\leqslant S\Big\}$$

<u>vérifie</u>

$$\omega(E) \gg (S/d)^{\ell/d} .$$

Démonstration du lemme 4.1.

Supposons que la conclusion ne soit pas vérifiée. D'un lemme de zéros de Masser [M] on déduit qu'il existe deux entiers r et k vérifiant $1 \leqslant r \leqslant d$, $1 \leqslant k \leqslant \ell$, $\ell r + kd > \ell d$, il existe des éléments $\lambda^{(1)}, \ldots, \lambda^{(r)}$ de $\mathbb{Z}^d$, linéairement indépendants sur $\mathbb{Z}$, avec $\lambda^{(\rho)} = (\lambda_1^{(\rho)}, \ldots, \lambda_d^{(\rho)})$, $(1 \leqslant \rho \leqslant r)$, et enfin des éléments $h^{(1)}, \ldots, h^{(k)}$ de $\mathbb{Z}^\ell$, linéairement indépendants sur $\mathbb{Z}$, avec $h^{(\varkappa)} = (h_1^{(\varkappa)}, \ldots, h_\ell^{(\varkappa)})$, $(1 \leqslant \varkappa \leqslant k)$, vérifiant

$$\prod_{i=1}^{d} \prod_{j=1}^{\ell} e^{\lambda_i^{(\rho)} h_j^{(\varkappa)} z_{ij}} = 1 , \qquad (1 \leqslant \rho \leqslant r , 1 \leqslant \varkappa \leqslant k).$$

De plus, le raffinement de ce résultat donné par Masser et Wüstholz dans [M-W] permet de choisir ces $\lambda^{(\rho)}$ et ces $h^{(\varkappa)}$ de telle sorte que

$$\max_{1 \leqslant i \leqslant d} |\lambda_i^{(\rho)}| \leqslant (S/d)^{\ell\rho/d} , \quad \max_{1 \leqslant j \leqslant \ell} |h_j^{(\varkappa)}| \leqslant (S/d)^{\ell r/d}$$

pour $1 \leqslant \rho \leqslant r$, $1 \leqslant \varkappa \leqslant k$.

Pour $1 \leqslant \rho \leqslant r$ et $1 \leqslant \varkappa \leqslant k$, notons

$$x'_\rho = \sum_{i=1}^{d} \lambda_i^{(\rho)} x_i , \quad y'_\varkappa = \sum_{j=1}^{\ell} h_j^{(\varkappa)} y_j ,$$

et

$$m_{\rho,\varkappa} = \frac{1}{2i\pi} \sum_{i=1}^{d} \sum_{j=1}^{\ell} \lambda_i^{(\rho)} h_j^{(\varkappa)} z_{ij} .$$

Ainsi $m_{\rho,\varkappa} \in \mathbb{Z}$, et $|m_{\rho,\varkappa}| \leqslant S^{2\ell}$ pour $S \geqslant S_0(\varepsilon)$ suffisamment grand. Soit s le rang de la matrice $(m_{\rho,\varkappa})_{1 \leqslant \rho \leqslant r, 1 \leqslant \varkappa \leqslant k}$. On suppose que la numérotation des $\lambda^{(\rho)}$ et des $h^{(\varkappa)}$ a été choisie de telle sorte que le déterminant de la matrice $(m_{\rho,\varkappa})_{1 \leqslant \rho,\varkappa \leqslant s}$ ne soit pas nul.

Il existe des entiers rationnels $a_{\sigma,\varkappa}$, $(0 \leqslant \sigma \leqslant s , 1 \leqslant \varkappa \leqslant k)$ avec $a_{0,\varkappa} \neq 0$, $(1 \leqslant \varkappa \leqslant k)$, et

$$a_{0,\varkappa} m_{\rho,\varkappa} = \sum_{\sigma=1}^{s} a_{\sigma,\varkappa} m_{\rho,\sigma} , \qquad (1 \leqslant \rho \leqslant r , 1 \leqslant \varkappa \leqslant k).$$

On peut choisir pour $a_{\sigma,\varkappa}$ des mineurs de la matrice $(m_{\rho,\varkappa})_{1 \leqslant \rho \leqslant r, 1 \leqslant \varkappa \leqslant k}$, donc

$$|a_{\sigma,\varkappa}| \leqslant s! \, S^{2\ell s} \leqslant d! \, S^{2\ell d} ,$$

pour $0 \leqslant \sigma \leqslant s$, $1 \leqslant \varkappa \leqslant k$. Pour $\sigma < \varkappa \leqslant k$, on définit

$$y''_\varkappa = a_{o\varkappa}\, y'_\varkappa - \sum_{\sigma=1}^{s} a_{\sigma,\varkappa}\, y'_\sigma \ .$$

Ainsi pour $1 \leqslant \rho \leqslant r$ et $s < \varkappa \leqslant k$ on a

$$\langle x'_\rho, y''_\varkappa\rangle = a_{o\varkappa}\ \langle x'_\rho, y'_j\rangle - \sum_{\sigma=1}^{s} a_{\sigma\varkappa}\ \langle x'_\rho, y'_\sigma\rangle \ ,$$

donc

$$|\langle x'_\rho, y''_\varkappa\rangle| < \exp(-\tfrac{1}{2}S^{\varepsilon}) \ .$$

L'hypothèse (1.1) implique alors $\langle x'_\rho, y''_\varkappa\rangle = 0$, pour $1 \leqslant \rho \leqslant r$, $s < \varkappa \leqslant k$. On termine la démonstration comme celle du lemme 5.4 de [Wal 1] pour obtenir une contradiction.

§5. La fonction auxiliaire.

Soient d et n deux entiers, $d > n \geqslant 1$, $x_1,\ldots,x_d$ des éléments de $\mathbb{C}^n$, c_0 un nombre réel suffisamment grand, T_0 un nombre réel encore plus grand. Pour chaque $T \geqslant T_0$ on définit N, S, D, U par

$$T = N^{\ell+d} \ , \ S = N^d \ , \ D = c_0^{-1}\ N^{\ell} \ , \ U^{n+1}\ c_0^{d+2} = N^{\ell d+\ell+d} \ .$$

Enfin soit $V \geqslant 25\ c_0\ T$.

Pour $t = (t_1,\ldots,t_n) \in \mathbb{C}^n$ on note $|t| = \max_{1\leqslant i\leqslant n} |t_i|$.

PROPOSITION 5.1. <u>Il existe un polynôme non nul</u> $F \in \mathbb{Z}[X_1,\ldots,X_d]$ <u>de degré</u> $\leqslant D$ <u>en chaque</u> X_i <u>et de taille</u> $\leqslant c_0^{-1}\ T$, <u>tel que la fonction</u>

$$\Phi(z_1,\ldots,z_d) = F(e^{z_1},\ldots,e^{z_d})$$

<u>vérifie la propriété suivante : pour tout</u> $z = (z_1,\ldots,z_d) \in \mathbb{C}^d$ <u>tel qu'il existe</u> $t = (t_1,\ldots,t_n) \in \mathbb{C}^n$ <u>avec</u>

$$|t| \leqslant c_0\ S \ , \ \underline{\text{et}} \ \max_{1\leqslant i\leqslant d} |z_i - \langle x_i, t\rangle| \leqslant e^{-V} \ ,$$

<u>on a</u>

$$|\Phi(z)| \leqslant e^{-U} + e^{-\frac{23}{24}V} \ .$$

<u>Démonstration</u>.

On construit le polynôme F de telle sorte que la fonction

$$\psi(t) = \Phi(\langle x_1,t\rangle,\ldots,\langle x_d,t\rangle)$$

satisfasse

$$|\psi(t)| \leqslant e^{-U}$$

pour tout $t \in \mathbb{C}^n$ vérifiant $|t| \leqslant c_0 S$. Il suffit d'appliquer le théorème 3.1 de [Wal 1] aux fonctions

$$\exp \langle \sum_{i=1}^{d} \lambda_i x_i , t \rangle , \qquad (0 \leqslant \lambda_i \leqslant D),$$

avec $L = ([D]+1)^d$, $r = c_0 S$, $R = er$ (le paramètre noté S dans [Wal 1] §3 est ici $c_0^{-1} T$).

Maintenant soit $t \in \mathbb{C}^n$, avec $|t| \leqslant c_0 S$, et soit $z = (z_1,\dots,z_d) \in \mathbb{C}^d$ avec

$$\max_{1 \leqslant i \leqslant d} |z_i - \langle x_i , t \rangle| \leqslant e^{-V} .$$

On va vérifier

$$|\Phi(z) - \psi(t)| \leqslant e^{-\frac{23}{24}V} .$$

On écrit

$$F(X_1,\dots,X_d) = \sum_{\lambda} p_\lambda \prod_{i=1}^{d} X_i^{\lambda_i} ,$$

avec $\lambda = (\lambda_1,\dots,\lambda_d)$, $0 \leqslant \lambda_i \leqslant D$. Alors

$$\Phi(z) - \psi(t) = \sum_{\lambda} p_\lambda \, (e^{u_\lambda} - e^{v_\lambda})$$

où

$$u_\lambda = \sum_{i=1}^{d} \lambda_i z_i \quad , \quad v_\lambda = \langle \sum_{i=1}^{d} \lambda_i x_i , t \rangle .$$

Mais pour u et v dans $\mathbb{C}$ on a

$$|e^u - e^v| \leqslant |e^v| . e^{|u-v|} . |u-v| .$$

D'où

$$|\Phi(z) - \psi(t)| \leqslant L . e^{c_0 T} . e^{-V} \leqslant e^{-\frac{23}{24}V} .$$

§6. Fin de la démonstration de la proposition 5.1.

On va d'abord montrer, sous les hypothèses de la proposition 3.1, l'inégalité :

(6.1) $$J(\theta_1,\dots,\theta_t) \geqslant (\ell d+\ell+d)/(n+1)(\ell+d) .$$

Comme $J(\theta_1,\ldots,\theta_t) \geq 1$, cela revient à choisir un nombre réel η vérifiant

$$1 \leq \eta < (\ell d+\ell+d)/(n+1)(\ell+d)$$

puis à construire, pour chaque $T \geq T_o$ et chaque $(\tilde{\theta}_1,\ldots,\tilde{\theta}_t) \in \mathbb{C}^t$ vérifiant

$$\max_{1\leq i\leq t} |\theta_i-\tilde{\theta}_i| < \exp(-2T^\eta) ,$$

un polynôme non nul $P \in \mathbb{Z}[X_1,\ldots,X_t]$ de taille $\leq T$ tel que

$$0 < |P(\tilde{\theta}_1,\ldots,\tilde{\theta}_t)| < \exp(-T^\eta) .$$

Posons $\varepsilon = 1/14$. On commence par choisir comme au §3 la racine $\tilde{\theta}_{t+1}$ de $B(\tilde{\theta}_1,\ldots,\tilde{\theta}_t,X)$ qui vérifie

$$|\theta_{t+1}-\tilde{\theta}_{t+1}| < \exp(-(2-\varepsilon)T^\eta) ,$$

et à en déduire $\tilde{\gamma}_{ij} \in \tilde{K}^\times$ avec

$$|\tilde{\gamma}_{ij}-\gamma_{ij}| < \exp(-(2-2\varepsilon)T^\eta) .$$

Reprenons le polynôme $F \in \mathbb{Z}[X_1,\ldots,X_d]$ construit à la proposition 5.1. L'hypothèse (3.2) montre qu'il existe $(h_1,\ldots,h_\ell) \in \mathbb{Z}^\ell$, $0 \leq h_j \leq S$, tel que le nombre

$$\xi = F(\tilde{\gamma}_1^{h_1}\ldots\tilde{\gamma}_\ell^{h_\ell})$$

ne soit pas nul. On choisit $z_{ij} \in \mathbb{C}$, $(1 \leq i \leq d\ , 1 \leq j \leq \ell)$ vérifiant $\exp z_{ij} = \tilde{\gamma}_{ij}$ et

$$|z_{ij} - \langle x_i,y_j\rangle| \leq \exp(-(2-3\varepsilon)T^\eta) .$$

Ainsi en posant

$$z_i = \sum_{j=1}^{\ell} h_j z_{ij} , \quad (1 \leq i \leq d) , \quad z = (z_1,\ldots,z_d), \ t = \sum_{j=1}^{\ell} h_j y_j ,$$

on trouve $\xi = \Phi(z)$, et

$$|t| \leq c_o S \quad \text{et} \quad \max_{1\leq i\leq d} |z_i - \langle x_i,t\rangle| \leq \exp(-(2-4\varepsilon)T^\eta) .$$

La proposition 5.1, avec $V = (2-4\varepsilon)T^\eta$, implique

$$0 < |\xi| < \exp(-(2-5\varepsilon)T^\eta) .$$

On définit $Q \in \mathbb{Z}[X_1,\ldots,X_{t+1}]$ par

$$Q = \sum_{\lambda} p_{\lambda} \prod_{i=1}^{d} \prod_{j=1}^{\ell} a_{ij}^{\lambda_i h_j} b_{ij}^{[D][S]-\lambda_i h_j} .$$

Ainsi

$$Q(\tilde\theta_1,\ldots,\tilde\theta_{t+1}) = \xi \prod_{i=1}^{d} \prod_{j=1}^{\ell} b_{ij}(\tilde\theta_1,\ldots,\tilde\theta_{t+1})^{[D][S]} ,$$

et donc

$$0 < |Q(\tilde\theta_1,\ldots,\tilde\theta_{t+1})| \leqslant \exp(-(2-6\varepsilon)T^{\eta}) .$$

D'autre part

$$t(Q) \leqslant c_0^{1/2} DS = T/c_0^{1/2} .$$

Posons $R(X) = Q(\tilde\theta_1,\ldots,\tilde\theta_t,X)$. Si le polynôme R est constant, on prend $P(X_1,\ldots,X_t) = Q(X_1,\ldots,X_t,0)$. Si $\deg R \geqslant 1$, on considère le semi-résultant de $R(X)$ et de $B(\tilde\theta_1,\ldots,\tilde\theta_t,X)$; on trouve ainsi (cf. [W-Z] lemmes 2.3 et 2.4) un polynôme non nul $P \in \mathbb{Z}[X_1,\ldots,X_t]$, vérifiant $t(P) \leqslant T$ et

$$0 < |P(\tilde\theta_1,\ldots,\tilde\theta_t)| \leqslant \exp(-(2-7\varepsilon)T^{\eta}) .$$

On en déduit (6.1).

Pour terminer la démonstration de la proposition 3.1 on utilise l'astuce de Landau-Philippon [P1] : on applique (6.1) aux sous-groupes X^k et Y^k de $(\mathbb{C}^n)^k$ pour k entier positif. L'hypothèse (3.2) étant stable par puissance cartésienne, on trouve

$$J(\theta_1,\ldots,\theta_t) \geqslant (k\ell d+\ell+d)/(kn+1)(\ell+d) ,$$

puis on fait tendre k vers l'infini.

§7. Compléments.

La démonstration que nous venons de présenter permet d'étudier, plus généralement, les sous-groupes à n paramètres de groupes algébriques définis sur une extension de $\mathbb{Q}$ de type fini. Nous nous contenterons, ici, de donner une généralisation en plusieurs variables du théorème principal de [M-W].

Soit $\wp$ une fonction elliptique de Weierstrass d'invariants g_2, g_3, n'ayant pas de multiplication complexe, et soient $X = \mathbb{Z}x_1+\ldots+\mathbb{Z}x_d$ et $Y = \mathbb{Z}y_1+\ldots+\mathbb{Z}y_\ell$ deux sous-groupes de type fini de $\mathbb{C}^n$ vérifiant la condition technique (1.1). Soit K le sous-corps de obtenu en adjoignant à $\mathbb{Q}$ les nombres g_2, g_3, et les valeurs de $\wp$

aux points $\langle x_i, y_j \rangle$, $(1 \leqslant i \leqslant d\ ,\ 1 \leqslant j \leqslant \ell)$ qui ne sont pas pôles de $\mathcal{P}$. Enfin soit t le degré de transcendance de K sur $\mathbb{Q}$.

THÉORÈME 7.1. <u>Si</u> $\mu(X)+\mu(Y) \neq 0$, <u>alors</u>

$$2^t \geqslant \mu(X)\mu(Y)/(2\mu(X)+\mu(Y)) .$$

Bibliographie

[C] G.V. ČUDNOVSKII.- Some analytic methods in the theory of transcendental numbers. Inst. Math., Ukr S.S.R. Acad. Sci., Preprints IM 74-8 et 74-9, Kiev 1974.

[E1] R. ENDELL.- Zur algebraischen Unabhängigkeit gewisser Werte der Exponentialfunktion (nach Chudnovsky). Diplomarbeit, Düsseldorf, 1981.

[E2] R. ENDELL.- Zur algebraischen Unabhängigkeit gewisser Werte der Exponentialfunktion, 1983, Noordwijkerhout, même volume.

[M] D.W. MASSER.- On polynomials and exponential polynomials in several complex variables. Invent. Math., 63 (1981), 81-95.

[M-W] D.W. MASSER and G. WÜSTHOLZ.- Fields of large transcendence degree generated by values of elliptic functions. Invent. Math., 72 (1983), 407-464.

[N] Ju.V. NESTERENKO.- On the algebraical independence of algebraic numbers to algebraic powers, in : "Approximations diophantiennes et nombres transcendants". Luminy 1982, Progress in Math. 31, Birkhäuser (1983), 199-220.

[P1] P. PHILIPPON.- Indépendance algébrique des valeurs de fonctions exponentielles p-adiques. J. reine angew. Math., 329 (1981), 42-51.

[P2] P. PHILIPPON.- Pour une théorie de l'indépendance algébrique. Thèse, Orsay, 1983 (à paraître).

[R] E. REYSSAT.- Un critère d'indépendance algébrique. J. reine angew. Math., 329 (1981), 66-81.

[Wal 1] M. WALDSCHMIDT.- Transcendance et exponentielles en plusieurs variables. Invent. Math., 63 (1981), 97-127.

[Wal 2] M. WALDSCHMIDT.- Algebraic independence of transcendental numbers. Gel'fond's method and its developments, à paraître.

[W-Z] M. WALDSCHMIDT et ZHU Yao Chen.- Une généralisation en plusieurs variables d'un critère de transcendance de Gel'fond, C. R. Acad. Sc. Paris, Ser. I, 297 (1983), 229-232.

[War] P. WARKENTIN.- Algebraische Unabhängigkeit gewisser p-adischer Zahlen. Diplomarbeit, Freiburg i. Br., 1978.

Michel WALDSCHMIDT
Institut Henri Poincaré
11, rue Pierre et Marie Curie
75231 PARIS CEDEX 05

RECENT PROGRESS IN TRANSCENDENCE THEORY

G. Wüstholz

1. Introduction

In the last few years, methods from commutative algebra, algebraic geometry, complex analysis of several variables, and even cohomology theory have been used to solve problems in transcendence theory which had long been regarded as inaccessible. One of the central objects is the so-called zero-estimates, or more generally multiplicity-estimates, on certain algebraic objects. Using these estimates and the techniques developed to obtain them, many open problems in transcendence theory have been solved. On the other hand, many new problems have arisen, and it seems that transcendence theory has finally become a theory. In this article, we would like to describe this development, and for this we have to begin with a short description of the theory of multiplicity-estimates. Let us begin with two very elementary examples to describe what we mean by multiplicity-estimates.

Example 1. Let $a_1,\dots,a_l$ be pairwise different complex numbers and $n_1,\dots,n_l$ positive integers. Then it is of course always possible to construct a non-zero polynomial $F(X)$ in one variable of degree

$$n \geq n_1+\dots+n_l$$

such that F vanishes at the points $a_1,\dots,a_l$ to order at least $n_1,\dots,n_l$. On the other hand it is well-known that a polynomial $F(X)$ of degree n that has $n+1$ zeroes (counted with multiplicities) is necessarily the zero-polynomial.

This is a trivial example of what we mean by multiplicity estimate It generalizes to several variables in the following way.

Example 2. Let $P_1,\dots,P_l$ be pairwise different points in $\mathbb{C}^d$, $n_1,\dots,$ positive integers. We want to construct a non-zero polynomial $F(X_1,\dots,X_d)$ of a given degree D that vanishes at $P_1,\dots,P_l$ with multiplicity $n_1,\dots,n_l$ respectively at least. This is possible as soon

$$N = \binom{n_1+d-1}{d} + \dots + \binom{n_l+d-1}{d} < \binom{D+d}{d}.$$

On the other hand it is not clear at all that the second implication in Example 1 should hold. It is in general not true in the case $d \geq 2$ that a polynomial $F(X_1,\dots,X_d)$ in d variables and of degree D that vanishes at all these points with multiplicity $n_1,\dots,n_l$ is necessarily the zero polynomial if only

$$N \geq \binom{D+d}{d}$$

holds.

These two examples illustrate that the situation in several variables is much more complicated than in one variable. Nevertheless situations like this have been studied quite extensively by many authors (Waldschmidt [Wa1], Chudnovsky [Ch1], Demailly [D], Masser [Ma1], [Ma2], Wüstholz [Wü1], Philippon [Ph1], Bombieri [Bo], Esnault-Vieweg [E-V1], [E-V2]) under very different points of view. It has turned out that the answer for such types of questions depends very much on the distribution of the points in question. Fortunately in most cases arising in transcendence theory we are concerned with a much more special situation which we want to describe now.

2. Zero estimates on group varieties

We consider a commutative algebraic group (group variety) G defined over the field of complex numbers, for example $\mathbb{G}_a$, the additive group, $\mathbb{G}_m$ the multiplicative group, E an elliptic curve or an arbitrary abelian variety. An arbitrary algebraic group G is then obtained by forming products or so-called extensions. Such an algebraic group G is a quasi-projective variety and can be embedded in some projective space in a nice way (see [Se] and [Fa-Wü])

$$G \hookrightarrow \mathbb{P}^N$$

for some N. For this one first compactifies G to $\overline{G}$ in such a way that G operates on $\overline{G}$ and after this embeds $\overline{G}$ into $\mathbb{P}^N$ by means of a certain very ample divisor. We denote the dimension of G by d. Then G replaces the space $\mathbb{C}^d$ of Example 2.

Next we take a finite set $X \subset G$ of elements of G that contains the neutral element of G. Then for integers $r \geq 1$ we define the sets rX as

$$X + \dots + X \text{ (r-times).}$$

Then dX replaces the set of points $\{P_1,\dots,P_l\}$ in Example 2. Now we define for integers r with $1 \le r \le d$ the numbers $Q_r(X)$ as

$$Q_r(X) = \begin{cases} |X| & \text{if G has no algebraic subgroup of codimension r} \\ \min\limits_{\substack{H \subset G \\ H \text{ algebraic} \\ \operatorname{cod} H = r \\ H \text{ connected}}} |X+H/H| & \text{otherwise.} \end{cases}$$

The numbers $Q_r(X)$ measure the distribution of the set X in G. Then Example 1 and Example 2 generalize in the following way.

Theorem 1 ([Ma-Wü1], [Ma-Wü2]). Let $P(X_0,\ldots,X_N)$ be a homogeneous polynomial of degree at most D that vanishes on dX. Then if

$$Q_r(X) > \deg G \cdot D^r \qquad (1 \le r \le d)$$

the polynomial P vanishes on all of G.

Here $\deg G$ is the degree of G in $\mathbb{P}^N$.

Remarks. 1. A geometric proof of a weaker version of Theorem 1 was given by Moreau ([Mo1] and [Mo2]). Unfortunately this proof does not seem to generalize to zeroes counted with multiplicities.
2. Various extensions and modifications of this result have been proved now. So for example there exists a version which also takes into account different degrees ([Ma-Wü2], see also [Wü4]). Another modification are the so-called "zero estimates with knobs on" used to prove results on large transcendence degree (see section 5 and [Ma-Wü3]). They also appear very useful for obtaining transcendence measures and measures for algebraic independence.

3. Multiplicity estimates on group varieties

As in the two examples at the beginning we now allow the zeroes to have multiplicities. We can do this even more generally than in these examples. Instead of taking into account multiplicities in all directions we take multiplicities only with respect to a certain subset of all directions. First results in this direction were obtained by Nesterenko [Ne1] and by Brownawell and Masser [Br-Ma], [Br1], [Br2]. But we are considering the following general situation.

Let $A \subseteq G$ be an analytic subgroup of G and let $a = \dim A$ be the dimension of A. This is the image of an analytic homomorphism

$$\Phi: \mathbb{C}^a \to G(\mathbb{C})$$

from the complex a-space to the complex Lie-group $G(\mathbb{C})$. This is the set of complex valued points of G. (It would be more precise but not more illuminating to work here with the notion of group schemes.) Such an analytic subgroup A provides us via its Lie-algebra with derivations $\Delta_1,\ldots,\Delta_a$ on G. This enables us to define the order of vanishing in the following way. We say that a homogeneous polynomial $P(X_0,\ldots,X_N)$ vanishes to order at least T along A (or Φ) at a point g in G with $x_0(g) \neq 0$ if

$$\Delta_1^{t_1} \ldots \Delta_a^{t_a} P(1,x_1/x_0,\ldots,x_N/x_0)(g) = 0$$

for all non-negative integers $t_1,\ldots,t_a$ with $t_1+\ldots+t_a < T$. Here $x_0,\ldots,x_N$ are the restrictions of the coordinate functions $X_0,\ldots,X_N$ on $\mathbb{P}^N$ to G.

Obvious examples force us in the same way as with the points to measure the "distribution" of the derivations corresponding to A. For this we define integers $\tau(V)$ in the following way. For algebraic subvarieties V of G with $V \cap A \neq \emptyset$ we define

$$\tau(V) = \operatorname{cod}_A V \cap A$$

where cod_A denotes the codimension in A. Then for integers r with $1 \leq r \leq d$ we put

$$\tau_r = \min_{\operatorname{cod} V = r} \tau(V)$$

and we can state the following fundamental result.

Theorem 2 ([Wü2]). There exists a positive constant c with the following property. Let $P(X_0,\ldots,X_N)$ be a homogeneous polynomial of degree at most equal to D that vanishes in dX along A with multiplicity at least T. Then if

$$(T/n)^{\tau_r} \cdot Q_r(X) > (cD)^r \qquad (1 \leq r \leq n)$$

the polynomial P vanishes on all of G.

Remark. It can be shown that Theorem 1 is best possible up to a numerical constant. It is very likely that Theorem 2 has the same property. It would be interesting to verify this.

We now illustrate our result with an explicit example.

Example 3. Let E be an elliptic curve and $K = (\mathrm{End}\ E)\otimes\mathbb{Q}$ the field of complex multiplications, where End E is the ring of endomorphisms of E. Associated with E is a Weierstraß elliptic function $\wp(z)$. Then let $1, x_1, \ldots, x_n$ be complex numbers which generate over K a vecto space of dimension $n+1$. This implies, as one verifies without any difficulties, that the functions

$$\wp(z_1),\ \ldots\ ,\ \wp(z_n),\ \wp(x_1 z_1+\ldots+x_n z_n)$$

are algebraically independent over $\mathbb{C}$ or even over $\mathbb{C}(z_1,\ldots,z_n)$. This also follows from a general result of Brownawell and Kubota [Br-Ku]. Now we take a polynomial $P(X_1,\ldots,X_{n+1})$ of degree D and define the function $\Phi(z_1,\ldots,z_n)$ by

$$\Phi(z_1,\ldots,z_n) = P(\wp(z_1),\ldots,\wp(z_n),\wp(x_1 z_1+\ldots+x_n z_n)).$$

This function is identically zero only if the polynomial P is the zer polynomial. Let now $w = (w_1,\ldots,w_n)$ be a point in $\mathbb{C}^n$ such that Φ is defined for all non-zero integer multiples of w and suppose that for some integers $S \geq 1$, $T \geq cD$ we have

$$\left(\frac{\partial}{\partial z_1}\right)^{t_1}\ldots\left(\frac{\partial}{\partial z_n}\right)^{t_n}\Phi(sw) = 0 \qquad (1 \leq s \leq S,\ 0 \leq t_1+\ldots+t_n < T).$$

Then we have the following corollary to Theorem 2.

Corollary. If $(T/n)^n(S/n) > (cD)^{n+1}$ then the polynomial P is the zero polynomial.

Here the coefficients τ_r, as well as the n other inequalities, do not appea anymore. The reason for this is that the conditions on the numbers $x_1,\ldots,x_n$ imply that $\tau_r \geq r$ for $1 \leq r \leq n$ and $\tau_{n+1} = n$. It then follows easily that the missing inequalities are consequences of the given one. What this example also shows is that the conditions of

Theorem 2, which seem to be very abstract, can be verified in all given situations without much difficulty. It reduces to a problem in linear algebra.

4. Small transcendence degree

In 1949 Gel'fond developed a new method to prove algebraic independence of two numbers out of a certain given set of numbers. The central point in his method was a general criterion for algebraic independence of two complex numbers, called "Gel'fond's criterion". In practice this criterion was until recently only applicable to numbers connected with the exponential function. The most striking result here was Gel'fond's result on the algebraic independence of the numbers

$$\alpha^{\beta},\ \alpha^{\beta^2}$$

where $\alpha \neq 0,1$ is algebraic and β cubic over the rationals.

One of the motivations for developing the theory of zero and multiplicity estimates was to prove the elliptic analogues of these types of results. In the same way as in the exponential case (for a complete account of this see [Wa2]) it is possible to embed this kind of results in a more general context. Here we will content ourselves with giving the elliptic analogue of Gel'fond's result. The general result can be found in [Ma-Wü4].

For this let $\wp(z)$ be a Weierstraß elliptic function with algebraic invariants g_2 and g_3. Denote as before by K its field of complex multiplications and denote by d its degree over $\mathbb{Q}$. Then it is known that $d = 1,2$. Let then β be an algebraic number of degree $l = 5-d$ over K and u a complex number such that $\wp(u)$ is algebraic. Then we have the following result.

<u>Theorem 3</u> ([Ma-Wü4]). At least two of the numbers

$$\wp(\beta u),\ldots,\wp(\beta^{l-1}u)$$

are algebraically independent (over $\mathbb{Q}$).

As an immediate consequence of this theorem we obtain in the case $d=2$ the elliptic analogue of Gel'fond's result.

The proof of this and related results will be given in [Ma-Wü5].

5. Large transcendence degree

The method initiated by Gel'fond was considerably extended by several authors to obtain algebraic independence of more than two numbers connected with the exponential function (Smelev [Sm], Waldschmidt [Wa2]) but these were still small transcendence degree results. It was G.V. Chudnovsky who then developed a method for proving the first result on large transcendence degree in this context ([Ch2]). He prov that the field generated over $\mathbb{Q}$ by numbers of the form

$$u_1,\ldots,u_n,v_1,\ldots,v_m,e^{u_1v_1},\ldots,e^{u_nv_m}$$

has transcendence degree which tends to infinity if m and n tend to infinity. Here one has to impose certain measures of linear independenc on the $u_1,\ldots,u_n$ and $v_1,\ldots,v_m$. A number of authors have developed Chudnovsky's ideas (Reyssat [R], Philippon [Ph2], Endell [E], Nesterenko [N2]) and this area of research is at the moment very active (Waldschmidt [Wa3], Zhu Yao Chen [Wa-Z] and Endell). A very natural and interesting question of course is to extend this sort of result to arbitrary algebraic groups. The main obstacle for doing this was the lack of the lemma of Tijdeman [T] in the general situation but since we have the zero estimates and the multiplicity estimates and both also "with knobs on" it is now possible to consider the general situation. The first step in this direction is due to D.W. Masser and the author [Ma-Wü3] where one elliptic curve is considered and an elliptic analogu of Chudnovsky's result is obtained.

In order to state it let E be an elliptic curve defined over the field of algebraic numbers and $\wp(z)$ an associated Weierstraß elliptic function with algebraic invariants g_2 and g_3. We assume that E has no complex multiplication. Then let $u_1,\ldots,u_n$ and $v_1,\ldots,v_m$ be compl numbers linearly independent over the rationals respectively such that there exists a positive real number κ with the property that

$$|s_1u_1+\ldots+s_nu_n| > \exp(-S^\kappa)$$

and

$$|t_1v_1+\ldots+t_mv_m| > \exp(-T^\kappa)$$

for all integers $s_1,\ldots,s_n,t_1,\ldots,t_m$ with $S = |s_1|+\ldots+|s_n|$ and $T = |t_1|+\ldots+|t_m|$ sufficiently large. Then we have the following result.

Theorem 4 ([Ma-Wü3]). Suppose that the integers m and n satisfy

$$mn \geq \{2^{k+1}(k+7)+4\kappa\}(m+2n)$$

for some integer $k \geq 0$. Then the transcendence degree of the field generated over the rationals by the numbers

$$\wp(u_i v_j) \qquad (1 \leq i \leq n;\ 1 \leq j \leq m)$$

is at least equal to k.

This result is of course only relatively weak and there might be some chance to improve it slightly using the result of Waldschmidt and Zhu Yao Chen already mentioned. Of course one can also ask for measures of algebraic independence. The general problem in this direction is the following one.

Problem. Let G be a commutative algebraic group defined over a finitely generated subfield K of the field of complex numbers and let Γ be a finitely generated subgroup of the tangent space $T(G)$ of G at the neutral element of G. Determine the transcendence degree of the smallest field L contained in the field of complex numbers such that

$$\exp_G(\Gamma) \subseteq G(L).$$

Of course in many special cases the answer is trivial. But in general this problem seems to be very difficult. For example it contains Schanuel's conjecture with $G = \mathbb{G}_a{}^n \times \mathbb{G}_m{}^n$ and $\Gamma = \mathbb{Z}\cdot(x_1,\ldots,x_n,x_1,\ldots,x_n)$.

5. Results of Lindemann's type

In 1882 Lindemann [Li] proved his famous theorem which is still one of the most beautiful results in transcendence theory. This result solved the old Greek problem of squaring the circle. The theorem says that if $\alpha_1,\ldots,\alpha_n$ are pairwise different algebraic numbers then the numbers

$$e^{\alpha_1},\ldots,e^{\alpha_n}$$

are linearly independent over the field of algebraic numbers. It is an immediate consequence of this result that if $\beta_1,\ldots,\beta_m$ are $\mathbb{Q}$-linearly independent algebraic numbers then the numbers

$$e^{\beta_1},\ldots,e^{\beta_m}$$

are algebraically independent over $\mathbb{Q}$. This result created a big theory developed by C.L. Siegel [Si1] in 1929 and later by Shidlovsky [Sh] and his school, namely the theory of so-called E- and G-functions. Her one is able to prove the algebraic independence of certain values of the E- and G-functions.

It seemed hopeless to obtain results of this sharpness for other classes of functions. But recently the newly developed theory of zero and multiplicity estimates together with effective algebraic constructions made it possible to prove in the complex multiplication case the abelian analogue of Lindemann's result on algebraic independence. The general result is very complicated to state. Therefore we restrict ourselves for simplicity to the elliptic case. For this let $\wp(z)$ be as usual a Weierstraß elliptic function with algebraic invariants g_2 and g_3. Furthermore we assume that the associated elliptic curve has comple multiplication. As usual we denote by K the associated imaginary quadratic field.

Theorem 5 ([Wü3], [Wü4]). Let $\beta_1,\ldots,\beta_m$ be algebraic numbers linearly independent over K. Then the numbers

$$\wp(\beta_1),\ldots,\wp(\beta_m)$$

are algebraically independent over $\mathbb{Q}$.

A slightly different proof of this result was given by Philippon [Ph4] using some ideas of the author. We should perhaps remark that our proof also gives the best possible (up to a ε) measure for algebraic independence. Let ε be a positive number.

Theorem 6 (see [Wü4]). Under the same hypothesis as before let $0 \neq P(X_1,\ldots,X_m)$ be a polynomial with integer coefficients of degree $d(P)$ and height $H(P)$. Then

$$\log |P(\wp(\beta_1),\ldots,\wp(\beta_m))| > -c\, d(P)^{m+\varepsilon} \log H(P)$$

where c is an effectively computable positive constant.

These results were proved for $m = 2,3$ by G.V.Chudnovsky [Ch3].

A somewhat related problem was posed by Gel'fond and by Schneider. Suppose that $\alpha \neq 0,1$ is an algebraic number and $\beta_1,\ldots,\beta_m$ are algebraic numbers that are linearly independent over $\mathbb{Q}$. Then the problem is to show that the numbers

$$\alpha^{\beta_1},\ldots,\alpha^{\beta_m}$$

are algebraically independent.

We do not know much on this problem. We know only in special situations from the result of Chudnovsky that the transcendence degree of the field generated by numbers of this type tends to infinity with m. And at present we do not know how to improve this in a significant way. Another special case which we know is the already mentioned result of Gel'fond.

Surprisingly the situation is quite different in the case of abelian functions. Here we know very sharp (but not yet best possible) results. They were proved by Philippon [Ph3] again relying on multiplicity estimates on group varieties. Here we will state only the simplest case of a more general result, and only the case of complex multiplication, which yields the best result. For this let $\wp(z)$ again be a Weierstraß elliptic function with algebraic g_2 and g_3. Let K have its usual meaning. Furthermore let F be an algebraic extension of K of degree m and $\beta_1,\ldots,\beta_m$ be a K-basis of F. Finally let u be a complex number such that $\wp(\beta_i u)$ is finite for $1 \leq i \leq m$. Then the following holds.

<u>Theorem 7</u> ([Ph3]). At least $\frac{m}{2} - 1$ of the numbers

$$\wp(\beta_1 u),\ldots,\wp(\beta_m u)$$

are algebraically independent.

Both Theorem 5 and Theorem 7 give a partial answer to the problem stated at the end of the last section. We end this section with the following

<u>Conjecture</u>. Let $E_1,\ldots,E_n$ be pairwise non-isogenous elliptic curves defined over the field of algebraic numbers and let $G = \mathbb{G}_m \times E_1 \times \ldots \times E_n$. Let $u \in T(G)(\mathbb{Q})$ have non-zero components . Then the dimension of

the Zariski closure X of $\exp_G(u)$ with respect to the $\overline{\mathbb{Q}}$- topology is equal to n+1. In other words $e, \wp_1(1), \ldots, \wp_n(1)$ are algebraically independent .

7. Analytic subgroups of algebraic groups

Since A.Baker had proved his famous theorem on linear forms in logarithms (see for this [Ba1]) a great number of authors have tried to prove analoguous results for elliptic and more generally abelian logarithms. These studies were initiated by Masser [Ma3] and further developed by Masser, Lang [L] and Coates and Lang [C-L]. It turned out that the crucial point here was a general multiplicity estimate which was not available. Therefore the authors had to use adhoc argumen to overcome these difficulties in very special situations. In addition the results that could be derived were rather weak except in the elliptic case with complex multiplication (Masser and Anderson). More recently Bertrand and Masser [B-M] succeeded to treat the elliptic case completely using a quite different method. Very surprisingly but in som sense quite naturally they could apply the criterion of Schneider and Lang in this situation. Nevertheless their approach is not very satisfactory for different reasons.

Since the necessary multiplicity estimates are now available (Theo rem 2) we obtain a completely general and universal result that cove all problems in this field.

In order to state it let G be as usual a commutative algebraic group defined over $\overline{\mathbb{Q}}$ and $A \subset G$ an analytic subgroup defined over $\overline{\mathbb{Q}}$ By this we mean that A is defined in the tangent space T(G) of G at the neutral element, which is in a natural way a $\overline{\mathbb{Q}}$-vector space, by $\overline{\mathbb{Q}}$-subspace T(A). Then the exponential map $\exp_G\colon T(G) \to G$ induces a local diffeomorphism between T(A) and A (or more precisely between their complex valued points $T(A)(\mathbb{C})$ and $A(\mathbb{C})$). Then we have the following result.

Theorem 8 ([Wü5]). Suppose that A contains a non-trivial algebraic point, i.e. $A(\overline{\mathbb{Q}}) \neq 0$. Then there exists an algebraic subgroup H of G with the following properties:

(i) H is defined over $\overline{\mathbb{Q}}$,

(ii) $\dim H > 0$,

(iii) $H \subseteq A$.

Remark. 1. It is obvious that the existence of such a subgroup H of G implies that $A(\overline{\mathbb{Q}}) \neq 0$. Hence the condition is also necessary. 2. As was pointed out by Gabber the theorem can also be stated as $\overline{A(\overline{\mathbb{Q}})} \subseteq A$. Here the second bar denotes the Zariski closure with respect to the $\overline{\mathbb{Q}}$-topology. In other words the Zariski closure (with respect to $\overline{\mathbb{Q}}$) of $A(\overline{\mathbb{Q}})$ is contained in A.

It is not very difficult to deduce Baker's theorem (in its qualitative version) from this general result. But we prefer to give the following more general result conjectured by Waldschmidt.

For this let $n,m \geq 1$ be integers and $\alpha_1,\ldots,\alpha_n$ be algebraic numbers such that $\log\alpha_1,\ldots,\log\alpha_n$ are $\mathbb{Q}$-linearly independent (for some fixed choice of the logarithms). Let further $\wp(z)$ be a Weierstraß elliptic function with algebraic g_2 and g_3 and $u_1,\ldots,u_m$ be complex numbers that are K-linearly independent ($K = (\mathrm{End}\ E)\otimes\mathbb{Q}$) and satisfy $\wp(u_1),\ldots,\wp(u_m) \in \overline{\mathbb{Q}}$. Finally let $\beta_1,\ldots,\beta_n$ and $\gamma_1,\ldots,\gamma_m$ be algebraic numbers not all zero.

Theorem 9 ([Wü6]). $\beta_1\log\alpha_1+\ldots+\beta_n\log\alpha_n+\gamma_1 u_1+\ldots+\gamma_m u_m \neq 0$.

Remark. The case $\gamma_1 = \ldots = \gamma_m = 0$ is Baker's theorem and for $\beta_1 = \ldots = \beta_n = 0$ we obtain the theorem of Bertrand and Masser.

A nice corollary of this result is the following result which was pointed out by Masser. Here we do not assume that $\log\alpha_1,\ldots,\log\alpha_n$ are $\mathbb{Q}$-linearly independent but that the numbers 1 and the non-zero elements among $\beta_1,\ldots,\beta_n$ are $\mathbb{Q}$-linearly independent.

Corollary. The complex number

$$\alpha_1^{\beta_1}\cdots\alpha_n^{\beta_n}\, e^{\gamma_1 u_1+\ldots+\gamma_m u_m}$$

is transcendental.

Of course it is not difficult to extend this result to abelian logarithms. Furthermore it is also possible to obtain quantitative versions of Theorem 8 nearly as precise as the bounds given by Baker for linear forms in classical logarithms. This applies for example to Siegel's theorem on integer points on curves and can be used to eliminate Roth's theorem, which is non-effective, in the proof of this theorem.

8. Periods of rational integrals.

Transcendence properties of periods of certain differential forms, especially elliptic and abelian, have been studied for a long time. The first result in this field was proved by Lindemann who obtained the transcendence of the non-zero periods of the differential form dx/x. Then Siegel [Si2] proved the transcendence of the periods of the ellitic differential $dx/\sqrt{4x^3-4x}$. Since then many authors (Schneider [Sch1], [Sch2], [Sch3], Baker [Ba2], [Ba3], Coates [Co1] , [Co2], [Co3] Masser [Ma3], [Ma4], [Ma5], Laurent [La1], [La2] and Bertrand [Be]) have obtained partial results. With the help of Theorem 8 it is now possible to prove a general result on arbitrary periods of rational lin integrals. This result covers all these results just refered to.

Let X be a smooth quasiprojective variety defined over $\overline{\mathbb{Q}}$ and ξ a closed holomorphic 1-form on X defined over $\overline{\mathbb{Q}}$. Suppose that γ is a closed path in X (in other words: γ represents an element in the first homology group $H_1(X,\mathbb{Z})$). Then we have the following result.

Theorem 10 ([Wü5]). The rational integral

$$\int_\gamma \xi$$

is either zero or transcendental.

One obtains immediately the following Corollary.

Corollary. The periods of an abelian integral (first, second and third kind) are either zero or transcendental.

At the present we can only deal with periods of 1-forms. It would be extremely interesting to extend this result as far as possible to arbitrary r-forms with $r \geq 2$. Of course it is not too hard to decide when the periods are zero. This can happen without the differential being exact or the path being homotopic to zero. This is done in [Wü 8] in the case of arbitrary elliptic integrals where the group generated by the polar divisor of the part of the third kind plays the essential rol Furthermore this is worked out in the case of abelian integrals of the second kind for products of elliptic curves ([Wü 7]). This solve a problem of Baker.

Of course it is possible to extend Theorem 10 to non-closed paths going from one algebraic point on X to another. One obtains then a re sult of the following type. Either the integral is transcendental or it is algebraic and one can determine why this is the case. But the details have yet to be worked out.

9. References

[Ba1] A.Baker, Transcendental number theory, Cambridge University Press, Cambridge (1975, 1979).

[Ba2] A.Baker, On the periods of the Weierstrass $\wp$-function, Symposia Math. IV, INDAM Rome, 1968 (Academic Press, London 1970), 155-174.

[Ba3] A.Baker, On the quasi-periods of the Weierstrass ζ-function, Göttinger Nachr. (1969), No.16, 145-157.

[Be] D.Bertrand, Endomorphismes de groupes algébriques; applications arithmétiques, Progr.Math.31, 1-46 (1983).

[Be-Ma] D.Bertrand, D.W.Masser, Formes linéaires d'integrales abéliennes, CRAS Paris, 290 (1980), 725-727.

[Bo] E.Bombieri, On the Thue-Siegel-Dyson theorem, Acta Math.148, 255-296 (1982).

[Br1] W.D.Brownawell, On the orders of zero of certain functions, Bull.Soc.Math. France, Mémoire 2 (1980), 5-20.

[Br2] W.D.Brownawell, Zero estimates for solutions of differential equations, Progr.Math.31, 67-94 (1983).

[Br-K] W.D.Brownawell, K.K.Kubota, The algebraic independence of Weierstrass functions and some related numbers, Acta Arith. 33 (1977), 111-149.

[Br-Ma] W.D.Brownawell, D.W.Masser, Multiplicity estimates for analytic functions, Duke Math.J.47 (1980), 273-295.

[Ch1] G.V.Chudnovsky, Singular points on complex hypersurfaces and multidimensional Schwarz Lemma, Progr.Math.12, 29-69 (1981).

[Ch2] G.V.Chudnovsky, Some analytic methods in the theory of transcendental numbers, Inst.of Math.,Ukr. SSSR Acad.Sci., Preprint IM 74-8 and 74-9, Kiev (1974).

[Ch3] G.V.Choodnovsky, Algebraic independence of the values of elliptic function at algebraic points, Inv.math.61 (1980), 267-290.

[Co1] J.Coates, Linear forms in the periods of the exponential and elliptic functions, Inv.math.12 (1971), 290-299.

[Co2] J.Coates, The transcendence of linear forms in $\omega_1, \omega_2, \eta_1, \eta_2$, $2\pi i$, Amer.J.Math.93 (1971), 385-397.

[Co3] J.Coates, Linear relations between $2\pi i$ and the periods of two elliptic curves, Diophantine approximation and its appl. cations (Acad.Press, London, 1973), 77-99.

[Co-L] J.Coates, S.Lang, Diophantine approximation on Abelian varieties with complex multiplications, Inv.math.34 (1976), 129-133.

[D] J.P.Demailly, Formules de Jensen en plusieurs variables et applications arithmétiques, Bull.Soc.Math. France 110 (1982) 75-102.

[E] R.Endell, Zur algebraischen Unabhängigkeit gewisser Werte der Exponentialfunktion, this volume.

[E -V1] H.Esnault, E.Vieweg, Sur une minoration du degré d'hypersurfaces s'annulant en certains points, Math.Ann.263, 75-86 (1983).

[E -V2] H.Esnault, E.Vieweg, Dyson's lemma for polynomials in several variables, preprint (1983).

[F-Wü] G.Faltings, G.Wüstholz, Einbettungen kommutativer algebraischer Gruppen und einige ihrer Eigenschaften, Manuskript.

[L] S.Lang, Diophantine approximation on abelian varieties with complex multiplication, Adv. in Math.17 (1975), 281-336.

[La1] M.Laurent, Transcendance de périodes d'integrales elliptiques, J. reine u. angew. Math. 316, 122-139 (1980).

[La2] M.Laurent, Transcendance de périodes d'integrales elliptiques II, J. reine u. angew. Math. 333, 144-161 (1982).

[Li] F.Lindemann, Über die Zahl π, Math.Ann. 20, 213-223 (1882)

[Ma1] D.W.Masser, A note on multiplicities of polynomials, Publ. Math.Univ.Paris VI (1981).

[Ma2] D.W.Masser, Interpolation on group varieties, Progr.Math.31 151-171 (1983).

[Ma3] D.W.Masser, Elliptic functions and transcendence, SLN 437 (1975).

[Ma4] D.W.Masser, Some vector spaces associated with two elliptic functions, Transcendence theory: advances and applications, Acad.Press, London, 101-119 (1977).

[Ma5] D.W.Masser, The transcendence of certain quasi-periods associated with Abelian functions, Comp.Math.35 (1977), 239-258.

[Ma-Wü1] D.W.Masser, G.Wüstholz, Zero estimates on group varieties I Inv.math.64 (1981), 489-516.

[Ma-Wü2] D.W.Masser, G.Wüstholz, Zero estimates on group varieties I preprint (1983).

[Ma-Wü3] D.W.Masser, G.Wüstholz, Fields of large transcendence degree generated by values of elliptic functions, Inv.math.72 (1983), 407-464.

[Ma-Wü4] D.W.Masser, G.Wüstholz, Algebraic independence properties of values of elliptic functions, London Math.Soc.Lécture Note 56 (1982), 360-363, Cambridge Univ. Press.

[Ma-Wü5] D.W.Masser, G.Wüstholz, Algebraic independence of values of elliptic functions, preprint (1983).

[Mo1] J.-C.Moreau, Démonstration géométrique de lemmes de zéros, I, Séminaire de Theorie des Nombres, Paris 1981-1982, to appear in Progr.Math.

[Mo2] J.-C. Moreau, Démonstration géométrique de lemmes de zéros, II, Progr.Math.31 (1983), 191-197.

[Ne1] Ju.V.Nesterenko, Bounds for the orders of zeros of functions in a special class and application to the theory of transcendental numbers, Izvestia Ser.math. 41 (1977), n°2, 253-284.

[Ne2] Ju.V.Nesterenko, On the algebraic independence of algebraic numbers to algebraic powers, Progr.Math.31 (1983), 199-220.

[Ph1] P.Philippon, Interpolation dans les espaces affines, Progr. Math.22 (1982), 221-236.

[Ph2] P.Philippon, Indépendance algébrique de valeurs de fonctions exponentielles p-adiques, J.reine u. angew. Math. 329 (1981), 42-51.

[Ph3] P.Philippon, Variétés abéliennes et indépendance algébrique I, Inv.math. 70 (1983), 259-318.

[Ph4] P.Philippon, Variétés abéliennes et indépendence algébrique II: un analogue abélien du théorème de Lindemann-Weierstraß, Inv.math. 72 (1983), 389-405.

[R] E.Reyssat, Un critère d'indépendance algébrique, J.Reine u. angew. Math. 329 (1981), 66-81.

[Sch1] Th.Schneider, Transzendenzuntersuchungen periodischer Funktionen I,II, J.reine u. angew. Math. 172 (1934), 65-74.

[Sch2] Th.Schneider, Arithmetische Untersuchungen elliptischer Integrale, Math.Ann. 113 (1937), 1-13.

[Sch3] Th.Schneider, Zur Theorie der Abelschen Funktionen und Integrale, J.reine u. angew. Math. 183 (1941), 110-128.

[Se] J.P.Serre, Quelques propriétés des groupes algébriques commutatifs, Astérisque 69-70 (1979), 191-202.

[Sh] A.V.Shidlovsky, On a criterion of algebraic independence, Izvestia Akad. Nauk SSSR 23 (1959), 35-66.

[Si1 C.L.Siegel, Über einige Anwendungen diophantischer Approximationen, Abh. Preuss. Akad. Wiss. 1 (1929).

[Si2] C.L.Siegel, Über die Perioden elliptischer Funktionen, J. reine u. angew. Math. 167 (1932), 62-69.

[Sm] A.A.Shmelev, On the question of the algebraic independence of algebraic powers of algebraic numbers, Mat.Zam.11 (1972) 635-644.

[T] R.Tijdeman, An auxiliary result in the theory of transcendental numbers, J. Number Theory 5 (1973), 80-94.

[Wa1] M.Waldschmidt, Nombres transcendants et groupes algébriques Astérisque 69-70 (1979).

[Wa2] M.Waldschmidt, Nombres transcendants, SLN 402 (1974).

[Wa3] M.Waldschmidt, Indépendance algébrique et exponentielles en plusieurs variables, to appear in this volume.

[Wa-Z] M.Waldschmidt, Zhu Yao Chen, Une genéralisation en plusieurs variables d'un critère de transcendance de Gel'fond, to appear in CRAS Paris.

[Wü1] G.Wüstholz, Nullstellenabschätzungen auf Varietäten, Progr. Math. 22 (1982), 359-362.

[Wü2] G.Wüstholz, Multiplicity estimates on group varieties, to appear.

[Wü3] G.Wüstholz, Sur l'analogue abélien du théorème de Lindemann CRAS Paris,Ser.I, 295 (1982), 35-37.

[Wü4] G.Wüstholz, Über das abelsche Analogon des Lindemannschen Satzes I, Inv.math.72 (1983), 363-388.

[Wü5] G.Wüstholz, Algebraic values of analytic homomorphisms of algebraic groups, to appear.

[Wü6] G.Wüstholz, Some remarks on a conjecture of Waldschmidt, Progr.Math. 31 (1983), 329-336.

[Wü7] G.Wüstholz, Zum Periodenproblem, to appear in Inv.math.

[Wü8] G.Wüstholz, Transzendenzeigenschaften von Perioden elliptischer Integrale, to appear in J. reine u. angew. Math.

G.Wüstholz
Max-Planck-Institut für Mathematik
Gottfried-Claren-Str. 26
D-5300 Bonn 3

Vol. 926: Geometric Techniques in Gauge Theories. Proceedings, 1981. Edited by R. Martini and E.M.de Jager. IX, 219 pages. 1982.

Vol. 927: Y. Z. Flicker, The Trace Formula and Base Change for GL (3). XII, 204 pages. 1982.

Vol. 928: Probability Measures on Groups. Proceedings 1981. Edited by H. Heyer. X, 477 pages. 1982.

Vol. 929: Ecole d'Eté de Probabilités de Saint-Flour X – 1980. Proceedings, 1980. Edited by P.L. Hennequin. X, 313 pages. 1982.

Vol. 930: P. Berthelot, L. Breen, et W. Messing, Théorie de Dieudonné Cristalline II. XI, 261 pages. 1982.

Vol. 931: D.M. Arnold, Finite Rank Torsion Free Abelian Groups and Rings. VII, 191 pages. 1982.

Vol. 932: Analytic Theory of Continued Fractions. Proceedings, 1981. Edited by W.B. Jones, W.J. Thron, and H. Waadeland. VI, 240 pages. 1982.

Vol. 933: Lie Algebras and Related Topics. Proceedings, 1981. Edited by D. Winter. VI, 236 pages. 1982.

Vol. 934: M. Sakai, Quadrature Domains. IV, 133 pages. 1982.

Vol. 935: R. Sot, Simple Morphisms in Algebraic Geometry. IV, 146 pages. 1982.

Vol. 936: S.M. Khaleelulla, Counterexamples in Topological Vector Spaces. XXI, 179 pages. 1982.

Vol. 937: E. Combet, Intégrales Exponentielles. VIII, 114 pages. 1982.

Vol. 938: Number Theory. Proceedings, 1981. Edited by K. Alladi. IX, 177 pages. 1982.

Vol. 939: Martingale Theory in Harmonic Analysis and Banach Spaces. Proceedings, 1981. Edited by J.-A. Chao and W.A. Woyczyński. VIII, 225 pages. 1982.

Vol. 940: S. Shelah, Proper Forcing. XXIX, 496 pages. 1982.

Vol. 941: A. Legrand, Homotopie des Espaces de Sections. VII, 132 pages. 1982.

Vol. 942: Theory and Applications of Singular Perturbations. Proceedings, 1981. Edited by W. Eckhaus and E.M. de Jager. V, 363 pages. 1982.

Vol. 943: V. Ancona, G. Tomassini, Modifications Analytiques. IV, 120 pages. 1982.

Vol. 944: Representations of Algebras. Workshop Proceedings, 1980. Edited by M. Auslander and E. Lluis. V, 258 pages. 1982.

Vol. 945: Measure Theory. Oberwolfach 1981, Proceedings. Edited by D. Kölzow and D. Maharam-Stone. XV, 431 pages. 1982.

Vol. 946: N. Spaltenstein, Classes Unipotentes et Sous-groupes de Borel. IX, 259 pages. 1982.

Vol. 947: Algebraic Threefolds. Proceedings, 1981. Edited by A. Conte. VII, 315 pages. 1982.

Vol. 948: Functional Analysis. Proceedings, 1981. Edited by D. Butković, H. Kraljević, and S. Kurepa. X, 239 pages. 1982.

Vol. 949: Harmonic Maps. Proceedings, 1980. Edited by R.J. Knill, M. Kalka and H.C.J. Sealey. V, 158 pages. 1982.

Vol. 950: Complex Analysis. Proceedings, 1980. Edited by J. Eells. V, 428 pages. 1982.

Vol. 951: Advances in Non-Commutative Ring Theory. Proceedings, 1981. Edited by P.J. Fleury. V, 142 pages. 1982.

Vol. 952: Combinatorial Mathematics IX. Proceedings, 1981. Edited by E. Billington, S. Oates-Williams, and A.P. Street. XI, 443 pages. 1982.

Vol. 953: Iterative Solution of Nonlinear Systems of Equations. Proceedings, 1982. Edited by R. Ansorge, Th. Meis, and W. Törnig. VII, 202 pages. 1982.

Vol. 954: S.G. Pandit, S.G. Deo, Differential Systems Involving Impulses. VII, 102 pages. 1982.

Vol. 955: G. Gierz, Bundles of Topological Vector Spaces and Their Duality. IV, 296 pages. 1982.

Vol. 956: Group Actions and Vector Fields. Proceedings, 1981. Edited by J.B. Carrell. V, 144 pages. 1982.

Vol. 957: Differential Equations. Proceedings, 1981. Edited by D.G. de Figueiredo. VIII, 301 pages. 1982.

Vol. 958: F.R. Beyl, J. Tappe, Group Extensions, Representations, and the Schur Multiplicator. IV, 278 pages. 1982.

Vol. 959: Géométrie Algébrique Réelle et Formes Quadratiques, Proceedings, 1981. Edité par J.-L. Colliot-Thélène, M. Coste, L. Mahé, et M.-F. Roy. X, 458 pages. 1982.

Vol. 960: Multigrid Methods. Proceedings, 1981. Edited by W. Hackbusch and U. Trottenberg. VII, 652 pages. 1982.

Vol. 961: Algebraic Geometry. Proceedings, 1981. Edited by J.M. Aroca, R. Buchweitz, M. Giusti, and M. Merle. X, 500 pages. 1982.

Vol. 962: Category Theory. Proceedings, 1981. Edited by K.H. Kamps, D. Pumplün, and W. Tholen, XV, 322 pages. 1982.

Vol. 963: R. Nottrot, Optimal Processes on Manifolds. VI, 124 pages. 1982.

Vol. 964: Ordinary and Partial Differential Equations. Proceedings, 1982. Edited by W.N. Everitt and B.D. Sleeman. XVIII, 726 pages. 1982.

Vol. 965: Topics in Numerical Analysis. Proceedings, 1981. Edited by P.R. Turner. IX, 202 pages. 1982.

Vol. 966: Algebraic K-Theory. Proceedings, 1980, Part I. Edited by R.K. Dennis. VIII, 407 pages. 1982.

Vol. 967: Algebraic K-Theory. Proceedings, 1980. Part II. VIII, 409 pages. 1982.

Vol. 968: Numerical Integration of Differential Equations and Large Linear Systems. Proceedings, 1980. Edited by J. Hinze. VI, 412 pages. 1982.

Vol. 969: Combinatorial Theory. Proceedings, 1982. Edited by D. Jungnickel and K. Vedder. V, 326 pages. 1982.

Vol. 970: Twistor Geometry and Non-Linear Systems. Proceedings, 1980. Edited by H.-D. Doebner and T.D. Palev. V, 216 pages. 1982.

Vol. 971: Kleinian Groups and Related Topics. Proceedings, 1981. Edited by D.M. Gallo and R.M. Porter. V, 117 pages. 1983.

Vol. 972: Nonlinear Filtering and Stochastic Control. Proceedings, 1981. Edited by S.K. Mitter and A. Moro. VIII, 297 pages. 1983.

Vol. 973: Matrix Pencils. Proceedings, 1982. Edited by B. Kågström and A. Ruhe. XI, 293 pages. 1983.

Vol. 974: A. Draux, Polynômes Orthogonaux Formels – Applications. VI, 625 pages. 1983.

Vol. 975: Radical Banach Algebras and Automatic Continuity. Proceedings, 1981. Edited by J.M. Bachar, W.G. Bade, P.C. Curtis Jr., H.G. Dales and M.P. Thomas. VIII, 470 pages. 1983.

Vol. 976: X. Fernique, P.W. Millar, D.W. Stroock, M. Weber, Ecole d'Eté de Probabilités de Saint-Flour XI – 1981. Edited by P.L. Hennequin. XI, 465 pages. 1983.

Vol. 977: T. Parthasarathy, On Global Univalence Theorems. VIII, 106 pages. 1983.

Vol. 978: J. Ławrynowicz, J. Krzyż, Quasiconformal Mappings in the Plane. VI, 177 pages. 1983.

Vol. 979: Mathematical Theories of Optimization. Proceedings, 1981. Edited by J.P. Cecconi and T. Zolezzi. V, 268 pages. 1983.

Vol. 980: L. Breen. Fonctions thêta et théorème du cube. XIII, 115 pages. 1983.

Vol. 981: Value Distribution Theory. Proceedings, 1981. Edited by I. Laine and S. Rickman. VIII, 245 pages. 1983.

Vol. 982: Stability Problems for Stochastic Models. Proceedings, 1982. Edited by V. V. Kalashnikov and V. M. Zolotarev. XVII, 295 pages. 1983.

Vol. 983: Nonstandard Analysis-Recent Developments. Edited by A. E. Hurd. V, 213 pages. 1983.

Vol. 984: A. Bove, J. E. Lewis, C. Parenti, Propagation of Singularities for Fuchsian Operators. IV, 161 pages. 1983.

Vol. 985: Asymptotic Analysis II. Edited by F. Verhulst. VI, 497 pages. 1983.

Vol. 986: Séminaire de Probabilités XVII 1981/82. Proceedings. Edited by J. Azéma and M. Yor. V, 512 pages. 1983.

Vol. 987: C. J. Bushnell, A. Fröhlich, Gauss Sums and p-adic Division Algebras. XI, 187 pages. 1983.

Vol. 988: J. Schwermer, Kohomologie arithmetisch definierter Gruppen und Eisensteinreihen. III, 170 pages. 1983.

Vol. 989: A. B. Mingarelli, Volterra-Stieltjes Integral Equations and Generalized Ordinary Differential Expressions. XIV, 318 pages. 1983.

Vol. 990: Probability in Banach Spaces IV. Proceedings, 1982. Edited by A. Beck and K. Jacobs. V, 234 pages. 1983.

Vol. 991: Banach Space Theory and its Applications. Proceedings, 1981. Edited by A. Pietsch, N. Popa and I. Singer. X, 302 pages. 1983.

Vol. 992: Harmonic Analysis, Proceedings, 1982. Edited by G. Mauceri, F. Ricci and G. Weiss. X, 449 pages. 1983.

Vol. 993: R. D. Bourgin, Geometric Aspects of Convex Sets with the Radon-Nikodým Property. XII, 474 pages. 1983.

Vol. 994: J.-L. Journé, Calderón-Zygmund Operators, Pseudo-Differential Operators and the Cauchy Integral of Calderón. VI, 129 pages. 1983.

Vol. 995: Banach Spaces, Harmonic Analysis, and Probability Theory. Proceedings, 1980–1981. Edited by R. C. Blei and S. J. Sidney. V, 173 pages. 1983.

Vol. 996: Invariant Theory. Proceedings, 1982. Edited by F. Gherardelli. V, 159 pages. 1983.

Vol. 997: Algebraic Geometry – Open Problems. Edited by C. Ciliberto, F. Ghione and F. Orecchia. VIII, 411 pages. 1983.

Vol. 998: Recent Developments in the Algebraic, Analytical, and Topological Theory of Semigroups. Proceedings, 1981. Edited by K. H. Hofmann, H. Jürgensen and H. J. Weinert. VI, 486 pages. 1983.

Vol. 999: C. Preston, Iterates of Maps on an Interval. VII, 205 pages. 1983.

Vol. 1000: H. Hopf, Differential Geometry in the Large. VII, 184 pages. 1983.

Vol. 1001: D. A. Hejhal, The Selberg Trace Formula for PSL(2, IR). Volume 2. VIII, 806 pages. 1983.

Vol. 1002: A. Edrei, E. B. Saff, R. S. Varga, Zeros of Sections of Power Series. VIII, 115 pages. 1983.

Vol. 1003: J. Schmets, Spaces of Vector-Valued Continuous Functions. VI, 117 pages. 1983.

Vol. 1004: Universal Algebra and Lattice Theory. Proceedings, 1982. Edited by R. S. Freese and O. C. Garcia. VI, 308 pages. 1983.

Vol. 1005: Numerical Methods. Proceedings, 1982. Edited by V. Pereyra and A. Reinoza. V, 296 pages. 1983.

Vol. 1006: Abelian Group Theory. Proceedings, 1982/83. Edited by R. Göbel, L. Lady and A. Mader. XVI, 771 pages. 1983.

Vol. 1007: Geometric Dynamics. Proceedings, 1981. Edited by J. Palis Jr. IX, 827 pages. 1983.

Vol. 1008: Algebraic Geometry. Proceedings, 1981. Edited by J. Dolgachev. V, 138 pages. 1983.

Vol. 1009: T. A. Chapman, Controlled Simple Homotopy Theory a Applications. III, 94 pages. 1983.

Vol. 1010: J.-E. Dies, Chaînes de Markov sur les permutations. 226 pages. 1983.

Vol. 1011: J. M. Sigal. Scattering Theory for Many-Body Quantu Mechanical Systems. IV, 132 pages. 1983.

Vol. 1012: S. Kantorovitz, Spectral Theory of Banach Space Op ators. V, 179 pages. 1983.

Vol. 1013: Complex Analysis – Fifth Romanian-Finnish Semin Part 1. Proceedings, 1981. Edited by C. Andreian Cazacu, N. Bob M. Jurchescu and I. Suciu. XX, 393 pages. 1983.

Vol. 1014: Complex Analysis – Fifth Romanian-Finnish Semin Part 2. Proceedings, 1981. Edited by C. Andreian Cazacu, N. Bob M. Jurchescu and I. Suciu. XX, 334 pages. 1983.

Vol. 1015: Equations différentielles et systèmes de Pfaff dans champ complexe – II. Seminar. Edited by R. Gérard et J. P. Ram V, 411 pages. 1983.

Vol. 1016: Algebraic Geometry. Proceedings, 1982. Edited by Raynaud and T. Shioda. VIII, 528 pages. 1983.

Vol. 1017: Equadiff 82. Proceedings, 1982. Edited by H. W. Knobl and K. Schmitt. XXIII, 666 pages. 1983.

Vol. 1018: Graph Theory, Łagów 1981. Proceedings, 1981. Edited M. Borowiecki, J. W. Kennedy and M. M. Sysło. X, 289 pages. 19

Vol. 1019: Cabal Seminar 79–81. Proceedings, 1979–81. Edited A. S. Kechris, D. A. Martin and Y. N. Moschovakis. V, 284 pages. 19

Vol. 1020: Non Commutative Harmonic Analysis and Lie Grou Proceedings, 1982. Edited by J. Carmona and M. Vergne. V, pages. 1983.

Vol. 1021: Probability Theory and Mathematical Statistics. F ceedings, 1982. Edited by K. Itô and J.V. Prokhorov. VIII, 747 pag 1983.

Vol. 1022: G. Gentili, S. Salamon and J.-P. Vigué. Geometry Semi "Luigi Bianchi", 1982. Edited by E. Vesentini. VI, 177 pages. 198

Vol. 1023: S. McAdam, Asymptotic Prime Divisors. IX, 118 pag 1983.

Vol. 1024: Lie Group Representations I. Proceedings, 1982–19 Edited by R. Herb, R. Lipsman and J. Rosenberg. IX, 369 pages. 19

Vol. 1025: D. Tanré, Homotopie Rationnelle: Modèles de Ch Quillen, Sullivan. X, 211 pages. 1983.

Vol. 1026: W. Plesken, Group Rings of Finite Groups Over p-a Integers. V, 151 pages. 1983.

Vol. 1027: M. Hasumi, Hardy Classes on Infinitely Connected mann Surfaces. XII, 280 pages. 1983.

Vol. 1028: Séminaire d'Analyse P. Lelong – P. Dolbeault – H. Sk Années 1981/1983. Edité par P. Lelong, P. Dolbeault et H. Sk VIII, 328 pages. 1983.

Vol. 1029: Séminaire d'Algèbre Paul Dubreil et Marie-Paule Mallia Proceedings, 1982. Edité par M.-P. Malliavin. V, 339 pages. 198

Vol. 1030: U. Christian, Selberg's Zeta-, L-, and Eisensteinser XII, 196 pages. 1983.

Vol. 1031: Dynamics and Processes. Proceedings, 1981. Edite Ph. Blanchard and L. Streit. IX, 213 pages. 1983.

Vol. 1032: Ordinary Differential Equations and Operators. ceedings, 1982. Edited by W. N. Everitt and R. T. Lewis. XV, 521 pa 1983.

Vol. 1033: Measure Theory and its Applications. Proceedings, 19 Edited by J. M. Belley, J. Dubois and P. Morales. XV, 317 pages. 19

Vol. 1034: J. Musielak, Orlicz Spaces and Modular Spaces 222 pages. 1983.